Boo

AN INTRODUCTION TO ABSTRACT ALGEBRA

AN INTRODUCTION TO ABSTRACT ALGEBRA

VOLUME 2

BY

F. M. HALL

Head of the Mathematics Faculty
Shrewsbury School

CAMBRIDGE
AT THE UNIVERSITY PRESS
1969

Published by the Syndics of the Cambridge University Press
Bentley House, 200 Euston Road, London N.W.1
American Branch: 32 East 57th Street, New York, N.Y. 10022

Standard Book Number: 521 5178 9 vol. 1
521 7055 4 vol. 2

ISBN 0521040554

Printed in Great Britain
at the University Printing House, Cambridge
(Brooke Crutchley, University Printer)

CONTENTS

11 BOOLEAN ALGEBRA

12 FURTHER RESULTS

PREFACE

This book is the second of a two volume work which attempts to give a broad introduction to the subject of abstract algebra. It assumes no previous knowledge of this work but does expect a fair amount of mathematical sophistication. Thus it is not intended to be used as a textbook at an elementary level in schools, as part of one of the 'modern mathematics' courses, but is aimed rather at a fairly intelligent sixth-former, who has been brought up on either traditional or modern syllabuses and who wishes to know something more about modern algebra, for example as a preparation for university work. This second volume in particular should be found useful by undergraduates, as giving a broad background before a study in more depth is undertaken, and it does in fact cover a great deal of work usually included in university first degree courses, though not all such work. Teachers in schools and students in teacher training colleges who wish to learn some abstract algebra as an aid and background to their teaching should also find parts of both volumes interesting and useful. Numerous books on the subject have been and are being written, but these are usually either too advanced, with a lack of motivation, for the beginner, or are written at a far less sophisticated level, for younger pupils. It is still not easy to find many which start from the beginning and lead up to substantial and important ideas gradually and in a fairly elementary manner.

Volume 1 dealt with various special sets, such as the integers, residue classes and polynomials, emphasising their structure before discussing the fundamental laws of algebra and finishing with three chapters on groups, carrying the theory as far as Lagrange's theorem.

The second volume is arranged so that it may be read independently of the first, and the first chapter gives a shortened account of the group theory which was included in volume 1. Thus it is not necessary for a student to have read volume 1 previously, and many university undergraduates may well use the present book by itself. A reader with a scant knowledge of structure would be advised however to look through at least part (if not all) of the first volume, and such a reader may omit chapter 1.

The first chapter is followed by essential work on group-homomorphisms, and we then consider rings, fields and integral domains. The work towards the end of chapter 4, and some in chapter 5, is quite advanced though explained as simply as possible. Chapter 6 continues the study of group theory by considering the vital concept of an invariant subgroup, and the next chapter introduces the related idea of an ideal in a ring. This chapter, and the following one, deals with quite deep properties, and the reader may well find them heavy going.

With chapter 9 we are on much simpler ground, and give a short account of the theory of vector spaces, from the point of view of abstract structure rather than as an adjunct to matrix theory, although we briefly introduce the idea of matrix as a representation of a linear transformation. Chapter 10 applies the vector space idea to the axiomatic foundations of geometry, and we conclude the formal work of the book by a fairly abstract account of Boolean algebra. The final chapter is informal, and indicates how some of the work may be further developed.

As in the first volume, I have been careful to give as much explanation as possible about the reasons for doing the various topics, and have explained the methods carefully. The difficulty which many students experience arises usually from their not understanding the motivation of the work as a whole: they do not see where it is leading. I have tried to avoid this sense of frustration by introducing new ideas carefully and sometimes, as in the case of invariant subgroups, at some length. The proofs are nearly always given in full detail and are selected for ease of understanding rather than conciseness. Concrete examples of new structures are given as much as possible, though inevitably they are usually drawn from mathematics itself, and are often necessarily rather more abstract than was the case in the first volume. But of course we meet again many of the particular sets that were studied in that volume, and they are now revealed in their true structural form (e.g. polynomials occur in many places in this volume).

The book may be read with little or no aid from a teacher, and each chapter except the last ends with a few worked exercises. The exercises themselves are, as in volume 1, divided into A and B: the first are quite straightforward and should be worked completely. The B exercises are very variable; some are fairly straightforward, others quite difficult and a few give extensions of the bookwork. The reader is not expected to be able to do all these exercises, at least not at a first reading.

I would like to thank a former pupil at Dulwich College, Mr J. R. Pratt, who read the manuscript and made many valuable suggestions. I am indebted to my colleague, Mr S. D. Baxter, who read the proofs; and am grateful to the Cambridge University Press for their help throughout all stages of the preparation of the book.

F.M.H.

The Schools
Shrewsbury
February 1969

1

GROUPS

1.1. Introduction—equivalence relations

In this first chapter we repeat briefly some of the basic work on group theory that was done in detail in volume 1. The reader who wishes to study the elementary theory in detail should read volume 1 and omit this chapter, but the student who possesses a fair amount of mathematical maturity and who wants to study other types of algebraic structure but who needs first to know the standard theorems and definitions of group theory may substitute this chapter, which will provide him with the knowledge he requires, though not with all the detailed examples and explanations that were given in the first volume.

We assume that the reader is familiar with set notation. We use the standard notation, but the following points should be noted.

Inclusion. We denote the fact that A is contained in B by $A \subseteq B$, reserving $A \subset B$ to mean $A \subseteq B$ but $A \neq B$.

The empty and the universal sets. The empty set is denoted by $\varnothing$, and we use no special notation for the universal set.

Complement and difference. The complement of A is written as A'; the difference $(A-B)$ means the set of elements in the universal set that are in A but not in B: it does not imply that $B \subseteq A$.

Sets in terms of their elements. The set containing elements $a, b, c, \ldots, k$ is denoted by $\{a, b, c, \ldots, k\}$. The set of elements $x_1, x_2, \ldots, x_n$ may also be written $\{x_i : i = 1, \ldots, n\}$ or even $\{x_i\}$. Sets defined by a property are denoted thus: $\{x : x \text{ is a triangle}\}$.

Equivalence relations

An important method of dividing a set into mutually exclusive subsets is by means of an equivalence relation, the result used being given in the following fundamental theorem.

Theorem 1.1.1. *The equivalence classes theorem.*

Suppose in a set S we have a relation R defined between certain pairs of elements, xRy meaning that x and y stand in the given relation R.

Suppose further that R has the following 3 *properties*:

(1) *It is reflexive*: *i.e. xRx for all $x \in S$.*

(2) *It is symmetric*: *i.e. $xRy \Rightarrow yRx$.*

(3) *It is transitive*: *i.e. xRy and $yRz \Rightarrow xRz$.*

Then R is an equivalence relation: *i.e. it divides S into mutually exclusive subsets so that every element of S is in one and only one subset, and so that two elements are in the same subset if and only if they stand in the relation R to one another.*

The subsets are called the *equivalence classes* defined by R.

Given any element x in S consider all the elements y such that xRy. These form a subset of S: let us call it A_x. We show first that two elements are in the same subset A_x if and only if they stand in the relation R to one another. Suppose yRz and $y \in A_x$. Then xRy and so xRz since R is transitive. Hence $z \in A_x$. Conversely if y and z are both in A_x we have xRy and xRz, i.e. yRx by the symmetric property and xRz: thus yRz by transitivity.

For each element x we now have a subset A_x, but these will not all be distinct. We will show that two such are either mutually exclusive or else identical. Suppose A_x and A_y both have an element z. Then xRz and yRz, and so zRy by the symmetric property; hence xRy by the transitive property. Now take any element w of A_x. Then xRw and since xRy we have yRx, giving yRw. So w is in A_y. Hence we have shown that if A_x and A_y have one element z in common, any element w of A_x is in A_y, i.e. $A_x \subseteq A_y$. Similarly $A_y \subseteq A_x$, and so the subsets A_x and A_y are identical.

Thus we have mutually exclusive subsets $A_{x_1}, A_{x_2}, \ldots$. Finally, by the reflexive property any element t is in one of the subsets, viz. A_t.

Hence S is divided into a set of mutually exclusive subsets as required.

Note that the subset A_x may equally well be described as A_y for any element y in it—the important things are the equivalence classes and not the individual elements.

1.2. Algebraic structures

In order to use the methods of algebra in a set we must have some process, such as addition or multiplication, connecting our elements of the set. A set which has one or more operations such as these is called an *algebraic structure.*

Our processes may be addition or multiplication, or both. Considered in isolation addition and multiplication are formally very similar in that they both combine pairs of elements and satisfy similar laws, differing only in notation and terminology: it is only when they are both present that a distinction arises, in connection with the Distributive Law and, consequent on this, in the impossibility of dividing by zero.

Our processes need not be of the addition and multiplication type. If we consider the set whose elements are all the subsets of a given set then we can define the intersection and union of two subsets, and such processes satisfy laws similar to our fundamental laws of algebra, but not quite the same. Again, we could form a new positive integer from two given positive integers x and y by raising x to the power y: but this is neither commutative nor associative and is not likely to be a fruitful idea as an algebraic operation.

It is possible that *three* (or more) elements need to be taken in order to define a new element. An example of this would be the formation of $\mathbf{a} \wedge (\mathbf{b} \wedge \mathbf{c})$ for the three-dimensional vectors $\mathbf{a}$, $\mathbf{b}$ and $\mathbf{c}$. We must expect the fundamental laws of combination for these more difficult processes to be correspondingly complicated. As is to be expected, by far the most fruitful ideas are those involving two elements only, and among these it is found that those which satisfy some of the ordinary fundamental laws of algebra are most useful in practice. This is natural, since these laws are precisely those which hold when we are dealing with ordinary numbers, which always remain one of the most fruitful sets in which to work. The processes of union and intersection are used in chapter 11, and in chapter 9 when we deal with vector spaces we introduce a rather different type of process, but on the whole we will be concerned in this and later chapters with forming sums and products.

1.3. Groups

We give at once the abstract definition of a group. For the ideas which lead us to adopt this definition, and for more discussion on its meaning and significance, see volume 1. The numbering of the laws refers to volume 1 also.

Abstract definition of a group

A set S of elements forms a group if to any two elements x and y of S taken in a particular order there is associated a unique third element of S, called their product and denoted by xy, which satisfies the following laws:

M2 *For any three elements* x, y *and* z, $(xy)z = x(yz)$.
M3 *There is an element called the* 'neutral element' *and denoted by* e *which has the property that* $xe = ex = x$ *for all elements* x.
M4 *Corresponding to each element* x *there is an element* x^{-1} *called the* 'inverse' *of* x *which has the property that*

$$xx^{-1} = x^{-1}x = e.$$

Thus given a set and a product we need five things to make it a group.

(*a*) The product must be unique. This is usually implicit in its definition.

(*b*) The product must be in the set. This again is usually obvious from the definition.

(*c*) The Associative Law M2 must hold. This is nearly always obvious, possibly needing a little thought.

(*d*) There must be a neutral element. It is usually clear which element this must be and it can then easily be tested.

(*e*) There must be an inverse to *every* element. This again is not often difficult to identify.

It seems from the above that it is usually a straightforward matter to test whether a structure is a group or not, and this is generally the case.

Elementary consequences

The following elementary but important theorems are all easily proved from the definition, and the reader is referred to volume 1 (§10.6) for the detailed proofs (§11.3 for theorem 1.3.1).

Theorem 1.3.1. *The Associative Law extends to the product of more than three elements, i.e. the product $x_1 x_2 x_3 \dots x_n$ is independent of the order in which we perform the multiplications of pairs, provided that we keep the x's in the same relative position.*

Theorem 1.3.2. *The uniqueness of the neutral element.*

There cannot be two different elements each having the property possessed by e.

Theorem 1.3.3. *The uniqueness of the inverse.*

For any x there cannot be two different elements each with the property possessed by x^{-1}.

Theorem 1.3.4. *The inverse of e is e.*

Theorem 1.3.5. $(x^{-1})^{-1} = x$.

Theorem 1.3.6. *The Cancellation Law.*

$$xy = xz \Rightarrow y = z.$$

$$yx = zx \Rightarrow y = z.$$

Note. We need to prove both parts, since we do not assume commutativity.

Theorem 1.3.7. *The equation $ax = b$ has the unique solution $x = a^{-1}b$. The equation $xa = b$ has the unique solution $x = ba^{-1}$.*

The first is proved by multiplying both sides of the equation $ax = b$ *on the left* by a^{-1}, and the second by similar multiplication *on the right*. There is no question of excluding 0: in a group *every* element has an inverse.

The order of a group

If a group has a finite number of elements this number is called the *order* of the group, and the group is said to be of finite order. A group with an infinite number of elements is said to be of infinite order.

It is possible to have a group consisting of just one element e with $ee = e$. This group has order 1 and is sometimes known as the trivial group.

It might be expected that nearly all important groups were infinite, since the sets used in elementary algebra (such as the real or complex numbers or the integers are). This is not the case. While many infinite groups are extremely useful there is a large number of important finite ones, and these in fact are usually more interesting as groups. The reason is, roughly speaking, that in a finite group the structure must, after a certain stage, turn over upon itself and become intertwined: it cannot continue indefinitely along a straight course. Such intertwinings and foldings back may occur of course in infinite groups, but it is usually in the finite case that they are exhibited to the highest degree. Thus a great deal of the interest and usefulness of the subject lies in finite groups, so much so that many works deal with these almost exclusively.

Abelian groups

In our definition of a group we did not assume the truth of the Commutative Law, that $xy = yx$ for all x and y. The reason was that it is not essential to much of our algebra and we wish in fact to work with many structures within which it does not hold.

On the other hand many groups *are* commutative. Such are called *Abelian groups* (after the Norwegian mathematician N. H. Abel (1802–29) who anticipated some of the later work in group theory). They have of course properties which are not possessed by non-commutative groups, on the whole they are easier to work with, but in many ways they are not as interesting as the others.

It is usual, though not universal, to use the addition rather than the multiplication notation for Abelian groups. We then have the following four laws.

A1 *For any two elements x and y, $x+y = y+x$.*

(The Commutative Law.)

A2 *For any three elements x, y, and z,*

$$(x+y)+z = x+(y+z).$$

(The Associative Law.)

A3 *There is an element called the* 'zero', *and denoted by* 0 *which has the property that* $x+0 = 0+x = x$ *for all elements* x.

A4 *Corresponding to each element* x *there is an element* $-x$ *called the* 'negative' *of* x *which has the property that*

$$x+(-x) = (-x)+x = 0.$$

The elementary theorems 1.3.1–1.3.7 are likewise modified.

We will not often use the additive notation when dealing with groups, but in the case of other structures, such as rings and fields, which possess two basic operations, it will be in general use.

The multiplication table

For finite groups it is possible to write out all possible products of two elements in the form of a table. Suppose we have n elements. Then we take a table with n rows and n columns and place each element at the head of one row and one column, usually taking them in the same order for columns as for rows. In the space of the table which is the intersection of the row headed by x and the column headed by y we place the element xy. The table so formed has several properties and is particularly useful for groups of small order.

Abstract groups

The only thing that we need to know about a given group is the product of any pair of elements. A set of abstract symbols, representing elements and combining in a given way, is known as an *abstract group*, and it is with these that we work when investigating group properties, although we think of and use many concrete examples when dealing with groups. This is an example of our familiar mathematical idea of abstraction. We isolate certain aspects of our set (in this case the group aspect), work with a model which exhibits these aspects without the extraneous properties possessed by the original, and then apply our results to our given set.

The inverse of a product

Theorem 1.3.8. $(xy)^{-1} = y^{-1}x^{-1}$.

For
$$\begin{aligned}(y^{-1}x^{-1})(xy) &= y^{-1}(x^{-1}x)\,y \\ &= y^{-1}ey \\ &= y^{-1}y \\ &= e.\end{aligned}$$

Also
$$\begin{aligned}(xy)(y^{-1}x^{-1}) &= x(yy^{-1})\,x^{-1} \\ &= xex^{-1} \\ &= xx^{-1} \\ &= e.\end{aligned}$$

As an extension we have that $(xyz)^{-1} = z^{-1}y^{-1}x^{-1}$, proved in the same way. Similarly for any number of elements.

The above theorem is easy to prove, but the result is extremely important, being rather surprising and a common source of error. To find the inverse of a product of two or more elements we must take the product of the inverses *in the reverse order*.

Notice that for an Abelian group $y^{-1}x^{-1} = x^{-1}y^{-1}$ and the difficulty disappears.

Powers of an element

We define x^2 to mean xx, x^3 to mean xxx, and generally x^r, where r is any positive integer, to mean the product of r x's.

We now write x^{-r} for $(x^{-1})^r$, and we notice that this is the inverse of x^r, for

$$(x^r)^{-1} = (xx \ldots x)^{-1} = x^{-1}x^{-1} \ldots x^{-1} = (x^{-1})^r.$$

As in elementary algebra we define x^0 to be e, and we then have the index laws, for a proof of which see volume 1, p. 210.

Theorem 1.3.9. *The Index Laws*

If m, n are integers, positive negative or zero, we have

(i) $x^m x^n = x^{m+n}$; (ii) $(x^m)^n = x^{mn}$.

It is important to notice that powers of the same element always commute, since $x^m x^n = x^n x^m = x^{m+n}$.

Now let us consider the set of elements $e(= x^0)$, x, x^2, x^3, ..., where x is some element of our group. In an infinite group it is possible for all these to be different but in a group of finite order we must sooner or later have two which are the same element. Suppose $x^m = x^n$ where $m < n$. Then $x^{n-m} = x^n x^{-m} = e$.

Hence in a finite group some positive power of any element must equal the neutral element. If n is the least positive integer such that $x^n = e$ we say that n is the *order* of the element x. It easily follows that the n elements $e, x, x^2, \ldots, x^{n-1}$ are distinct.

It is obvious that $x^{rn+s} = x^s$ and that $x^m = e \Leftrightarrow m$ is a multiple of n, possibly the zero multiple. We see also that

$$x^{-1} = x^n x^{-1} = x^{n-1}, \quad x^{-2} = x^{n-2}, \text{ etc.},$$

and that $x^{-n} = e$, so that the order of x is the same as the order of x^{-1}.

The order of e is 1, and e is the only element of order 1.

The order of any element is not greater than the order of the group, since the n elements $e, x, \ldots, x^{n-1}$ are all distinct.

In an infinite group there may be no n such that $x^n = e$, in which case all elements $e, x^{\pm 1}, x^{\pm 2}, \ldots$ are distinct. In this case we say that x has infinite order. There may, however, be elements of finite order even in an infinite group.

1.4. Standard examples of groups

We now give a selection of groups which occur in practice or in other branches of mathematics. They include many which are important from the theoretical point of view and we draw upon these examples later in the book, where many of them appear again as other structures. The verification that a certain set is a group will usually be left to the reader: it is not usually difficult.

Examples from sets of numbers

The real numbers form a group under addition and also, if we except 0, under multiplication.

The complex numbers form a group under addition and, except 0, under multiplication.

The rationals form groups likewise, since the sum and product of two rationals are themselves rational.

The set of integers, positive, negative, or zero, clearly form a group under addition. This is an extremely important example, known as the *group of integers* or the *infinite cyclic group*, because of its similarity to the finite cyclic groups to be discussed later.

The set of all integers, positive, negative or zero, which are multiples of a fixed integer r form a group under addition, all such being isomorphic to the infinite cyclic group (for a discussion of *isomorphism* see chapter 2, §2.4).

The positive real numbers and the positive rationals both form groups under multiplication.

Residue classes: the cyclic groups

If n is a fixed positive integer, the set of all integers, positive, negative or zero, may be decomposed into equivalence classes such that any two integers are in the same class if and only if they are congruent modulo n, that is if and only if they leave the same remainder when divided by n. These classes are called *residue classes* modulo n. We can easily see that there are just n residue classes modulo n, a typical one consisting of all those integers that leave remainder r when divided by n, where r ranges over the values $0, 1, \ldots, (n-1)$. The class of the integers congruent to any of these r's is often denoted merely by r itself, when it is understood that we are working with residues. (The term 'residue class' is often shortened to 'residue'.)

Example. If $n = 4$ we obtain 4 residue classes modulo 4, namely 0, 1, 2, 3, where 0 consists of all integers divisible by 4, 1 all those that leave remainder 1 when divided by 4, and so on.

It is easily proved that we can add, subtract and multiply residues, in the following sense. If we add any member of the class r (working modulo some fixed n) to any member of the class s then we obtain a member of the class containing $r+s$, and the sum of the classes r and s is defined to be the class containing $r+s$ (note that this may not be the *class* $r+s$, since this may be greater than n, and we denote a class by its smallest positive member—the important thing is that all the sums are in the same residue class). Thus modulo 4, $2+3 = 1$. Similarly for subtraction, where $r-s$ is the class containing $r-s$, and multiplication, where rs is the class containing rs. Thus modulo 4, $2-3 = 3$ and $2.3 = 2$.

Division of residues is more complicated. It can be shown that r has an inverse if and only if r is prime to n. If r and n have a common factor then there is no inverse to r, and the equation $rx = s$ has no solution if s is not divisible by the H.C.F. of r and n, and has more than one solution if s is divisible by this H.C.F. Thus modulo 9, the residues 1, 2, 4, 5, 7, 8 have inverses, for example the inverse of 2 is 5 and that of 4 is 7, whereas 6 has no inverse: the equation $6x = 4$ has no solution, while the equation $6x = 3$ has solutions 2, 5 and 8.

Since it is always possible to add and subtract two residues, it is fairly clear that under addition the set of residues modulo n forms a group of order n, the neutral element being 0 and the inverse of r being $(n-r)$. This group is the *cyclic group* of order n, defined abstractly as follows.

A *cyclic group* is one which consists merely of powers of one of its elements, called a *generator*. If the generator x is of finite order n then the elements are $e, x, x^2, x^3, \dots x^{n-1}$, and of course $x^n = e$. All cyclic groups of order n are isomorphic, and we speak of *the* cyclic group of order n, denoting it by C_n. Note that this ensures that there exists at least one group of each finite order.

In the additive notation the elements are

$$e, x, 2x, 3x, \dots (n-1)x \quad \text{with} \quad nx = e.$$

Thus we see that the group of residues modulo n under addition is the cyclic group C_n, where 1 is a generator.

The cyclic groups are Abelian. We note the similarity with the infinite cyclic group, which consists of all powers of a generator which has infinite order (the integer 1).

Groups of residues under multiplication

Since division of residues is not always possible we cannot always form a group of residues under multiplication. But if we take the set of residues prime to n each will have an inverse in the set, and we still include the neutral element 1. Furthermore, if r and s are both prime to n so is rs, and so the product of any two elements is still in the set. Hence the residues prime to n form a group under multiplication.

If n is prime this group includes all elements except 0, i.e. its order is $(n-1)$. The order of the group, the number of numbers less than n prime to n, is denoted by $\phi(n)$ and is known as *Euler's function*.

These groups are interesting, though by no means so important as the cyclic groups, since their structure can be fairly complicated, and haphazard.

Groups of polynomials

All polynomials in a single variable x clearly form a group under addition, the neutral element being the polynomial 0 and the inverse of $P(x)$ being $-P(x)$.

In particular all polynomials of the nth degree form a group under addition, where we allow any coefficient or coefficients to be zero, including the coefficient of x^n. This group is isomorphic to the group of $(n+1)$-dimensional vectors. The coefficients may of course be the elements of any group (with the addition notation).

The set of rational functions, excluding 0, form a group under multiplication.

Vectors

n-dimensional vectors form a group under addition.

Permutations

If we have a set of distinct objects a *permutation* of them is a re-arrangement of them among themselves. Thus if we have n objects we label them $1, 2, 3, \ldots, n$, where no arithmetical meaning is to be attached to the integers—they are merely labels and express the fact that the objects are distinguishable from one another. A permutation then replaces the numbers (i.e. the objects) $1, \ldots, n$ by themselves in a different order. If it replaces i by a_i we write it in the form

$$\begin{pmatrix} 1 & 2 & 3 & \ldots & n \\ a_1 & a_2 & a_3 & \ldots & a_n \end{pmatrix}.$$

If we have two permutations on n objects we may define their product to be the permutation obtained by applying the first and following it by the second. Thus if x sends i into a_i, and y sends a_i into b_i, the product xy (in that order) is the permutation which sends i into b_i. This is of course a permutation. Symbolically

$$x = \begin{pmatrix} 1 & 2 & \dots & n \\ a_1 & a_2 & \dots & a_n \end{pmatrix}, \quad y = \begin{pmatrix} a_1 & a_2 & \dots & a_n \\ b_1 & b_2 & \dots & b_n \end{pmatrix},$$

$$xy = \begin{pmatrix} 1 & 2 & \dots & n \\ a_1 & a_2 & \dots & a_n \end{pmatrix}\begin{pmatrix} a_1 & a_2 & \dots & a_n \\ b_1 & b_2 & \dots & b_n \end{pmatrix} = \begin{pmatrix} 1 & 2 & \dots & n \\ b_1 & b_2 & \dots & b_n \end{pmatrix},$$

where we are using the fact that the top row in the permutation y may be written in any order, so long as we make the corresponding changes in the bottom row.

It is easily seen that with this definition of product the set of all $n!$ permutations on n objects forms a group. This is called the *symmetric group* of degree n, and is written as S_n. The neutral element is the identity permutation, which leaves each object unchanged, and if x is the permutation which sends i into a_i then its inverse is that which sends a_i into i. The symmetric groups is non-Abelian for $n \geqslant 3$.

The symmetric group of degree 1 has just 1 element and is the trivial group. That of degree 2 has two elements,

$$e = \begin{pmatrix} 1 & 2 \\ 1 & 2 \end{pmatrix} \quad \text{and} \quad a = \begin{pmatrix} 1 & 2 \\ 2 & 1 \end{pmatrix} \quad \text{with} \quad a^2 = e,$$

and is therefore isomorphic to the cyclic group of order 2, i.e. is the same abstract group. The symmetric group of degree 3 has 6 elements. Let us denote

$$\begin{pmatrix} 1 & 2 & 3 \\ 2 & 3 & 1 \end{pmatrix} \quad \text{and} \quad \begin{pmatrix} 1 & 2 & 3 \\ 1 & 3 & 2 \end{pmatrix}$$

by a and b respectively. Then the other elements can be seen to be a^2, ba, ba^2 and of course e, and it is easily verified that $a^3 = e$, $b^2 = e$ and $ab = ba^2$.

Even and odd permutations

Any permutation may be expressed as a product of transpositions, where a transposition is a permutation which interchanges two of the objects but leads the others unaltered. Thus

$$\begin{pmatrix} 1 & 2 & 3 & 4 \\ 1 & 4 & 3 & 2 \end{pmatrix}$$

is a transposition. The expression of a given permutation in this way will not be unique, but it can be proved (see volume 1, §12.6 for details) that there will always be either an even number of transpositions in the product, or else an odd number. Permutations requiring an even number of transpositions are called *even permutations*, those requiring an odd number are known as *odd permutations*. It follows that the product of two even or two odd permutations is even, whereas the product of an odd and an even is odd. Since the identity permutation is even, the inverse of an even one must be even, and so the set of even permutations on n objects forms a group, called the *Alternating Group of degree n*, and containing (as may be proved) exactly half the possible $n!$ permutations.

1.5. Transformation groups

An important type of group is that consisting of all possible transformations of a figure. We define a transformation to be a movement of space which leaves our figure occupying the same position, two movements which have the same effect being taken to be the same. The product of two transformations is defined to be that which has the same effect as when we apply one transformation and follow it by the second. Under this definition the set of all possible transformations on any figure always forms a group (product is associative), there is a neutral element (the identity transformation which leaves the figure unaltered) and there is an inverse to each transformation (that which has the reverse effect). Such groups give a measure and a description of the symmetry of our figure and provide a way of isolating the symmetry properties. They are particularly useful in the case of regular polygons, solids or figures of higher dimension: figures which possess some symmetry have non-trivial transformation groups, while for an object with no symmetry the group of transformations consists merely of the trivial group $\{e\}$.

Transformations of a one-faced polygon

Consider a regular n-gon which can move in its plane, but which we imagine cannot be turned over. The only transformations are rotations clockwise or anti-clockwise through multiples of $2\pi/n$ radians. If we call that which rotates through $2\pi/n$ clockwise a we see that $a, a^2, \ldots, a^{n-1}$ give different transformations, while $a^n = e$. Thus the group is the cyclic group C_n, a being a generator.

Transformations of a circle

The transformation group of a one-sided circle is the group of real numbers under addition, modulo 2π, where the number x corresponds to a rotation through the angle x clockwise, say.

Transformations of a two-faced polygon: the dihedral groups

We take a regular n-gon whose two faces are identical, and we allow movements in three dimensions, so that transformations which turn the polygon over are allowed. There are $2n$ possible transformations. n of these arise from rotations which do not turn the polygon over (as in the one-faced case), while the other n turn it over and rotate it. It is possible to obtain all the second set by a turning about some radius (i.e. a line joining the centre of the polygon to a vertex) fixed *in space* followed by a rotation in the plane of the figure. Let us call the transformation which rotates clockwise through $2\pi/n$, a, and let b be the element which turns the polygon over by a rotation through π about a fixed radius, which we will take for simplicity to lie up the page in our examples. The n elements which do not turn the polygon over are, as in the one-sided case, $e, a, a^2, \ldots, a^{n-1}$ with $a^n = e$. The other n elements are $b, ba, ba^2, \ldots, ba^{n-1}$. We also have the relation that $b^2 = e$. We can prove (see figure 1) that $ab = ba^{n-1}$, and the group is described completely by the three relations $a^n = e$, $b^2 = e$, $ab = ba^{n-1}$, where a and b are elements and b is not a power of a. The third relation enables us to put any product of a's and b's in one of the forms a^r or ba^r: in other words turning over first (if necessary) and then rotating.

The group is known as the *dihedral group of order* $2n$. It is non-Abelian for $n \geqslant 3$, and is denoted by D_{2n}.

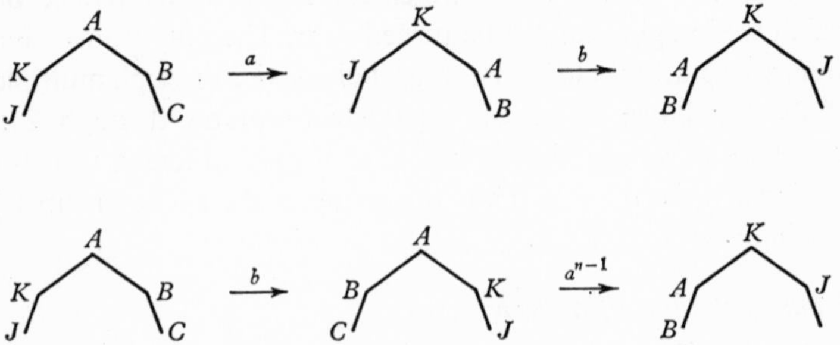

Fig. 1

The dihedral group of order 4 has elements e, a, b, $ab = c$, where $a^2 = b^2 = e$ and $ab = ba$. It is Abelian and every element has order 2 (except e of course). It is sometimes known as the *Vierergruppe*, and its multiplication table is given in figure 2.

	e	a	b	c
e	e	a	b	c
a	a	e	c	b
b	b	c	e	a
c	c	b	a	e

Fig. 2

The dihedral group of order 6 has the relations $a^3 = b^2 = e$ and $ab = ba^2$, and is thus identical to the symmetric group of degree 3. It is non-Abelian, and in fact is the smallest non-Abelian group. Its multiplication table is given in figure 3.

	e	a	a^2	b	ba	ba^2
e	e	a	a^2	b	ba	ba^2
a	a	a^2	e	ba^2	b	ba
a^2	a^2	e	a	ba	ba^2	b
b	b	ba	ba^2	e	a	a^2
ba	ba	ba^2	b	a^2	e	a
ba^2	ba^2	b	ba	a	a^2	e

Fig. 3

Transformation groups of the regular polyhedra

It was proved in volume 1 that the group of the regular tetrahedron has 12 elements and is the alternating group of degree 4 (i.e. the group of even permutations on 4 objects). It was also shown that the groups of the octahedron and cube are both the symmetric group of degree 4, having 24 elements, and it was stated without proof that the dodecahedron and icosahedron have transformation groups of 60 elements which are the alternating group of degree 5.

1.6. Direct Products

Suppose we have any two groups G and H and consider the set of all ordered pairs (x, y) where $x \in G$ and $y \in H$. Define a product of two pairs by $(x_1, y_1)(x_2, y_2) = (x_1x_2, y_1y_2)$, the products x_1x_2 and y_1y_2 being those defined in G and H respectively. Then under this product the set of ordered pairs forms a group. For the product is unique and in the set. It is associative since the products defined in G and H are: we see that

$$(x_1, y_1)(x_2, y_2)(x_3, y_3) = (x_1x_2x_3, y_1y_2y_3).$$

If e and f are the respective neutral elements of G and H it is easily seen that (e, f) acts as a neutral element in our set of pairs, while the inverse of (x, y) is given by (x^{-1}, y^{-1}).

The group we have defined is known as the *direct product* of G and H and is denoted by $G \times H$. It gives us a powerful method of forming new groups from known ones and, in more advanced work, of investigating the structure of a given group. In volume 1 we called this group the *direct sum* and wrote it as $G \oplus H$, but we now use the more usual notation.

G and H may be any groups, one or both finite, one or both Abelian. If they are both finite of orders m and n, the order of $G \times H$ is mn. $G \times H$ is Abelian if and only if both G and H are.

We may extend the idea to n groups $G_1, G_2, \ldots, G_n$, by defining the direct product $G_1 \times G_2 \times \ldots \times G_n$ to be the set of all ordered n-tuples $(g_1, g_2, \ldots, g_n)$, $g_i \in G_i$, with product defined as

$$(g_1, g_2, \ldots, g_n)(g_1', g_2', \ldots, g_n') = (g_1g_1', g_2g_2', \ldots, g_ng_n').$$

1.7. Generators

In describing the dihedral group of order 6 we have used the fact that all its elements may be expressed in terms of two of them, a and b, and that these satisfy the equations $a^3 = b^2 = e$ and $ab = ba^2$.

A set of elements of a group which have the property that *all* the elements may be expressed in terms of them or their inverses is called a *set of generators* for the group. If none of them can be expressed in terms of the others they are a *minimal set* of generators, and in this case we cannot omit any of them and still have a set of generators.

Any finite group possesses a set of generators, viz. the set of all elements of the group and this may be reduced to a minimal set by omitting any that are expressible in terms of the rest. The minimal set need not be unique: in the dihedral group D_6 both $\{a, b\}$ and $\{a, ba\}$ have the property and there are other pairs.

An infinite group may or may not possess a finite set of generators.

1.8. Subgroups

A non-empty subset H of a group G is a *subgroup* of G if:

(a) Given any ordered pair (h_1, h_2) of elements of H the product $h_1 h_2$ (as defined in G) is in H.

(b) The neutral element of G is in H.

(c) Given any element h in H the inverse h^{-1} (as defined in G) is in H.

Thus a subset of elements of G is a subgroup if and only if it forms a group, under the same group product.

Note that (b) is in fact superfluous, since if $h \in H$, $h^{-1} \in H$ by (c) and so $hh^{-1} = e \in H$ by (a).

The subgroups $\{e\}$ and G itself are called *trivial* subgroups: any others are *non-trivial* or *proper*.

Theorem 1.8.1. *A necessary and sufficient condition for a non-empty subset H to be a subgroup of G is that $gh^{-1} \in H$ for all $g, h \in H$.*

Necessary. If H is a subgroup $h^{-1} \in H$ and so $gh^{-1} \in H$.

Sufficient. If g is any element of H we have $gg^{-1} \in H$, i.e. $e \in H$. Hence $eg^{-1} = g^{-1} \in H$, i.e. H contains inverses. Thus if $g, h \in H$, so is h^{-1} and hence $g(h^{-1})^{-1} \in H$, i.e. $gh \in H$.

The subgroup of powers of an element

If g is an element of G the powers of g form a subset of G. This is always a subgroup, since $g^m . g^n = g^{m+n}$, and is in the set. $e = g^0$ is in it, as is $(g^n)^{-1} = g^{-n}$.

If g has infinite order the subgroup of its powers is the infinite cyclic group, while if g has finite order n the subgroup of powers of g is the finite cyclic group C_n, of order n. We see here the connection between the two uses of the word 'order', as applied to groups and to elements.

Subgroups generated by subsets of elements

If we have any subset of elements in a group we consider the set of all possible products formed by writing down finite strings of these elements or their inverses in any order, with repetitions if we wish. Thus considering the subset $\{g, h, k\}$ we would include elements such as $hk^{-1}k^{-1}gkkkh^{-1}g^{-1}$. If our elements commuted all such could be written in the form $g_1^{\lambda_1} g_2^{\lambda_2} \dots g_r^{\lambda_r}$ where our subset consists of $g_1, g_2, \dots, g_r$ and the λ_i's are positive, zero or negative integers; but for non-commutative elements more complicated expressions will arise. All products formed in the way described must be in the original group, though they will not all be different. The set of all such is a subset of the group, and is always a subgroup, as may be seen after a moment's thought. We call this subgroup the *subgroup generated by our subset*. It may of course be the whole group.

The subgroup generated by a subset is the smallest subgroup that contains that subset, and is the intersection of all subgroups that contain the subset.

Frobenius notation for subsets of a group

We denote subsets of the set of elements of a group G by capital letters, usually H, K, etc. In addition to the normal set structure connecting these subsets we may take the group product of any element of H with any of K. If we do this over all elements of H and K we obtain a collection of elements which is itself a subset of G, and we introduce the notation HK for this set. It may be called the *product* of H and K. Thus HK means the set of all elements hk for $h \in H$ and $k \in H$. Note that the same element may be given by two or more different pairs hk.

A subset of a group is often called a *complex*. Product of complexes is associative, since $H(KL) = (HK)L = \{hkl\}$, $h \in H, k \in K, l \in L$. It is not in general commutative but may be in special cases. It is important to realise that $HK = KH$ does *not* mean that all pairs of elements of H and K commute, i.e. that $hk = kh$ for all $h \in H$ and $k \in K$: it merely expresses the fact that if we form all products hk we obtain the same elements as if we form products kh, but they need not be given in the same order.

One of H and K may be a single element g. We then write gH to mean the set of elements $\{gh\}$ as h varies over H, or Hg to be the set $\{hg\}$.

Extending the notation we write H^{-1} to mean the set $\{h^{-1}\}$ for $h \in H$. We also denote HH by H^2 with obvious extensions for higher powers.

Theorem 1.8.2. *If H is a subgroup, $H^2 = H$ and $H^{-1} = H$.*

Since H is a subgroup, $hh' \in H$ if h and h' are, and so $H^2 \subseteq H$. But since $h = he$ and $e \in H$, $H \subseteq H^2$. Hence $H^2 = H$.

$h \in H \Rightarrow h^{-1} \in H$, therefore $H^{-1} \subseteq H$. Hence $(H^{-1})^{-1} \subseteq H^{-1}$, i.e. $H \subseteq H^{-1}$. Thus $H^{-1} = H$.

Theorem 1.8.3. *If H is non-empty, $H^2 = H$ and $H^{-1} = H$ together imply that H is a subgroup.*

If h, k are in H we have hk and h^{-1} are both in H, and so $hh^{-1} = e$ is in it. Hence H is a subgroup.

Theorem 1.8.4. *If G is finite and H non-empty*

$$H^2 \subseteq H \Rightarrow H^{-1} \subseteq H, \quad \textit{and} \quad H^2 \subseteq H$$

is sufficient to make H a subgroup.

For any $h \in H$, $h^2 \in H^2$ and so $\in H$, $h^3 \in H$, etc. Thus if the order of h is n, $h^{-1} = h^{n-1} \in H$, i.e. $H^{-1} \subseteq H$. Thus for any $h, h' \in H$, $(h')^{-1} \in H$ and so $h(h')^{-1}$ is, making H a subgroup.

So we have the result that for a finite group G a necessary and sufficient condition for a non-empty complex H to be a subgroup of G is that $H^2 \subseteq H$.

Theorem 1.8.5. *If H and K are subgroups so is $H \cap K$.*

If g and h are in $H \cap K$ they are both in H and both in K. Hence, since H and K are subgroups, gh^{-1} is in H and in K, and so in $H \cap K$. Also $H \cap K \ni e$ and so is non-empty. Hence $H \cap K$ is a subgroup by theorem 1.8.1.

Notice that $H \cup K$ need not be a subgroup, since if g and $h \in H \cup K$ there is no guarantee that they are both either in H or in K, and we can say nothing about gh^{-1}.

We can similarly show that the intersection of any finite number of subgroups is itself a subgroup.

1.9. Cosets and Lagrange's theorem

In the group of two-dimensional vectors a line through the origin forms a subgroup, but a line not through O does not. Such a line obviously bears some resemblance to one through O, particularly to its parallel, and all lines parallel to a given one, Γ, through O form a decomposition of the whole group of vectors into subsets similar to Γ though not subgroups. The condition for two vectors **a** and **b** to lie on the same line is that $(\mathbf{a}-\mathbf{b})$ is a vector on Γ. We have an analogous idea in vectors of three or more dimensions. In three dimensions we may decompose the whole group either into lines parallel to a line Γ through O, or into planes parallel to a plane π through O, and in either case the condition for two vectors to belong to the same element of the decomposition is that their difference is a vector of Γ or π.

A similar idea is seen in the group of integers under addition. The set of integers divisible by n forms a subgroup H. Now consider all integers which have remainder r when divided by n, that is the set $\{r+\alpha n\}$, for a variable integer α (i.e. the residue class r modulo n). This set is in some respects similar to H—it consists of the elements of H plus r and is in a sense 'parallel' to H. It is not a subgroup. As r takes successively the values $0, 1, \ldots, (n-1)$ we obtain a decomposition of the whole group of integers into a collection of these 'parallel' subsets, and the condition that two integers belong to the same subset is that their difference is divisible by n, i.e. is in H.

In both the above examples we took a subgroup and formed a decomposition of our whole group into subsets 'parallel' to H, two elements being in the same subset if their difference was in H. The idea may be generalised to any group and subgroup of it. The 'difference' becomes in the product notation the quotient of two elements g and k, gk^{-1} or $k^{-1}g$. We may take either of these, but will find it convenient to use the latter. Thus we will decompose our group G into subsets such that g and k are in the same one if $k^{-1}g \in H$. The subsets are known as *cosets*. To prove that we always have a true decomposition we use the idea of equivalence relations.

Thus if G is any group with H a subgroup we say that gRk if $k^{-1}g \in H$.

Theorem 1.9.1. *The relation defined above is an equivalence relation.*

Reflexive. We always have gRg since $g^{-1}g = e \in H$ since H is a subgroup.

Symmetric. If gRk then $k^{-1}g \in H$. Hence its inverse is in H, i.e. $g^{-1}k \in H$ and so kRg.

Transitive. If gRk and kRl then $k^{-1}g \in H$ and $l^{-1}k \in H$. Hence $(l^{-1}k)(k^{-1}g) = l^{-1}g \in H$ and so gRl.

Note that in the above we have used precisely those properties of H that make it a subgroup—for the work we are doing at the present it is vital that H is a subgroup.

By the equivalence classes theorem (theorem 1.1.1) we have a decomposition of G into a set of equivalence classes, mutually exclusive, such that two elements g and k are in the same class if and only if $k^{-1}g \in H$. These classes are called the *left cosets of G relative to H.*

We may consider cosets from a slightly different point of view. The coset containing a given element g, i.e. the equivalence class containing g, consists of those elements k such that

gRk, i.e. $k^{-1}g \in H$. But this is true if and only if $g^{-1}k \in H$, i.e. $k = gh$ for some $h \in H$. Hence the coset containing g is precisely the complex gH, the set $\{gh\}$ for $h \in H$. We thus have the following theorem.

Theorem 1.9.2. *The left cosets of G relative to H are the complexes gH, $g \in G$. Any left coset may be expressed in this form for any g in it. Any two cosets are either the same or have no element in common.*

We may similarly define *right cosets* of G relative to H as the equivalence classes under the relation given by $gk^{-1} \in H$. They are the complexes Hg.

Although in general, for non-Abelian groups, left and right cosets do not coincide their properties are the same and either may be used. It is advisable to restrict oneself to either one or the other, and so we will usually use left cosets, and will call them merely *cosets*.

The study of cosets leads immediately to Lagrange's theorem, which is one of the most fundamental in the whole subject of group theory. The study of a group is largely the study of its subgroups. To discover all subgroups of a given group is not a simple task, but we can limit our search in the finite case by applying Lagrange's theorem.

Theorem 1.9.3. *Lagrange's theorem.*

The order of any subgroup of a finite group is a factor of the order of the group.

Suppose H is a subgroup of G, where the order of G is n and that of H is m. Then we wish to prove that m is a factor of n.

The whole group G may be decomposed into cosets gH relative to H. The number of elements in each coset must be m, since the coset gH is formed by taking all elements of H and pre-multiplying them by g, and all the elements so formed are different by the Cancellation Law. But the cosets are completely disjoint. Hence if there are r cosets we have that $n = m.r$, and so m is a factor of n.

The number r is called the *index* of H in G.

Theorem 1.9.4. *The order of any element of a finite group is a factor of the order of the group.*

Consider the subgroup of powers of the element. Its order is equal to that of the element, and the result follows by Lagrange's theorem.

Theorem 1.9.5. *If x is any element of a finite group G, where G is of order n, $x^n = e$.*

Let x have order m. Then $n = m.r$ and $x^m = e$. Hence $x^n = (x^m)^r = e$.

Theorem 1.9.6. *The only group of order p, where p is prime, is the cyclic group of order p.*

Take any non-neutral element x. The order of x must be a factor of p, by theorem 1.9.4. But p is prime, and so x has order p. Thus the subgroup generated by x must be the whole group, which is therefore cyclic.

Note that we have proved the fairly obvious fact that any non-neutral element is a generator.

Worked exercises

1. Suppose a, b are elements of a group and that $a^5 = e$, $b^4 = e$, $ab = ba^3$. Prove (i) $a^2b = ba$; (ii) $ab^3 = b^3a^2$.

(i) $a^2b = a.ab = aba^3 = ba^3a^3 = ba^6 = ba$, since $a^5 = e$.

(ii) Multiplying both sides of (i) by b^3 both in front and behind, we obtain $b^3a^2b^4 = b^4ab^3$.

Hence $b^3a^2 = ab^3$ as required, since $b^4 = e$.

2. Identify the group of residues modulo 16 prime to 16 under multiplication.

The residues prime to 16 are 1, 3, 5, 7, 9, 11, 13, 15. Let $a = 3$. Then $a^2 = 9$, $a^3 = 11$, $a^4 = 1$. Let $b = 7$. Then $b^2 = 1$, $ab = 5$, $a^2b = 15$, $a^3b = 13$. Hence the 8 elements are e, a, a^2, a^3, b, ab, a^2b, a^3b with $a^4 = b^2 = e$ and of course $ab = ba$ since the group is Abelian.

Thus the group is isomorphic to $C_4 \times C_2$.

3. List all subgroups of $C_2 \times C_2 \times C_2$.

Let the generators be a, b, c. The order of any subgroup other than the trivial ones must be 2, or 4, since it is a factor of 8, the order of the whole group. The subgroups of order 2 must consist of e and an element of order 2. But every element has this order, and so we obtain seven subgroups each consisting of e and one other element.

Consider the subgroups of order 4. Since every element has order 2, all such subgroups must be isomorphic to the Vierergruppe (the only other group of order 4 was proved in volume 1 to be C_4, which contains elements of order 4), and it is easy to see that there are just seven such, generated by a, b; a, c; b, c; a, bc; b, ca; c, ab; ab, bc.

Exercises 1A

Are the relations in **1–3** equivalence relations? If not, state which laws are not satisfied; if they are give the equivalence classes.

1. In the set of real numbers aRb means $a \leqslant b$.

2. In the set of real numbers aRb means $a-b$ is an integer.

3. In the set of all circles in a plane aRb means a and b meet in at least one real point.

Which of the sets **4–6** are groups under the given operations? If the set is a group give the neutral element and the inverses; if it is not state which laws are not satisfied.

4. All numbers of the form 2^n where n is an integer, positive, negative or zero, under multiplication.

5. All vectors of the form either $(0, y)$ or $(x, 0)$ under vector addition.

6. The elements 0 and 1 with product defined as $0.0 = 0, 0.1 = 0, 1.0 = 0, 1.1 = 1$.

7. Construct the multiplication table for the group of residues modulo 6 under addition, and give the orders and inverses of each element.

8. If, in a group, $b = x^{-1}ax$ show that $b^2 = x^{-1}a^2x$ and hence show that a and b have the same order.

9. Prove that the elements xy and yx in a group have the same order.

10. Give all the single generators, and the orders of all other elements, of the groups C_5, C_8, C_{12}.

11. Give the multiplication tables for the following groups of residues (except 0) under multiplication and identify the groups: (i) modulo 7, (ii) modulo 11.

12. Identify the following groups of residues modulo n prime to n under multiplication:

$$\text{(i) } n = 8, \quad \text{(ii) } n = 10, \quad \text{(iii) } n = 14.$$

13. For D_{12} prove the following from the relations $a^6 = b^2 = e$, $ab = ba^5$ and illustrate by figures showing the transformations:

$$\text{(i) } (ba^2)^2 = e, \quad \text{(ii) } aba = b, \quad \text{(iii) } ba^3b = a^3.$$

14. In S_4, if
$$a = \begin{pmatrix} 1 & 2 & 3 & 4 \\ 2 & 3 & 4 & 1 \end{pmatrix}, \quad b = \begin{pmatrix} 1 & 2 & 3 & 4 \\ 1 & 3 & 2 & 4 \end{pmatrix},$$
find (i) ab, (ii) ba, (iii) a^2b, (iv) $(ab)^2$, (v) aba^3.

15. Express the following as products of transpositions and hence determine whether they are even or odd:
$$\text{(i)} \begin{pmatrix} 1 & 2 & 3 & 4 \\ 2 & 1 & 4 & 3 \end{pmatrix}, \quad \text{(ii)} \begin{pmatrix} 1 & 2 & 3 & 4 & 5 \\ 3 & 4 & 2 & 1 & 5 \end{pmatrix}, \quad \text{(iii)} \begin{pmatrix} 1 & 2 & 3 & 4 & 5 & 6 \\ 4 & 6 & 3 & 5 & 1 & 2 \end{pmatrix}.$$

16. Prove that an odd permutation must have even order.

17. What are the transformation groups of the following: (i) a rhombus; (ii) a rectangular box; (iii) a prism whose cross-section is a regular pentagon?

18. Give the multiplication tables for $C_4 \times C_2$, $C_2 \times C_2 \times C_2$. Is either of these isomorphic to C_8?

19. Prove that $C_5 \times C_2$ is isomorphic to C_{10}.

20. Give a minimal set of generators for $C_4 \times C_2$.

21. Is the set of all transformations of a cube which leave one, or both, of two given vertices fixed a subgroup of the group of transformations of the cube?

22. List all the subgroups of D_6 and give their indices.

23. Repeat **22** for $C_3 \times C_2$.

24. Show that if H is a subgroup of G, then $\{g^{-1}Hg\}$, for any $g \in G$, is a subgroup isomorphic to H but not necessarily distinct from it.

25. Prove that for any complexes H, K, L of G, $H(K \cap L) \subseteq HK \cap HL$. Give an example where the strict inclusion is true.

26. Give the left and right cosets of S_4 relative to the subgroup
$$\left\{\begin{pmatrix} 1 & 2 & 3 & 4 \\ 1 & 2 & 3 & 4 \end{pmatrix}, \begin{pmatrix} 1 & 2 & 3 & 4 \\ 2 & 1 & 4 & 3 \end{pmatrix}, \begin{pmatrix} 1 & 2 & 3 & 4 \\ 3 & 4 & 1 & 2 \end{pmatrix}, \begin{pmatrix} 1 & 2 & 3 & 4 \\ 4 & 3 & 2 & 1 \end{pmatrix}\right\}.$$

27. Show by giving a counter example that if H, K, L are complexes of a group G, $HK = HL$ does not necessarily imply that $K = L$. Prove that $gK = gL \Rightarrow K = L$.

28. If H is a subgroup and K a complex with $K \subseteq H$, show that
$$HK = H = KH.$$
Does this result hold if H is not a subgroup?

29. Prove that if H and K are subgroups whose orders are r and s where r and s are co-prime then $H \cap K = \{e\}$.

Exercises 1B

1. Show that if R is a relation such that (i) $aRa \; \forall \; a$ and (ii) aRb and $bRc \Rightarrow cRa$, then R is an equivalence relation.

2. In the set of points (x, y) in a plane let $(a, b)\, R(c, d)$ mean that

$$a+d = b+c.$$

Prove that R is an equivalence relation and find its equivalence classes.

3. Suppose we have a set S in which a unique product is defined which is associative. Suppose also that $\exists$ a left neutral element (i.e. an element e such that $ex = x \; \forall \; x \in S$) and a left inverse of every element x (i.e. an element y such that $yx = e$). Prove that e is also a right neutral element, y is a right inverse and that S is a group under the given product.

4. Suppose we have a *finite* set S with an associative product defined. Suppose also that the Cancellation Law (theorem 1.3.6) holds. Prove that M3 and M4 are true and hence that S is a group.

5. Prove that if x, y, xy are all of order 2 then $xy = yx$. Hence show that if every non-neutral element of a group is of order 2 the group is Abelian.

6. In a group of even order prove that there is at least one element of order 2. [*Hint*. Consider pairs consisting of elements with their inverses.]

7. Find a generator of the group of residues modulo 23 under multiplication. Hence show that the group is C_{22} and find an element of order 2 and one of order 11.

8. Consider an infinite straight line with points marked along it at unit intervals. Let a be the transformation which moves all points of space 1 unit to the right and let b rotate it though π about a point of space corresponding to one of the marked points. Show that the group of transformations which leave the marked points occupying the same positions is generated by a and b with the properties (i) a has infinite order; (ii) $b^2 = e$; (iii) $aba = b$. Prove from (iii) that $a^r b a^r = b$ and verify this geometrically. Why is the group not $Z \times C_2$, where Z is the infinite cyclic group?

9. If we have 2 regular tetrahedra $ABCD$ and $A'B'C'D'$ interlocking symmetrically (so that AB and $C'D'$ intersect and bisect each other, and similarly for the other pairs of sides) the resulting solid is called a *stella octangular*. Show that its transformation group is isomorphic to S_4.

10. If p and q are co-prime prove that $C_p \times C_q$ is isomorphic to C_{pq}.

11. If p and q are not co-prime prove that the order of every element of $C_p \times C_q$ is a factor of the L.C.M. of p and q, and deduce that $C_p \times C_q$ is not isomorphic to C_{pq}.

12. Show that the group of two-dimensional vectors may be expressed as the direct product of *any* two lines through O, and that the group of three-dimensional vectors may be expressed either as the direct product of any three non-coplanar lines through O, or as the direct product of any plane through O and any line through O not in that plane.

13. By use of Lagrange's theorem and its corollaries prove Fermat's theorem that if p is prime and x is prime to p then $x^{p-1} \equiv 1 \pmod{p}$. Prove further that if n is not prime, but x is prime to n, then $x^{\phi(n)} \equiv 1 \pmod{n}$, where $\phi(n)$ is Euler's function (the number of numbers less than n prime to n).

14. If A_i $(i = 1, \ldots, r)$ are subgroups of a group G, show that $B = \bigcap_{i=1}^{n} A_i$ is the greatest common subgroup of $A_1, \ldots, A_r$ in the following sense:

(i) B is a subgroup of each A_i, proper or improper but not empty;

(ii) any common subgroup of the A_i's is a subgroup of B.

[Note the analogy with H.C.F.]

15. In the group of integers let H_r denote the subgroup generated by r, where r is a positive integer. Prove

(i) $H_r H_s$ is generated by r and s and is H_λ where λ is the H.C.F. of r and s.

(ii) $H_r \cap H_s = H_\mu$ where μ is the L.C.M. of r and s.

State and prove analogous results in the cyclic group of order n, and in this case show further that the index of H_λ is the L.C.M. of the indices of H_r and H_s, while the index of H_μ is their H.C.F.

16. In an Abelian group G let the subgroups H and K be generated by the sets of elements $\{h_1, \ldots, h_r\}$ and $\{k_1, \ldots, k_s\}$ respectively. Show that HK is a subgroup and is generated by $\{h_1, \ldots, h_r, k_1, \ldots, k_s\}$, which need not be a minimal set.

17. Suppose H, K are subgroups of a group G, of orders r and s, respectively. Let $L = H \cap K$ have order d. Prove that if the cosets of H relative to L are $h_1 L, \ldots, h_\lambda L$ then $HK = h_1 K \cup \ldots \cup h_\lambda K$. Show further that no two of the $h_i K$'s have elements in common and deduce that the number of elements in the complex HK is rs/d.

2

HOMOMORPHISMS OF GROUPS

2.1. Mappings

This section is a brief résumé of part of chapter 9 of volume 1, and the reader who has that volume to hand should re-read the chapter in question. On the other hand, the inclusion of the material again here makes the succeeding work on homomorphisms self-contained.

A *mapping* is an extension of the idea of a function to general sets. The function $y = f(x)$ associates a number y to each number x, where the 'numbers' x and y are usually real but may be complex, or indeed restricted to the integers or other subsets of the complex numbers. To generalise this idea, a *mapping* of the set A into the set B associates a unique element b (in B) to each element a in A. The sets A and B may be the same or different. For a mapping to be properly defined we must have a unique b associated with *each* element of A. The elements b may be all different or not, so that two or more elements of A may give rise to the same element of B. But we must not have two b's corresponding to the same a.

Notation

A is called the *object space* of the mapping, B is the *image space*, a is the *object* whose *image* is b, and we say that we have a mapping of A *into* B.

If the mapping is denoted by θ, we write $\theta: A \to B$, and we say that $a\theta = b$. (Some writers use the notation $\theta(a)$, but our method is often more convenient, particularly in more advanced work.) We sometimes say that $a \to b$.

Inverse mappings

By analogy with inverse functions (e.g. sin and $\sin^{-1}$, exp and log), the inverse of θ: $A \to B$ is a mapping $\phi: B \to A$ such that if $a\theta = b$, $b\phi = a$ and if $b\phi = a$, $a\theta = b$. Such a mapping exists only when every element of B is the image of one and only one element of A. We thus need two conditions on θ.

(1) Each element of B must be the image of some $a \in A$, i.e. the images of the elements of A must together cover the whole of B. If this is the case we say that the mapping θ is *onto* B. (Distinguish carefully between the terms *into* and *onto*: all mappings are into the image space, but are onto only if the images cover the whole of the image space.)

(2) Any element of B must be the image of only one element of A (i.e. if $a\theta = a'\theta$ then $a = a'$). A mapping with this property is said to be 1–1 (one-to-one).

If both these conditions exist then θ has an inverse mapping, viz. that mapping of B into A which sends an element of B into the element of A which gave rise to it under θ. The inverse mapping is denoted by θ^{-1}, and is of course also onto and 1–1. The inverse of θ^{-1} is θ.

The set of all the images of a subset A' of A, under θ, is a subset of B and is called the image of A', being denoted by $A'\theta$. The set $A\theta$ is called the image of the mapping (*not* the same as the image space, which is B): it is a subset of B, and $A\theta = B$ is the condition that θ is onto.

A mapping which is onto and 1–1 gives a pairing off between the elements of A and B. This is sometimes known as a 1–1 correspondence, and is of course possible for finite sets only if they have the same number of elements.

Products of mappings

If $\theta: A \to B$ and $\phi: B \to C$ are such that $a\theta = b$ and $b\phi = c$, then the mapping of A into C which sends a into c is called the *product* of θ and ϕ and is denoted by $\theta\phi$. The idea is analogous to that of a function of a function. We may similarly have the product of any number of mappings, provided that the object space of each is the same as the image space of the preceding one, or contains it as a subset. The Associative Law is true for these products.

The mapping of A into A which sends each element into itself is called the identity mapping and is denoted by I, or sometimes

I_A. Then if $\theta: A \to B$ is onto and 1–1, so that θ^{-1} exists, we see that $\theta\theta^{-1} = I_A$, $\theta^{-1}\theta = I_B$.

Theorem 2.1.1. *If* θ *and* ϕ *are both onto and* 1–1, *so is* $\theta\phi$, *and* $(\theta\phi)^{-1} = \phi^{-1}\theta^{-1}$.

This is fairly obvious, the latter part because if $a\theta\phi = c$ then $c\phi^{-1}\theta^{-1} = a\theta\phi\phi^{-1}\theta^{-1} = a$ by pairing off from the inside outwards. A full proof is given on p. 166 of volume 1.

Involutions

If θ is a 1–1 correspondence between elements of A and itself such that $\theta = \theta^{-1}$, then θ is said to be an involution. The condition is equivalent to $\theta^2 = I$.

2.2. Mappings which preserve structure: homomorphisms

The idea of a mapping is a very general one. In order to make it a useful concept in any particular context we need to select certain mappings and concern ourselves with these exclusively; only thus can we obtain useful theorems. Different selections are made in different contexts. In elementary algebra and analysis we concern ourselves more or less exclusively with a few important functions: polynomials, circular functions, logarithmic, exponential and hyperbolic functions between them cater for most of our needs, though other functions are developed for special use. Similarly in abstract algebra we wish to select certain types of mapping as being more significant. Since abstract algebra is concerned chiefly with structure (operations on elements of our set, such as sum, product, union, intersection) it is natural to select mappings which preserve some or all of the structure in which we are interested. A few general examples will clarify this.

Example 1. Let θ be a mapping of the set of two-dimensional vectors into itself defined by $(x, y)\,\theta = (x-y, x+y)$. (This in fact gives a rotation through $\frac{1}{4}\pi$ anti-clockwise together with a magnification by $\sqrt{2}$, if we consider the vectors geometrically.)

Let $$\mathbf{a} = (a_1, a_2) \quad \text{and} \quad \mathbf{b} = (b_1, b_2).$$

Then $\mathbf{a}+\mathbf{b} = (a_1+b_1, a_2+b_2)$ and under θ this goes to

$$(a_1+b_1-a_2-b_2,\ a_1+b_1+a_2+b_2).$$

Now if we add the images of **a** and **b**, i.e. $(a_1 - a_2, a_1 + a_2)$ and $(b_1 - b_2, b_1 + b_2)$ we obtain the same result. Hence the image of the sum of any two vectors is the same as the sum of their images or, in symbols, $(\mathbf{a}+\mathbf{b})\,\theta = \mathbf{a}\theta + \mathbf{b}\theta$. Thus the sum structure is unaltered by the mapping: i.e. if two vectors sum to a third, their images sum to the image of that third vector.

Not only is the sum structure unaltered, but the scalar multiple structure is also not affected. For the image of $\lambda\mathbf{a}$ is $(\lambda a_1 - \lambda a_2, \lambda a_1 + \lambda a_2)$, and this is $\lambda(a_1 - a_2, a_1 + a_2)$, i.e. is λ times the image of **a**. Symbolically $(\lambda\mathbf{a})\,\theta = \lambda(\mathbf{a}\theta)$.

Thus θ preserves both the sum and scalar multiple structure. The same is in fact true for any mapping of the form

$$(x, y) \to (\alpha_{11}x + \alpha_{12}y, \alpha_{21}x + \alpha_{22}y).$$

Example 2. Let θ be a mapping of the complete set of integers into the set of residues modulo n, where n is a fixed positive integer, such that any integer r is mapped into the residue class containing r (i.e. into the class r', where r' is the remainder when r is divided by n).

Then it is clear from our knowledge of residues (see volume 1, chapter 6) that the sum of two integers is mapped into the sum of their images ('sum' is here used, for residue classes, in the sense of the group operation, so that multiples of n are discarded throughout), and the product of two integers has as image the product of their images. Notice that the set of multiples of n is mapped into the zero residue, and two integers (if they differ by a multiple of n) have the same image, so that the mapping is certainly not 1–1, though it *is* onto.

Example 3. If θ: the set of all non-zero polynomials $\to$ the positive integers maps a polynomial into its degree, then the product of two polynomials maps into the sum of their images: structure is similar, provided we connect the product of polynomials with the sum of integers.

Example 4. The mapping of complex numbers into two-dimensional vectors which maps $a + bi$ into (a, b) clearly preserves both the sum and scalar multiple structure.

Example 5. The mapping of the real numbers into themselves given by $a \to 2a$ preserves sums but not products

$$(2(a+b) = 2a+2b \quad \text{but} \quad 2ab \neq (2a)(2b)).$$

Similarly the mapping $a \to a^2$ preserves products but not sums.

The above examples give mappings which preserve some or all of the structure of the sets in question. This type of mapping is the one with which we are chiefly concerned in abstract algebra, and such mappings are called *homomorphisms*. A homomorphism will preserve the structure with which we are dealing: thus in example 1 both the addition and scalar multiple structures are unaltered, in example 2 sum and product are both unaffected, while the mappings in example 5 each preserve one of sum and product.

In the present chapter we are dealing with groups, and so it is the group product that must be preserved. A mapping which does this is called a *group-homomorphism*, or often simply a *homomorphism* if it is clear that groups are being used. We thus have the following formal definition.

Definition of group-homomorphism

If G and H are groups, then $\theta : G \to H$ is a group-homomorphism if for any two elements $g, g' \in G$, $(gg')\,\theta = (g\theta)\,(g'\theta)$.

Note that the product gg' is the group product defined in G, while $(g\theta)(g'\theta)$ is the group product in H, since $g\theta$ and $g'\theta$ are both elements of H. The condition is that the image of the product gg' (defined in G) is the same element as the product (in H) of the images of g and g'.

Examples of group-homomorphisms

Several examples and types will be given in §2.5, but we give one or two here to help fix the ideas.

Example 1. The mapping in example 1 on p. 31, where both G and H are the group of two-dimensional vectors under addition. This preserves the group product (i.e. vector addition) and thus is a homomorphism. The fact that it also preserves scalar multiplication is irrelevant here, but important if we are considering the whole structure of vector sets.

Example 2. If G is the group of integers under addition and H the group of residues modulo n under addition of residues then $\theta: r \rightarrow$ the residue class containing r is a homomorphism.

Example 3. G is the group of complex numbers under addition and H is the group of two-dimensional vectors under addition, with $\theta: a+bi \rightarrow (a, b)$.

Example 4. G and H are both the group of integers under addition and $\theta: a \rightarrow 2a$. Similarly if G and H are both the group of reals except 0 under multiplication and $\theta: a \rightarrow a^2$.

Example 5. As in the first example of 4 above, where H is the group of even integers under addition.

Example 6. Let G be the cyclic group of order 3, generator a, and let H be the symmetric group of degree 4.

Consider the mapping θ which sends

$$e \text{ into } \begin{pmatrix} 1 & 2 & 3 & 4 \\ 1 & 2 & 3 & 4 \end{pmatrix}, \quad a \text{ into } \begin{pmatrix} 1 & 2 & 3 & 4 \\ 2 & 3 & 1 & 4 \end{pmatrix} \quad \text{and } a^2 \text{ into } \begin{pmatrix} 1 & 2 & 3 & 4 \\ 3 & 1 & 2 & 4 \end{pmatrix}.$$

It may easily be verified that θ is a homomorphism.

The idea of homomorphism is an extremely important one. It gives us a method of linking two groups together, so that we may learn more about the properties of one by knowing those of the other. In the present chapter we will not be able to pursue the subject very deeply, but in the more advanced theory of groups homomorphisms are fundamental in their importance: we shall obtain a glimpse of their use in chapter 6, when we deal with invariant subgroups. We shall also study homomorphisms in connection with other types of algebraic structure: in fact when considering any new structure the algebraist, after the elementary work has been done, first explores the substructures (subgroups in the case of groups), and then the homomorphisms (those mappings which preserve the combinations between elements of his structure).

In the definition of homomorphism (we will omit the word 'group-', since we are now exclusively concerned with group-homomorphisms) we merely required that our mapping preserved product. If we look at our examples we notice that in all cases the neutral element of G is mapped into the neutral element of H, and inverses are mapped into inverses. Thus in example 1, $(0, 0) \to (0-0, 0+0) = (0, 0)$ and

$$-(x, y) = (-x, -y) \to (-x+y, -x-y) = -(x-y, x+y).$$

This is true generally, as is easily proved below.

Theorem 2.2.1. *If G and H are groups with neutral elements e and f, and if $\theta: G \to H$ is a homomorphism, then $e\theta = f$.*

If g is any element of G,

$$eg = g.$$

Hence $$(eg)\theta = g\theta.$$

But since θ is a homomorphism,

$$(eg)\theta = (e\theta)(g\theta).$$

Thus $(e\theta)(g\theta) = g\theta$, and so the element $g\theta$ of H is unchanged when multiplied by $e\theta$.

Hence $e\theta$ is f, the neutral element of H.

(For $(e\theta)(g\theta)(g\theta)^{-1} = (g\theta)(g\theta)^{-1} = f$ and so $e\theta = f$.)

Theorem 2.2.2. *If $g\theta = h$, $g^{-1}\theta = h^{-1}$.*

$gg^{-1} = e$ and so $(g\theta)(g^{-1}\theta) = e\theta = f$ by theorem 2.2.1.

Hence $g^{-1}\theta$ is the inverse of $h = g\theta$, i.e. $g^{-1}\theta = h^{-1}$.

(For the reader who wishes every proof to be completed in detail, we have $f = (g\theta)(g^{-1}\theta) = h(g^{-1}\theta)$ and so

$$h^{-1} = h^{-1}h(g^{-1}\theta) = g^{-1}\theta.)$$

If G is a finite group, then clearly $G\theta$ is finite, since it cannot contain more elements than G (although it may contain less). This does not of course mean that H is finite. For example, let G be the cyclic group of order 2, elements e and a, and let H be the group of rotations of a circle (an infinite group—see p. 15). Then the mapping that maps e into the zero rotation, and a into a rotation through π clockwise is evidently a homomorphism (since it maps $a^2 = e$ into the zero rotation, and this is equivalent to the product of two rotations through π).

The converse, that if $G\theta$ is finite then G is finite, is not true. For consider example 2 above, as a counter-example.

If G is Abelian, then so is $G\theta$. For if $g\theta$ and $g'\theta$ are any two elements of $G\theta$, $(g\theta)(g'\theta) = (gg')\,\theta = (g'g)\,\theta$ (since G is Abelian), $= (g'\theta)(g\theta)$. But of course H need not be Abelian, as we can see from the counter-example in example 6. It is also possible for H to be Abelian with G non-Abelian. (The mapping between S_n and C_2 which maps the even permutations into e and the odd ones into the other element of C_2 is a case in point. That this is a homomorphism may easily be proved: see exercises 2A, no. 10.)

2.3. The kernel and the image of a homomorphism

We have seen that in any homomorphism the neutral element e of G maps into the neutral element f of H. In some cases (examples 1, 3, 4, 5 and 6 of §2.2, for instance) this is the only element whose image is f. But often there are other elements of G which map into f. For example, in example 2 of §2.2 all multiples of n have image 0, the zero residue. The set of such elements, a subset of G, is called the *kernel* of the homomorphism, and is an extremely important subset. In the example of the residues given we see that the kernel, consisting of multiples of n, is not only a subset of the set of integers, but is also a *subgroup*, and this is true generally.

Theorem 2.3.1. *The kernel of any homomorphism is a subgroup of G.*

Let g and g' be any two elements in the kernel of $\theta: G \to H$. Thus $g\theta = g'\theta = f$, the neutral element of H.
Hence

$$(gg'^{-1})\,\theta = (g\theta)\,(g'^{-1}\theta) = (g\theta)\,(g'\theta)^{-1} \text{ by theorem 2.2.2}$$
$$= ff^{-1} = f \quad \text{and so } gg'^{-1} \text{ is in the kernel.}$$

Also $e \in$ kernel (theorem 2.2.1).
Hence the kernel is a subgroup (theorem 1.8.1).

Let us now see why the kernel is so important. In our example 2, the mapping of the integers into the group of residues modulo n, let us consider which elements map into a particular residue r. The answer of course is the set of integers which have r as remainder when divided by n, i.e. the set $\{r+\alpha n\}$. But this is a coset (in fact the coset containing r) of the group of integers relative to the subgroup of multiples of n, i.e. relative to the kernel. Thus each coset relative to the kernel, in this case, maps into just one element, and this is true generally. We state and prove this important theorem, in two forms, below.

Theorem 2.3.2. *Let $\theta: G \to H$ be a homomorphism with kernel K.*

(i) The element g' has the same image as a given element g if and only if $g' \in gK$, the coset containing g.

(ii) Two elements g and g' have the same image under θ if and only if they are in the same coset relative to K.

(i) Suppose $g' \in gK$, i.e. let $g' = gk$ for some $k \in K$.

Then $g'\theta = (g\theta)(k\theta) = (g\theta)f$, since $k \in K$, the kernel of θ.

Thus $g'\theta = g\theta$ and so g' has the same image as g.

Conversely let $g'\theta = g\theta$.

Then $(g'\theta)(g\theta)^{-1} = f$ and so $g'g^{-1} = k$, an element of K, since $(g'g^{-1})\theta = (g'\theta)(g\theta)^{-1}$.

Hence $g' = gk$ and so is in the coset gK.

(ii) $g\theta = g'\theta \Leftrightarrow (gg'^{-1})\theta = f \Leftrightarrow gg'^{-1} \in K$, as above. But this is true if and only if g and g' are in the same coset.

Note 1. The two parts of the theorem state the same thing, and are proved in much the same way, the difference being that each refers more directly to one of our two approaches to the idea of cosets (see volume 1, p. 277).

Note 2. We use above left cosets, but the work equally well applies with right cosets instead—the left and right cosets in fact always coincide for kernels: this point is pursued in chapter 6.

The kernel is an important subgroup of the object group. An almost equally important subgroup of the image group is the set of the images of all elements of G. This is clearly a subset of H, and is easily proved to be a *subgroup*. It is called the *image of the homomorphism*, and is of course denoted by $G\theta$, so that $G\theta = \{g\theta : g \in G\}$.

Theorem 2.3.3. *The image $G\theta$ is a subgroup of H.*

Let any two elements of $G\theta$ be $g\theta$ and $g'\theta$ (they must be of this form for some g and g' in G, since otherwise they would not be in the image $G\theta$).

Then $(g\theta)(g'\theta)^{-1} = (gg'^{-1})\theta$ and so is the image of gg'^{-1}, and hence is in $G\theta$.

Also $f \in G\theta$ by theorem 2.2.1.

Hence $G\theta$ is a subgroup, by theorem 1.8.1.

Changing the image and object spaces

Suppose we are given a homomorphism $\theta: G \to H$. Then the image space H may be extended at will without affecting the homomorphism. For example, a homomorphism into the real numbers may equally well be considered as one into the complex numbers. But H may not be compressed arbitrarily—this may only be done provided we do not thereby omit any of the image $G\theta$.

The object space G may be compressed to still give us a homomorphism. Thus we may consider that part of θ which maps any subgroup of G: for example a homomorphism of the complex numbers gives us also a homomorphism of the real numbers (we sometimes say that it 'induces' the second homomorphism). But the object space may not be extended, at any rate not without some further definition to give the complete homomorphism.

Thus any homomorphism induces one from a compressed object space but not an extended object space, and one to an extended image space but not to a compressed image space.

Notice the rather rough duality between the object and image spaces. Another example of this duality is that the kernel is a subgroup of the object space, while the image is a subgroup of the image space. This duality runs through much of our work on homomorphisms, but must not be examined too critically or forced too far.

2.4. Types of homomorphism

When dealing with general mappings in §2.1 we were interested in the existence of inverse mappings, and were led to consider mappings which were onto, and those which were 1–1. The same is true when we restrict ourselves to homomorphisms, and in this case the special types have technical names, which are worth remembering.

Epimorphisms

A homomorphism which is onto the image space is called an *epimorphism*: the condition of course is that $G\theta = H$: the image of the homomorphism is not merely a subgroup of H but is the whole group H.

Any homomorphism may be made into an epimorphism by compressing the image space into the subgroup $G\theta$. Thus in example 4 of §2.2 the homomorphism $\theta: a \to 2a$ is not an epimorphism when we consider the image space as being the group of integers, but may be made into one if we take the image space as being the group of *even* integers.

Monomorphisms

A homomorphism that is 1–1 is known as a *monomorphism*. Thus in a monomorphism different objects always have different images: in other words $g\theta = g'\theta \Rightarrow g = g'$.

The simplest way of testing for monomorphisms is by considering the kernel. This is always a subgroup of the object space G, and of course always contains the neutral element e of G by theorem 2.2.1. The condition for a monomorphism is that the kernel is the trivial subgroup consisting of e alone.

Theorem 2.4.1. *$\theta: G \to H$ is a monomorphism if and only if the kernel of θ is $\{e\}$.*

Only if. Suppose θ is a monomorphism and suppose the kernel contains an element $k \neq e$. Then $k\theta = e\theta = f$, the neutral element of H, and hence since θ is a monomorphism $k = e$, which contradicts our hypothesis. Hence the kernel is $\{e\}$.

If. Suppose the kernel is $\{e\}$ and let $g\theta = g'\theta$. Then

$$(gg'^{-1})\,\theta = (g\theta)\,(g'\theta)^{-1} = f,$$

and so gg'^{-1} is in the kernel and thus must be e. Hence $g = g'$ and so θ is a monomorphism.

Isomorphisms

For a homomorphism to have an inverse it must be both onto and 1–1: such a homomorphism is called an *isomorphism*. Thus an isomorphism is a homomorphism in which *every* element of the image space is the image of one and only one element of the object space. As a mnemonic 'epi-+mono- = iso-'.

If θ is an isomorphism we may define the inverse homomorphism $\theta^{-1}: H \to G$ by $h\theta^{-1} = g$ where g is the object whose image under θ is h (i.e. $g\theta = h$). g is of course unique, and always exists, by the isomorphism hypothesis. It is fairly obvious, but needs proving, that θ^{-1} *is* a homomorphism.

Theorem 2.4.2. *If $\theta: G \to H$ is an isomorphism, the mapping $\theta^{-1}: H \to G$ defined by $h\theta^{-1} = g$ where $g\theta = h$ is a homomorphism.*

Let $h\theta^{-1} = g$ and $h'\theta^{-1} = g'$, so that $g\theta = h$ and $g'\theta = h'$. Then $(gg')\,\theta = hh'$ and so $(hh')\,\theta^{-1} = gg' = (h\theta^{-1})\,(h'\theta^{-1})$.

θ^{-1} is in fact also an isomorphism: it is onto G since all elements of G have an image under θ in H, and it is 1–1 since no element of G can give rise to two different elements of H under θ, and so two elements of H cannot map under θ^{-1} into the same element of G.

An isomorphism gives a 1–1 correspondence (a pairing off) between the elements of G and H, and this correspondence preserves the group structure. If G and H possess an isomorphism then they are to all intents and purposes identical groups, *considered purely as groups*: provided we deal only with group theoretic properties it does not matter whether we work with G or H. Such groups are called *isomorphic groups*, and we write $G \cong H$. They will sometimes be spoken loosely of as being the *same* group (considered as abstract groups). Note that if we say $G \cong H$ then we are implying that an isomorphism exists between them: in many cases there will be more than one such isomorphism, and the groups will be isomorphic in several ways. (See example 4 below.)

The ideas of isomorphism and isomorphic groups are exceedingly important: by showing that a certain group is isomorphic to a known one we can immediately discover its group properties. Some examples are given below, but we first prove a trivial theorem about monomorphisms.

Theorem 2.4.3. *If θ is a monomorphism, then $G \cong G\theta$.*

If θ is considered as a homomorphism between G and $G\theta$ it is onto, and is 1–1 by hypothesis. Hence it is an isomorphism and so $G \cong G\theta$.

Examples of isomorphisms

Example 1. The mapping of the complex numbers under addition into the group of two-dimensional vectors under addition defined by $x+iy \to (x, y)$ gives an isomorphism. We also have an isomorphism between the group of displacements in a plane (product being given by successive displacements) and the group of ordered pairs of real numbers, and either of these groups may be identified with the group of two-dimensional vectors.

Example 2. The polynomials with real coefficients of degree $< n$ are isomorphic to the group of n-dimensional vectors, both groups under addition, by an obvious isomorphism:

$$\sum_{i=0}^{n-1} a_i x^i \to (a_i).$$

Example 3. The subgroup of complex numbers of the form $a+i0$ is isomorphic to the group of real numbers, both under addition, by the isomorphism $a+i0 \to a$. This isomorphism is used to identify the complex numbers of this form with the reals, when we define complex numbers by means of number pairs.

Example 4. The group of residues modulo 5 under multiplication (excluding 0) is isomorphic to the group of residues modulo 4 under addition by the isomorphism $1 \to 0$, $2 \to 1$, $3 \to 3$ and $4 \to 2$. Another isomorphism between these two groups is given by $1 \to 0$, $2 \to 3$, $3 \to 1$ and $4 \to 2$. (This example is considered in more detail on p. 205 of volume 1.)

Example 5. The homomorphism of the group of integers under addition into the group of even integers under addition given by $a \to 2a$. Both groups are (or *are isomorphic to* to be more precise) the infinite cyclic group.

Example 6. The group of rotations of a regular one-faced polygon of n sides is isomorphic to the group of residues modulo n under addition, an isomorphism being given by a rotation through $r(2\pi/n)$ clockwise mapping into the residue r.

(Another isomorphism is given by the corresponding anti-clockwise rotations.) The groups, considered as abstract groups, are the same: both are the cyclic group of order n.

Homomorphisms with the same object and image spaces

A homomorphism which has identical object and image spaces is often called an *endomorphism*. An example is the mapping of two-dimensional vectors into itself given by $(x, y) \to (x-y, x+y)$. Another example is the mapping of the integers into itself given by $a \to 2a$, *provided* we consider the image as being the whole group of integers: if we think of this mapping as being an isomorphism as in example 5 above then it is no longer strictly an endomorphism.

An endomorphism may of course not be onto, or 1–1. If it is also an isomorphism then it is called an *automorphism*. Thus an automorphism is an isomorphism of a group into itself: an example in the group of residues modulo 5 under multiplication is given by $1 \to 1$, $2 \to 3$, $3 \to 2$ and $4 \to 4$.

2.5. Examples of homomorphisms

The identity homomorphism

The mapping which sends any element of a group G into itself is clearly a homomorphism, and is called the identity homomorphism in G. It is often denoted by I_G, and is of course an automorphism of G.

The trivial homomorphism

The mapping $\theta: G \to H$ which sends every element of G into the neutral element f of H preserves products, since

$$(g\theta)(g'\theta) = ff = f = (gg')\,\theta.$$

It is called the trivial homomorphism. Its kernel is the whole group G and its image is the trivial subgroup $\{f\}$ of H. Note that with this we have the greatest possible kernel and least possible image, whereas in any isomorphism the kernel is the least possible (namely $\{e\}$) and image greatest, being $\simeq G$.

Homomorphisms from the infinite cyclic group

We have seen that the mapping of the integer r into the residue class containing r modulo n gives a homomorphism of the group of integers (the infinite cyclic group) into the group of residues modulo n under addition (the cyclic group of order n). It is an epimorphism but not of course a monomorphism, the kernel being the subgroup of multiples of n.

If a is any element of any group H we may map the integer r into a^r, and this will give us a homomorphism of the infinite cyclic group into the group H. The image is the subgroup of powers of a and the kernel is the subgroup $\{kn\}$ where n is the order of a in H: if a has infinite order the mapping is a monomorphism.

Endomorphisms of the infinite cyclic group

If we map the integer 1 into s, then we obtain an endomorphism if we map any integer r into rs. Thus corresponding to each integer s we have an endomorphism. All these are monomorphic except the trivial one (for which $s = 0$) and the typical one gives an isomorphism between the group of integers and the subgroup of multiples of s. Thus this subgroup is isomorphic to the infinite cyclic group. The only two such mappings which are automorphisms are those where $s = 1$ or -1, since only in these cases is the image the *whole* group of integers.

Homomorphisms between finite cyclic groups

Suppose we wish to have a homomorphism of C_n (generator a) into C_m (generator b). Let $a \to b^r$. Then $e = a^n$ must map into b^{nr}. But for the mapping to be homomorphic e must go into f, the neutral element of C_m, and so b^{nr} must be f, i.e. $nr = km$ for some integer k. Various special cases follow.

(1) If $n = m$ we have an endomorphism of C_n into itself when $a \to a^r$ for any r. Thus we have n endomorphisms. Of these one is the trivial endomorphism ($r = 0$), and we obtain an automorphism if and only if r is prime to n, so that a^r is a generator. Thus if n is prime we obtain $(n-1)$ automorphisms.

(2) If n and m are co-prime we must have $r = 0$ or m (so that

$k = 0$ or n). Both cases give us the trivial homomorphism, and so this is the only one in this case.

(3) If the H.C.F. of n and m is h, where $n = h\nu$ and $m = h\mu$, we have $\nu r = \mu k$ and so r must be a multiple of μ. For each such multiple we obtain a homomorphism.

Homomorphisms of vectors

It may easily be verified that the mapping

$$(x, y) \rightarrow (\alpha_{11}x+\alpha_{12}y, \alpha_{21}x+\alpha_{22}y)$$

preserves the sum of vectors, and hence is a homomorphism of the group of two-dimensional vectors into itself (in fact it also preserves scalar multiples, but we are not concerned with this here).

The kernel consists of all (x, y), where x and y are solutions of

$$\alpha_{11}x+\alpha_{12}y = 0,$$
$$\alpha_{21}x+\alpha_{22}y = 0.$$

Thus $(0, 0)$ is the kernel (as it must be, since it is the neutral element), and is the only vector in the kernel unless

$$\alpha_{11}\alpha_{22}-\alpha_{12}\alpha_{21} = 0, \text{ or } \begin{vmatrix} \alpha_{11} & \alpha_{12} \\ \alpha_{21} & \alpha_{22} \end{vmatrix} = 0.$$

Thus if
$$\begin{vmatrix} \alpha_{11} & \alpha_{12} \\ \alpha_{21} & \alpha_{22} \end{vmatrix} \neq 0$$

the kernel consists of $(0, 0)$ alone and the homomorphism is 1–1. In this case it is also onto, since the equations

$$\alpha_{11}x+\alpha_{12}y = X,$$
$$\alpha_{21}x+\alpha_{22}y = Y$$

have a unique solution. Thus if the determinant is not zero we have an automorphism of the group of vectors into itself.

Such an automorphism is called a *non-singular linear transformation*, and the reader familiar with matrix theory will recognise the condition on the determinant as being the condition that the matrix of coefficients is non-singular. The theory is easily extended to vectors in n-dimensions. In the non-singular case the inverse automorphism is given by the inverse matrix.

Direct Products—projections

The mapping which, considered geometrically, projects two-dimensional vectors onto a line through the origin, is given analytically by $(x, y) \to (x, 0)$ where for simplicity we consider the image space as being the x-axis. Since

$$(x_1, y_1)+(x_2, y_2) = (x_1+x_2, y_1+y_2)$$

and projects into $(x_1+x_2, 0)$, which is the sum of the projections of the summands, this projection is a homomorphism. It is in fact an endomorphism of the group of two-dimensional vectors into itself (a singular linear transformation), but may also be considered as an epimorphism of the group of two-dimensional vectors onto the group of scalars, by identifying (isomorphically) $(x, 0)$ with the scalar x. The idea works equally well from the group of vectors of n dimensions onto the group of vectors of m dimensions, where $m < n$, by mapping the surplus co-ordinates into 0 and omitting them (thus $(x, y, z, t) \to (x, y)$ gives a projection from 4 to 2 dimensions).

This work may be generalised to direct products, of which vectors form one example. If G is the direct product $A \times B$, typical element (a, b), the set $(a, 0)$ forms a subgroup of G isomorphic to A, and we consider the mapping $(a, b) \to (a, 0)$. This is an endomorphism by the same reasoning as in the case of vectors above, and gives an epimorphism of $G = A \times B$ onto A if we identify $(a, 0)$ with a.

Injections

If H is a subgroup of G the mapping $h \to h$, where the object h is considered as an element of H and the image h is considered as an element of G, gives a monomorphism of H into G. It is called the *injection* of $H \to G$, and is the identity automorphism restricted to H as the object space. It is particularly useful when we work with sequences of homomorphisms in more advanced work, as a means of transferring from H to the larger group G.

Inner automorphisms

Theorem 2.5.1. *The mapping of G into itself given by $g \to x^{-1}gx$, where x is any fixed element of G, is an automorphism.*

For $g \to x^{-1}gx$ and $g' \to x^{-1}g'x$, and the product of these images is $x^{-1}gxx^{-1}g'x = x^{-1}gg'x$, the image of gg'.

Hence the mapping preserves the product structure and so is an endomorphism. To show that it is an automorphism we must show that any element of G is the image of one and only one element.

But any element h may be written as $x^{-1}(xhx^{-1})\,x$ and so is the image of xhx^{-1}, and this object is unique since if g and g' have the same image we must have $x^{-1}gx = x^{-1}g'x$ and so $g = g'$ by pre- and post-multiplying by x and x^{-1} respectively.

If G is Abelian all such automorphisms are the identity automorphism since $x^{-1}gx = g$. However, if G is non-Abelian they will not all be the identity. Such automorphisms are called *inner automorphisms* and form a very important type of automorphism in the non-Abelian case. There is one corresponding to each element x of G, but they will not all be different. Not all automorphisms are of this type—the others are sometimes called *outer automorphisms.*

Some isomorphisms

Various isomorphisms were shown in volume 1, to which the reader is referred for details.

The symmetric group of degree 3 is isomorphic to the dihedral group of order 6 (the group of the equilateral triangle), the isomorphism being given by a permutation of (1 2 3) corresponding to a transformation of the triangle obtained by numbering the *positions which the vertices occupy* and transforming each vertex into the position numbered as in the image under the permutation. Thus

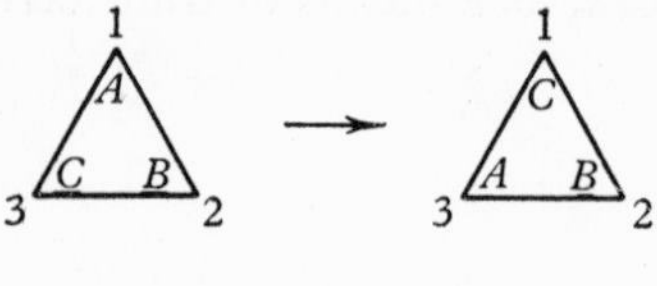

corresponds to
$$\begin{pmatrix} 1 & 2 & 3 \\ 3 & 2 & 1 \end{pmatrix}$$
(see volume 1, p. 234).

Each of these two groups is also isomorphic to the group of cross-ratios (see volume 1, p. 245).

The tetrahedral group is isomorphic to the alternating group of degree 4 (i.e. to the group of *even* permutations on four numbers). An isomorphism is obtained in a similar manner to the above (using the four vertices of a tetrahedron instead of the three vertices of a triangle) and noting that only the even permutations give rise to a corresponding transformation of the tetrahedron (the odd ones give mirror images of the 12 positions). (For details see volume 1, p. 240.)

It was also shown that the octahedral group is isomorphic to the symmetric group of degree 4, and it was stated without proof that the dodecahedral group is isomorphic to the alternating group of degree 5.

The group of the rectangle (the dihedral group of order 4) is isomorphic to $C_2 \times C_2$, both being examples of the Vierergruppe. (See volume 1, pp. 229 and 249.)

2.6. Sets of homomorphisms and their algebra

In §2.1 we mentioned products and inverses of mappings. When our mappings are homomorphisms, we will see that any products and inverses are also homomorphisms, and thus we can obtain a certain amount of algebra by combining homomorphisms. We are here treating a homomorphism as an element of some given set of mappings, and we must not confuse this element with elements of the various object and image spaces. We will see also that it is only in certain cases that products and inverses may be formed.

Products of homomorphisms

We recall that if $\theta: A \to B$ and $\phi: B \to C$ are mappings where $a\theta = b$ and $b\phi = c$, then the mapping of $A \to C$ which sends a into c is called the product of θ and ϕ and is denoted by $\theta\phi$. If the sets are groups (which we will denote by G, H and K to make the distinction clearer) and θ and ϕ are homomorphisms, then $\theta\phi$ is also a homomorphism. For θ preserves the structure from G to H and ϕ preserves it from H to K, and so $\theta\phi$ will preserve the group structure from G to K. We give a formal proof below.

Theorem 2.6.1. *If $\theta: G \to H$ and $\phi: H \to K$ are homomorphisms, so is the mapping $\theta\phi: G \to K$.*

If g and g' are any two elements of G, and if $g\theta = h, g'\theta = h'$, $h\phi = k, h'\phi = k'$, we know that $(gg')\,\theta = hh'$ and $(hh')\,\phi = kk'$ by the homomorphism property.

Hence
$$(gg')\,\theta\phi = (hh')\,\phi = kk'.$$

But $k = g\theta\phi$ and $k' = g'\theta\phi$ and so $(gg')\,\theta\phi = (g\theta\phi)\,(g'\theta\phi)$, showing that $\theta\phi$ is a homomorphism.

We may similarly prove that the product of any number of homomorphisms is a homomorphism. Such a product exists only when the object space of each is the same as the image space of the one preceding (or possibly contains the image space as a subgroup, although this case is better avoided, as will be discussed later). It is very convenient to show the various homomorphisms diagrammatically as in figure 4.

$$G \xrightarrow{\theta} H \xrightarrow{\phi} K \xrightarrow{\psi} M$$

Fig. 4

The inverse of a homomorphism

We have seen that for a mapping to have an inverse it must be both onto and 1–1: thus in the homomorphism case the mapping must be both epimorphic and monomorphic—in other words it must be an isomorphism. We showed in theorem 2.4.2 that the inverse mapping is necessarily a homomorphism, in fact an isomorphism.

By theorem 2.1.1 applied to the case of homomorphisms, if $\theta: G \to H$ and $\phi: H \to K$ are both isomorphisms, so is $\theta\phi: G \to K$, and $(\theta\phi)^{-1} = \phi^{-1}\,\theta^{-1}$.

The identity isomorphism

If $\theta: G \to H$ is a homomorphism, it is immediate that
$$\theta I_H = I_G\theta = \theta$$
and if θ is an isomorphism that
$$\theta\theta^{-1} = I_G, \quad \theta^{-1}\theta = I_H.$$

Changing the object and image spaces

For $\theta\phi$ to exist the image space of θ should be the same as the object space of ϕ. Let us denote the image space of θ by H and the object space of ϕ by H'. Then $H = H'$ is sufficient to ensure the existence of $\theta\phi$. We may however be able to define $\theta\phi$ when $H \neq H'$.

For example suppose H is a subgroup of H'. Then every image of an element of G under θ is in H and so is in H', and thus may be mapped into K, so that $\theta\phi$ is well-defined. In this case it is much more convenient if we introduce a third mapping, the injection i of H into H' as defined in §2.5, and obtain a sequence of three homomorphisms $G \xrightarrow{\theta} H \xrightarrow{i} H' \xrightarrow{\phi} K$ where now each image space is the same as the succeeding object space.

$\theta\phi$ is also defined when H' is a subgroup of H *provided* it contains $G\theta$. In this case it is often useful to replace H by its subgroup H' (this does not affect θ).

The sum of two homomorphisms

Suppose θ and θ' are two homomorphisms from $G \to H$, with $g\theta = h$ and $g\theta' = h'$. Consider the mapping of $G \to H$ defined by $g \to hh'$, where hh' is the group product in H. We will call this the 'sum' of θ and θ' and denote it by $\theta + \theta'$. (The term 'product' may seem more natural, but has already been used. In fact sum is an appropriate word, as we will see shortly.) $\theta + \theta'$ is always a mapping, but is not necessarily a homomorphism. We will investigate the conditions which ensure that it preserves the structure of G.

For $\theta + \theta'$ to be homomorphic we must have

$$(g\gamma)(\theta + \theta') = g(\theta + \theta').\gamma(\theta + \theta')$$

for any two elements g and γ in G.

Suppose $g\theta = h$ and $g\theta' = h'$ so that $g(\theta + \theta') = hh'$, and $\gamma\theta = \eta$ and $\gamma\theta' = \eta'$ so that $\gamma(\theta + \theta') = \eta\eta'$.

Also $(g\gamma)\theta = h\eta$ and $(g\gamma)\theta' = h'\eta'$ since both θ and θ' are homomorphisms, and so $(g\gamma)(\theta + \theta') = h\eta h'\eta'$.

Hence our condition is that $h\eta h'\eta' = hh'\eta\eta'$, i.e. that $\eta h' = h'\eta$.

Thus $\theta+\theta'$ is a homomorphism if any image of θ, i.e. any element of $G\theta$, commutes with any element of $G\theta'$, and this is certainly true if H is Abelian. In this case we may write the group product in H in the additive notation and the notation $\theta+\theta'$ is quite natural. We have proved the following theorem.

Theorem 2.6.2. *If θ and θ' are two homomorphisms from G into H, where H is Abelian, then the mapping $\theta+\theta'$ defined by*

$$g(\theta+\theta') = g\theta+g\theta'$$

is a homomorphism.

2.7. Groups of automorphisms

We have seen that under certain conditions homomorphisms may be combined to form a product, that there are sometimes inverses, and that identity homomorphisms exist with properties analogous to the neutral element of a group. It is natural to ask whether we cannot form groups of homomorphisms, treating a homomorphism as an element.

For a set of homomorphisms to form a group it must be possible to form inverses from all of them, and so they must all be isomorphisms. It must also be possible to multiply *any* pair, i.e. the image space of any must be the same as the object space of any, and so they must clearly all be endomorphisms of the same group into itself. Thus, combining these two conditions, they must all be automorphisms of the same group. Any such set of automorphisms, containing the identity automorphism (as a neutral element), and such that the product of any two and the inverse of any one are each in the set, forms a group. In particular, we obtain an important group by considering *all* the automorphisms of any given group G. (Product is obviously associative.) This group is known as the *group of automorphisms of* G. It may or may not be Abelian; even if G itself is, its group of automorphisms need not be.

Theorem 2.7.1. *If G is finite and of order n, its group of automorphisms is isomorphic to a subgroup of S_n, the symmetric group of degree n.*

Any automorphism is a 1–1 correspondence between the

elements of G and themselves, i.e. is a permutation of the n elements, and so corresponds to an element of S_n (no two automorphisms corresponding to the same permutation).

The product of two automorphisms, obtained by applying first one then the other, is formed in exactly the same way as the product of permutations.

This gives the required isomorphism.

In general the subgroup will be a proper one, since by no means all permutations of the elements of G are automorphisms. It will of course usually be a very small subgroup, whereas S_n is a large group for all values of n except the few lowest.

Since the neutral element e always stays unchanged the group of automorphisms is in fact isomorphic to a subgroup of S_{n-1}, given by the permutations of the other $(n-1)$ elements.

Worked exercises

1. Prove that the mapping of the real numbers under addition onto the rotation group of the circle which sends x into a rotation through x radians gives a homomorphism. Give the kernel and image and state its type.

The real number $x+y$ is mapped into the rotation through $x+y$ radians, which is the 'product' (in the rotation group) of rotations through x radians and y radians, working modulo 2π in this group. Hence the mapping is a homomorphism.

The kernel consists of all real numbers which map into the identity rotation, i.e. of all multiples of 2π. These of course form a subgroup of the group of reals, the subgroup being isomorphic to the infinite cyclic group.

The image is the whole group of rotations.

Thus the mapping is an epimorphism.

2. Identify the group of automorphisms of the symmetric group S_3.

	I	X	U	Y	V	W
e	e	e	e	e	e	e
a	a	a	a	a^2	a^2	a^2
a^2	a^2	a^2	a^2	a	a	a
b	b	ba	ba^2	b	ba	ba^2
ba	ba	ba^2	b	ba^2	b	ba
ba^2	ba^2	b	ba	ba	ba^2	b

Fig. 5

The first step is to find all the automorphisms. Let the elements of S_3 be e, a, a^2, b, ba, ba^2 where $a^3 = b^2 = e$ and $ab = ba^2$. Then any homomorphism is completely determined if we know the images of a and b.

Since the image of $a^3 = e$ must be e, the image of a must have order 3, and similarly that of b must have order 2 (neither can go into e since we are concerned with automorphisms, which are 1–1, and e has image e).

Hence the image of a is either a or a^2 and that of b is b, ba or ba^2, since all these latter have order 2. The images of the other elements are forced, and we obtain 6 homomorphisms as shown in figure 5 above, which are all seen to be automorphisms and give us the complete set of automorphisms. (The figure gives the images of the elements of S_3 under each automorphism.)

We must now identify the group of automorphisms. By inspecting the table we see that $X^2 = U$ and $X^3 = I$, that $Y^2 = I$ and $YX = V$, $YX^2 = W$, while $XY = YX^2$. Hence the group is the symmetric group of degree 3.

Note that in this case the group of automorphisms is the same as the original group. This is of course by no means always the case.

3. Prove that the mapping $\theta: g \to g^{-1}$ of G into itself is always 1–1 and onto, and is an automorphism if and only if G is Abelian.

1–1. If $g\theta = h\theta$ we have $g^{-1} = h^{-1}$ and so $g = h$.

Onto. Any element k is the image of k^{-1}, and so is in the image of the mapping.

Hence θ is an automorphism if and only if it is a *homomorphism*, and this is the case if and only if $(gh)\,\theta = (g\theta)\,(h\theta)$, i.e. if and only if

$$(gh)^{-1} = (g^{-1})\,(h^{-1}).$$

But $(gh)^{-1} = h^{-1}g^{-1}$ and so the condition is that $g^{-1}h^{-1} = h^{-1}g^{-1}$ for all g and h in G. Thus the condition is that G is Abelian. (Any two inverses commute if and only if any two elements commute, since as elements range over G so do inverses.)

4. Find all possible images (as abstract groups) of the Vierergruppe $C_2 \times C_2$ under homomorphisms.

Let the elements of $V = C_2 \times C_2$ be e, a, b, ab. Except for e, these all have order 2 and so their images under any homomorphism have order 2 (or 1).

Suppose first that all elements go into the neutral elements f of the image group. Then the homomorphic image is the trivial group $\{f\}$.

Suppose a maps into the neutral element, and b into x say, an element of order 2. Then ab maps into x also and the image consists of f and x, and is C_2.

If a maps into a non-neutral element x and b into f then the image is C_2 as before. Now suppose a and b map into the same element x. Then ab goes into $x^2 = f$ and we again obtain C_2.

Finally suppose a maps into x and b into y, a different element. Then ab goes into xy and of course x and y both have order 2. We obtain an isomorphism and the image is the Vierergruppe.

Hence the only possible images are $\{f\}$, C_2 and $C_2 \times C_2$.

Exercises 2A

In these exercises we will write Z for the group of integers under addition (the infinite cyclic group), R for the group of real numbers under addition, C for the complex numbers under addition, and V_n for the group of n-dimensional vectors. Other standard notations, such as C_n, D_{2n} and S_n are also used.

Prove that the mappings in **1–17** give homomorphisms. In each case give the kernel and image, and state the type of homomorphism (i.e. monomorphism, automorphism, etc.).

1. $\theta: Z \times Z \to C$ defined by $(m, n)\,\theta = m+in$.

2. $\theta: R \times R \to C$ defined by $(x, y)\,\theta = x+iy$.

3. $\theta: R \times R \to C$ defined by $(x, y)\,\theta = iy$.

4. $\theta: V_2 \to V_2$ defined by $(x, y)\,\theta = (x-y, x+y)$.

5. $\theta: V_3 \to V_3$ defined by $(x, y, z)\,\theta = (x, x+y, x+y+z)$.

6. $\theta: V_2 \to V_3$ defined by $(x, y)\,\theta = (x-y, x+y)$.

7. $\theta: V_3 \to V_2$ defined by $(x, y, z)\,\theta = (x-y, x+y)$.

8. $\theta: Z \to$ complex numbers (except 0) under multiplication defined by $r\theta = e^{2\pi ir/n}$ for some positive integer n.

9. $\theta: R \to V_2$ defined by $x\theta = (x, x)$.

10. $\theta: S_n \to$ the group consisting of -1 and $+1$ under multiplication defined by $P\theta = \zeta(P)$ where P is any permutation in S_n and $\zeta(P) = +1$ if P is even and -1 if P is odd.

11. $\theta: D_{2n} \to$ the group consisting of -1 and $+1$ under multiplication defined by $x\theta = +1$ if x does not turn the polygon over, and $x\theta = -1$ if x turns the polygon over.

12. θ: group of residues modulo 10 and prime to 10 under multiplication $\to C_4$, defined by $3\theta = a$ and $3^n\theta = a^n$ for any integer n, where a is a generator of C_4.

13. As in **12** but with $3\theta = a^2$ and $3^n\theta = a^{2n}$.

14. As in **12** but $\to C_8$ with $3\theta = a^2$ and $3^n\theta = a^{2n}$, where a is a generator of C_8.

15. $\theta: C_{10} \to C_{10}$ defined by $a\theta = a^4$ and $a^n\theta = a^{4n}$, where a is a generator.

16. $\theta: C_{12} \to C_8$ defined by $a\theta = b^2$ and $a^n\theta = b^{2n}$, where a is a generator of C_{12} and b a generator of C_8.

17. $\theta: C_{12} \to C_4$ defined by $a\theta = b^3$ and $a^n\theta = b^{3n}$, where a is a generator of C_{12} and b a generator of C_4.

18. Give an isomorphism between the group of residues modulo 9 and prime to 9 under multiplication and C_6.

19. Repeat **18** for residues modulo 20 and prime to 20 under multiplication, and $C_4 \times C_2$.

20. List all the endomorphisms of $C_6 \to C_6$, giving the kernels and images in each case. Which are automorphisms?

21. Repeat **20** for the homomorphisms of $C_6 \to C_8$. Why can none of these be isomorphisms? State which are monomorphisms and which epimorphisms.

22. Prove that $\theta: V_2 \to V_2$ defined by $(x, y)\,\theta = (\alpha_{11}x+\alpha_{12}y, \alpha_{21}x+\alpha_{22}y)$ is a homomorphism.

23. If θ, ϕ, ψ are isomorphisms prove that $\theta\phi\psi$ is an isomorphism and that $(\theta\phi\psi)^{-1} = \psi^{-1}\phi^{-1}\theta^{-1}$.

24. By considering $A \xrightarrow{i} A \times B \xrightarrow{p} A$, where i is an injection and p the projection sending (a, b) into a, prove that the product of two homomorphisms may be an isomorphism even if neither of the two is.

25. If $\theta\phi$ is an isomorphism prove that ϕ must be an epimorphism and θ must be a monomorphism.

26. If $\theta\phi$ is an epimorphism show that ϕ must be an epimorphism,

27. If $\theta\phi$ is a monomorphism show that θ must be a monomorphism.

28. Prove that if $\theta: C_n \to H$ then the image $C_n\theta$ is always cyclic, and that its order must be a factor of n.

Identify the groups of automorphisms of the groups in **29–32**.

29. The infinite cyclic group.

30. C_5.

31. C_6.

32. The Vierergruppe $C_2 \times C_2$.

33. Prove theorem 2.4.1 by using theorem 2.3.2.

34. Prove that there is no non-trivial homomorphism of $C_n \to$ the infinite cyclic group.

35. Prove that if G is non-Abelian then not all the inner automorphisms are the identity automorphism.

36. Prove that the inner automorphism $g \to x^{-1}gx$ is the identity automorphism if and only if x is in the centre of the group. (The centre is the set of elements that commute with all elements of the group.)

37. θ and θ' are homomorphisms from $G \to H$ where G and H are Abelian, so that $\theta+\theta'$ is defined. Show that if θ and θ' are both isomorphisms then $\theta+\theta'$ is not necessarily an isomorphism.

(Consider a case where $g\theta' = (g\theta)^{-1}$.)

If θ and θ' are both monomorphisms, need $\theta+\theta'$ be a monomorphism?

If θ and θ' are both epimorphisms, need $\theta+\theta'$ be an epimorphism?

38. With the same notation as in **37**, if $\theta+\theta'$ is an isomorphism, need θ and θ' both be isomorphisms?

Exercises 2B

1. If λ, μ are co-prime, we know from §2.5 that $\theta: C_{h\nu} \to C_{h\mu}$ defined by $a\theta = b^r$, a and b being generators, is a homomorphism if and only if $r = \lambda\mu$ for some λ. Prove conditions for θ to be (i) a monomorphism, (ii) an epimorphism.

2. List the 6 inner automorphisms of S_3 into itself, giving the images of all elements in each case. Which of these automorphisms are the same?

3. Repeat **2** for D_8.

4. Show that the group of automorphisms of C_n is isomorphic to the group of residues modulo n prime to n under multiplication.

5. Show that the group of automorphisms of the group of rational numbers under addition is isomorphic to the group of rationals (except 0) under multiplication.

6. Let the homomorphism $\theta: G \to H$ be given by $g\theta = h$. Prove that the mapping $\phi: G \to H$ given by $g\phi = h^{-1}$ is a homomorphism if and only if $G\theta$ is Abelian.

7. Prove that the set of inner automorphisms of G is a *subgroup* of the group of automorphisms of G, not merely a subset.

8. If the inner automorphism of G defined by $g \to x^{-1}gx$ is denoted by T_x, show that the mapping $x \to T_x$ from G into the group of automorphisms of G is a homomorphism, and that its kernel is the centre of G.

9. Show that the homomorphisms of an Abelian group G into another Abelian group H form a group under sum of homomorphisms as defined in §2.6. What is the neutral element?

10. If $G = H = C_n$ in **9** show that the group of homomorphisms is also C_n.

11. If $G = C_n$, $H = C_m$ in **9**, where the H.C.F. of n and m is h, show that the group of homomorphisms is C_h.

12. A sequence $\theta_i: G_i \to G_{i+1}$ of homomorphisms is said to be *exact* if the kernel of any θ_{i+1} is the image of θ_i. Prove that $\theta_i\theta_{i+1}$ is the trivial homomorphism.

3

RINGS

3.1. Definition of a ring

We have seen that in a group we have a set together with an operation which obeys certain rules analogous to the rules of ordinary multiplication, or addition. It is immaterial whether we call this operation addition or multiplication, except that it is conventional to use the product notation in the general case, reserving the sum notation for Abelian groups.

It often happens that we have a set which contains two different operations, and that these operations are linked in some way. The obvious example is in the set of real numbers where we have addition and multiplication, which are linked by the Distributive Law. Subtraction arises from addition (by the introduction of negatives) and division from multiplication (by the introduction of inverses), but the Distributive Law forms the only real connection between sum and product, and distinguishes between them since the dual Distributive Law (that $x+yz = (x+y)(x+z)$, obtained by interchanging sum and product) is not satisfied.

The set of reals is by no means the only set containing two operations. Other examples are given below.

Example 1. The set of integers admits of addition (and subtraction) and multiplication (but not division).

Example 2. The set of polynomials in x (of any degree, with real coefficients) admits of addition and multiplication, including subtraction but not division in general, though it may do so in special cases.

Example 3. The set of residues modulo any integer n also admits of addition, subtraction and multiplication, but of division only if n is prime.

Example 4. The even integers may be added or multiplied to give an even integer.

Example 5. The quaternions may be manipulated according

to all four elementary operations. A quaternion is a 'number' of the form $a+bi+cj+dk$, where two quaternions are added as though a quaternion was in the form of a vector (a, b, c, d) and multiplication is done as though i, j and k were ordinary numbers with

$$i^2 = j^2 = k^2 = -1, \quad ij = -ji = k, \quad jk = -kj = i,$$

$$ki = -ik = j.$$

In all the above examples, and in many more, some of which will be studied later in this chapter, we have two processes present, akin to ordinary addition and multiplication. Because of the wealth of such examples and the algebra that can be abstracted from them it is found useful to define a *ring* as a set with addition, subtraction and multiplication, but not division. Note that when addition and multiplication are both present there is a definite distinction between them, caused by the Distributive Law.

Addition. For a set to be a ring we must first ensure that it possesses an additive structure and, to enable much algebra to be done with it, this structure must satisfy the group axioms. Thus our set must be a group under addition. It must further be an Abelian group, the reason for this being given later in this section.

Multiplication. The set, in order to be a ring, must also possess a multiplicative structure. This must be Associative (for the theory to be fruitful) but division need not be ensured (i.e. there need be no inverses, as in examples 1–4 above, nor need there be a unity, or neutral element for multiplication, as in example 4). If there is also division we have a less general structure, known as a *field* (or *skew field* if not commutative) which is the subject of chapter 4.

The Distributive Law. It is often possible to have a set with both addition and multiplication, but which are not connected in any way. Thus on any group we could impose a second group operation having no connection with the first. But for the two operations to give us any more structure than purely that of a group they must be linked, and the Distributive Law provides the obvious linkage, as it does in the case of numbers.

Abstract definition of a ring

A set R of elements forms a ring if to any two elements x and y of R taken in a particular order there are associated two elements of R, called their sum and product and denoted by $x+y$ and xy respectively, both being unique, and such that the following laws are satisfied:

A1 *For any two elements* x *and* y, $x+y = y+x$.
A2 *For any three elements* x, y *and* z, $(x+y)+z = x+(y+z)$.
A3 *There is an element called the zero and denoted by* 0 *which has the property that* $x+0 = 0+x = x$ *for all elements* x.
A4 *Corresponding to each element* x *there is an element* $-x$ *called the negative of* x *which has the property that*

$$x+-x = -x+x = 0.$$

M2 *For any three elements* x, y *and* z, $(xy)z = x(yz)$.
D1 *For any three elements* x, y *and* z,

$$(x+y)z = xz+yz, \quad x(y+z) = xy+xz.$$

The numbers of the laws refer to chapter 10 of volume 1.

Note that we must have both forms of the Distributive Law since multiplication is not assumed commutative.

The need for A1

In group theory there is no need for the operation to be commutative. As soon as we introduce a second operation and D1, however, A1 becomes essential for any useful algebra. The reason appears if we try to extend D1 in an obvious way.

Consider $(x+y)(z+w)$. This can be thought of in two ways.

(*a*) $(x+y)(z+w) = x(z+w)+y(z+w)$ by the first part of D1,
$= xz+xw+yz+yw$ by the second part of D1;

(*b*) $(x+y)(z+w) = (x+y)z+(x+y)w$ by the second part of D1,
$= xz+yz+xw+yw$ by the first part of D1.

Thus we have two natural expressions for the expansion, which are equal only if the elements xw and yz commute under addition, and so A1 is necessary to prevent formidable difficulties arising. But note that there is no such need to assume commutativity of multiplication.

Clearly rings are more restricted than groups, since while all rings are groups by no means all groups are rings. Thus our results are more restricted in application, but on the other hand there is more algebra available since we assume more axioms initially when dealing with rings.

It turns out in fact that there is comparatively little *elementary* ring theory. Unlike the theory of groups, which gives us such an important theorem as Lagrange's Theorem at a very early stage, rings do not give us any startling results at an elementary level. They are important at a more advanced stage (the advanced theory depends on the concept of an 'ideal' and a very little of this type of work is dealt with in chapter 7) but at present we must be content with very straightforward work on the subject.

3.2. Elementary results

A ring is always an Abelian group under addition, and hence all the group theorems apply to this part of its structure. In particular the following elementary results are true, the proofs of which are given in volume 1, chapter 10, and will not be repeated here.

Theorem 3.2.1. *The Associative Law extends to the sum of more than three elements, i.e. the sum* $x_1+x_2+x_3+\ldots+x_n$ *is independent of the order in which we perform the addition of pairs, and is also independent of the order in which the* x_i*'s are written, since the Commutative Law holds.*

Theorem 3.2.2. *The uniqueness of zero.*

There cannot be two different elements each having the zero property.

Theorem 3.2.3. *The uniqueness of the negative.*

For any x *there cannot be two different elements each with the property possessed by* $-x$.

Theorem 3.2.4. *The negative of* 0 *is* 0.

Theorem 3.2.5. *The negative of* $-x$ *is* x.

Theorem 3.2.6. *The Cancellation Law of addition.*

$$x+y = x+z \Rightarrow y = z.$$

(Note that here we need only one Cancellation Law, since addition is commutative.)

Theorem 3.2.7. *The equation* $x+a = b$ *has the unique solution* $x = b-a$. (*Which is the same as* $-a+b$ *of course.*)

The Associative Law of multiplication may be extended to more than three elements in the same way, and so we have the following theorem.

Theorem 3.2.8. *The product* $x_1 x_2 x_3 \ldots x_n$ *is independent of the order in which we perform the multiplication of pairs, provided in this case that we keep the* x_i*'s in the same order* (*since multiplication is not necessarily commutative*).

Because of the Distributive Law we have the following theorems, which were also proved in volume 1, chapter 10.

Theorem 3.2.9. *The Distributive Law may be extended so that*

$$(x+y+z)w = xw+yw+zw,$$

$$x(y+z+w) = xy+xz+xw, \text{ etc.}$$

Theorem 3.2.10. $x0 = 0x = 0$ *for any* x.

Theorem 3.2.11. $(-x)y = -(xy), \quad x(-y) = -(xy),$

$$(-x)(-y) = xy.$$

Finally we have a theorem about the negative of $x+y$. Note that because of the commutativity we do not need to reverse the order of x and y, as was the case for groups.

Theorem 3.2.12. $-(x+y) = -x+-y$: *i.e. the negative of* $x+y$ *is the sum of the negatives of* x *and* y.

3.3. Basic types of rings

In §3.1 we gave the basic structure necessary for a set to be a ring. We note that there are very few multiplicative axioms —merely that product is associative—and we do not want to restrict our sets by imposing further postulates on them. On the other hand many rings possess additional structure, and this enables more algebra to be done.

Commutative rings

The most obvious additional axiom is that of commutativity of multiplication, that $xy = yx$ for all x and y in the ring. Such a ring is said to be a *commutative ring* (addition is of course always commutative). In the examples of §3.1 the first four are all commutative rings, but the ring of quaternions is not.

Rings with unity

A ring admits of multiplication but not necessarily division. For the latter to be possible we must have a neutral element, and also inverses, which are defined in terms of the neutral element. If both these are present the structure becomes a *field*, and such are studied in the next chapter. However, a ring may have a neutral element yet not possess inverses (e.g. the ring of integers). Such a ring is called a *ring with unity*. The neutral element is called the unity and is usually denoted by 1 (in the case of groups we use the symbol e but, partly because of the separate historical development of the two subjects, in ring theory 1 is customary), and has the property that $x.1 = 1.x = x$ for all elements x in the ring. It is clearly unique, if it exists, by the same proof as that of the uniqueness of the zero. Distinguish between the *unity* and *units*, the latter term being used technically in a different sense (see chapter 5).

Examples 1, 2, 3 and 5 in §3.1 are all rings with unity, whereas 4 is not. Note that the quaternions have a unity but are not commutative.

While the above types of ring clearly have more properties than general rings, they are not in fact very different: the extra assumptions do not have far reaching effects. But if we take the next step, that of assuming the existence of inverses, we do obtain structures that are quite different in their properties: they of course possess all the ring properties, but the additional ones which they have because of the division possibilities are extremely important. As we have said, such structures are called fields and form the subject of chapter 4.

Of our examples in §3.1, only the quaternions and the residues modulo a prime have inverses. All except residues modulo a composite number have the property, however, of the Cancellation Law of multiplication—that either

$$xy = xz \quad \text{or} \quad yx = zx \Rightarrow y = z,$$

unless $x = 0$. While this is a consequence of the existence of inverses, it may nevertheless exist in a ring even though the latter is not a field. Rings with this property (and also with commutativity of multiplication and a unity) are called *integral domains*. These also exhibit many additional properties and are discussed in chapter 5.

3.4. The additive and multiplicative structures of a ring

A ring has two structures, the additive and multiplicative. These are not independent, because of the Distributive Law, and while the results of this section will on the whole refer to one or the other the connection will sometimes be made use of.

The additive structure

The additive structure is that of a group and so all the results of group theory apply to it, with the additional fact of commutativity being applicable.

Additive powers of an element

With the addition notation, the positive powers of an element x are x, $x+x$, $x+x+x$, etc. We write $x+x = 2x$, $x+x+x = 3x$, etc., where it is important to notice that nx does *not* mean the ring product of n and x: indeed there may be no such element n. nx is merely a shorthand for the sum of x with itself n times.

We may define the unit power $1x$ to be x, and the zero power $0x$ to be 0 (this is true also for the ring product of 0 and x) and the negative power $(-n)x$ to be $n(-x)$, i.e. the sum of the negative $-x$ with itself n times.

Theorem 3.4.1. *The Index Laws for addition.*

If m and n are integers, positive, negative or zero, then

(i) $(m+n)\,x = mx+nx$;

(ii) $(m-n)\,x = mx-nx$;

(iii) $(mn)x = m(nx)$.

The proofs of these, using the general multiplicative notation, were given in volume 1 (p. 210), and need not be reproduced here, being quite straightforward once we are clear about what is to be proved. It must be emphasized again that m and n are not elements of the ring and that we are not using the ring product at all here: (i) is not the Distributive Law but merely states that the sum of $(m+n)$ x's is the same as the sum of m of them added to the sum of n of them.

Theorem 3.4.2. *If n is an integer, positive, negative or zero, then $n(x+y) = nx+ny$.*

This is an immediate consequence of the definition of nx, and of the commutativity of addition.

To give an example of the above work we consider the ring of even integers. The positive powers of, say, the element 4 are 4, 8, 12, 16, ... and we can write 12, for example, as 3.4, where 4 is an element of the ring but 3 is not, and 3.4 stands as a shorthand for 4+4+4. The negative powers are of course -4, -8, -12, ... and the zero is 0.

Additive powers in a ring with unity

Suppose our ring has a unity, which we will call e for convenience here, though 1 is more usual. Then its powers are $2e$, $3e$, ...

Thereom 3.4.3. *In a ring with unity e, $nx = (ne)x$, where the product on the* R.H.S. *is the ring product.*

$$\begin{aligned} nx &= x+x+\ldots+x \; (n \text{ terms}) \\ &= ex+ex+\ldots+ex \\ &= (e+e+\ldots+e)x \text{ by the Distributive Law,} \\ &= (ne)x \text{ by definition of } ne. \end{aligned}$$

Thus in a ring with unity nx may be thought of as a product. If we call the unity 1, we may then denote ne by the number n and thus identify the number n with an element of the ring. This causes no confusion, since by theorem 3.4.3 nx, which may mean either the sum of n x's as before, or the product of the element n with x, has the same value whichever way it is thought of.

Additive order of an element

Since the additive structure is that of a group, any element x has an additive order, namely the least positive n for which $nx = 0$, if this exists, and the elements $x, 2x, 3x, \ldots (n-1)x$ are all distinct. (If no such n exists, the addition order is infinite.) Further, if the ring has finite order, n is a factor of this order, by theorem 1.9.4.

We can in general say nothing about the additive structure, except that under it the ring forms an Abelian group. For any Abelian group may be made into a ring, a zero-ring, as is shown below.

Theorem 3.4.4. *If R is any Abelian group and if we define a product in R by $xy = 0$ for any x and y in R, then under this product and the group addition R forms a ring.*

The Distributive Law and the Associative Law of multiplication are immediately seen to be true, since $0+0 = 0$.

Such rings are called *zero-rings*.

Thus for general rings the additive group is arbitrary. If, however, we consider only rings with unity the possible such groups are restricted, by the following theorem.

Theorem 3.4.5. *If R is a ring with a unity e, and the additive order of e is m, then the order of any element of R is a factor of m.*

By theorem 3.4.3 if x is any element,

$$\begin{aligned} mx &= (me)x \\ &= 0x, \text{ since } e \text{ has order } m, \\ &= 0. \end{aligned}$$

Hence the result follows.

If e has finite order, so that a finite m exists, m is called the *characteristic* of the ring. If e has infinite order we say that the characteristic is zero.

The characteristic is not a very important concept in ring theory. It is much more fruitful in the case of a field, as will be seen in the next chapter.

The multiplicative structure

In rings as such the multiplicative structure is by no means as rich as the additive. In general all we have is the Associative Law, and we can do very little algebra with this alone.

Multiplicative powers of an element

As in group theory, we write x^2 for xx, x^3 for xxx, and so on for positive powers of x. The unit power x^1 is x, but we cannot usefully define x^0, unless the ring has a unity, in which case of course x^0 is understood to be this unity. Nor can we define negative powers (unless there are inverses). These are severe restrictions, but within them we have the index laws applying as usual.

Theorem 3.4.6. *The Index Laws for multiplication.*

If m and n are positive *integers, then*

(i) $x^m x^n = x^{m+n}$;

(ii) $(x^m)^n = x^{mn}$.

The proof is as for groups, and is immediate from the definition.

Theorem 3.4.7. *If the elements x and y commute, so that $xy = yx$, then so do their positive powers: $x^m y^n = y^n x^m$. We also have $(xy)^n = x^n y^n$.*

The proof is simple, and will be omitted (see volume 1, p. 211).

Multiplicative order of an element

Since the multiplicative structure does not in general form a group, we must not expect that there always exists a positive integer n such that $x^n = 1$, even in finite rings. If such an n does exist then we may call it the multiplicative *order* of x (it is the order of the multiplicative subgroup generated by x), and we may say that x is a root of unity.

3.5. Examples of rings

We have already met most of the examples below as groups, and some of them are also fields and will be dealt with as such in the next chapter. In comparatively few cases is the chief structure the ring structure.

Rings of numbers

The sets of the rationals, the reals and the complex numbers all form rings under the usual processes. They are all commutative and have a unity. As we have already seen, the quaternions form a non-commutative ring with a unity.

The ring of integers and similar rings

The complete set of integers forms a ring, commutative and with a unity. We have already met this, and it is very easy to show that the necessary axioms are satisfied (as before the detailed proof that an example *is* a ring will be omitted). The ring of integers is a fairly important structure *as a ring*, since division is not always possible. It does however have the cancellation property of multiplication and is thus an integral domain.

The *Gaussian integers* (all complex numbers of the form $a+bi$, where a and b are integers) form a ring.

The set of numbers $a+b\sqrt{n}$, where a and b are integers and n is a fixed positive integer, form a ring under the usual rules. If n is a perfect square the ring is just the ring of integers. If n is not a perfect square, note that although the additive structures of the rings are isomorphic for all n (see p. 220 of volume 1), this is not true for the multiplicative structures. For

$$(a_1+b_1\sqrt{n})\,(a_2+b_2\sqrt{n}) = a_1a_2+b_1b_2n+(a_1b_2+b_1a_2)\sqrt{n},$$

and this depends for its form on n.

The set of integral multiples of a fixed positive integer n form a ring under the usual rules, since $an+bn = (a+b)n$ and $(an)(bn) = (abn)n$. These rings are not all the same (since the integer n enters into the multiplication 'coefficient') and are not the same as the ring of integers. They are commutative but have no unity (unless $n = 1$, when we have the ring of integers). In the case $n = 2$ we obtain the ring of even integers.

Rings of residues

For any positive integer n the set of residues modulo n admits of addition, subtraction and multiplication, and the Associative, Distributive and both Commutative Laws are clearly satisfied. Thus for each n we have a ring, commutative and with a unity. The additive structure of these rings is of course that of the corresponding cyclic group.

If n is prime then division exists and we have a field. If n is composite there is no general division and inverses do not in general exist (r in fact has an inverse if and only if it is prime to n); nor does the Cancellation Law of multiplication apply since $rx = ry \Rightarrow x = y$ only if r is prime to n. We thus have examples of finite rings which are not fields: such structures are useful in giving counter-examples.

Polynomials

Various sets of polynomials form important examples of rings and will be dealt with in some detail in the next section.

Rings of functions

Consider the set of all functions $f(x)$ defined for all real values of x, where $f(x)$ is real and unique. Then the sum and product of any two such functions exist and are unique, and, of course are themselves functions. It is easy to see that all the usual laws apply, and that the set of functions forms a ring. (Note that an element is a *function*, defined at all points x, and is not just an individual value of the function.)

The concept is a very general one, and we may obtain various rings by generalising in the following ways.

(*a*) In the above we took the object set of the function to be the set of real numbers—$f(x)$ was defined for all real x. We may however take *any* object set, so that $f(x)$ need be defined only for certain specified x, which need not even be real numbers, provided only that all functions of our set are defined over the same object set. Thus we may have functions defined over the integers (i.e. sequences), those defined for $0 \leqslant x \leqslant 1$, those defined for each element of a given group, ring or other set and so on. If the object set is finite we may represent a function $f(x)$ by an ordered set of images $(f_1, f_2, \ldots, f_n)$ where n is the number of elements in the object set.

(*b*) The values $f(x)$ were taken to be real (i.e. the image set was the set of real numbers). We may generalise the image set to be any ring.

(*c*) We may restrict the admissible functions, instead of allowing any. Thus where the image and object sets are both continuous sets of real numbers we may restrict our functions to be the continuous functions (over the given object set), or the differentiable functions, or indeed to any type provided the sum and product of two such are of the required type.

We thus obtain the generalised concept of a ring of mappings from any set into any ring. The ring obtained will be commutative if the image ring is (for then $fg = gf$), and will have a unity if the image ring has (viz. the mapping which has the value 1 at every point). It may seem at first sight that these mappings will form a field, but this is not so, since not only does 0 have no inverse (0 being the mapping which has the zero value at *every* point), but neither does any mapping that has the zero value at any one or more points. We do not even have the Cancellation Law, as is seen in the following example.

Example. Consider the ring of all continuous real-valued functions defined for all real values x. Define the following such functions:

$$f(x) = 0 \quad \text{for all} \quad x \leqslant 0 \quad \text{and} = x \quad \text{for all} \quad x > 0,$$

$$g(x) = x+2 \quad \text{for all} \quad x \leqslant 0 \quad \text{and} = 2 \quad \text{for all} \quad x > 0,$$

$$h(x) = -x+2 \quad \text{for all} \quad x \leqslant 0 \quad \text{and} = 2 \quad \text{for all} \quad x > 0.$$

Then fg and fh are both 0 for all $x \leqslant 0$ and are both $2x$ for all $x > 0$, hence $fg = fh$. But $f \neq 0$ and $g \neq h$ and so the Cancellation Law is not true in this case.

A ring from set theory

Consider the set of all possible subsets of a given set S. Then we saw (volume 1, p. 246), that these form an Abelian group under $A+B = (A \cup B)-(A \cap B)$.

Now define $AB = A \cap B$.

Then multiplication is commutative and associative, and the Distributive Law may easily be verified. Thus the set of subsets of S with these definitions of sum and product forms a commutative ring. It has a unity, the whole set S. Each element, except $\varnothing$, has additive order 2, since $A+A = \varnothing$. Since

$$A^2 = A^3 = \dots = A$$

no element except S is a root of unity, and no element except S has an inverse. (If A^{-1} exists then $A^2A^{-1} = AA^{-1} = S$, i.e. $A = S$.)

Direct sums

In the same way as we obtain a group, called the direct product, from any two given groups, so we may obtain a direct sum of any two rings, and this will itself be a ring. It gives us a method of obtaining new rings from old, and of investigating the structure of complicated rings in terms of simpler ones. The definition is a natural extension of that for groups.

Suppose we have any two rings R and S and consider the set of ordered pairs (x, y) where $x \in R$ and $y \in S$. We define the sum and product of two pairs by

$$(x_1, y_1)+(x_2, y_2) = (x_1+x_2, y_1+y_2),$$
$$(x_1, y_1).(x_2, y_2) = (x_1x_2, y_1y_2),$$

where the sum and product of the x's are those defined in R, and those of the y's are those defined in S. Then the sum and product of the pairs are unique and in the set, under this definition of sum the pairs form a group (the zero is $(0, 0)$ where the two 0's are the zeros in R and S and the negative of (x, y) is $(-x, -y)$), and product is associative since it is in R and S, and finally the Distributive Law holds since it does so in R and S. Hence the set of ordered pairs under these definitions of sum and product form a ring, called the *direct sum* of R and S and denoted by $R \oplus S$. (We use *direct sum* here, rather than *direct product*, since the group structures are Abelian, for which the additive notation is employed.)

If R and S are finite of orders m and n, then the order of $R \oplus S$ is mn. $R \oplus S$ is commutative if and only if both R and S

are, and has a unity if and only if both R and S have, the unity in this case being (e, f) where e and f are the unities of R and S respectively.

So far the theory is analogous to that for groups, but if we try to restrict our rings still further then we cease to obtain the expected results. Thus if R and S are fields (with inverses defined) it might be thought that $R \oplus S$ is a field, but this is not so: in fact $R \oplus S$ is *never* a field. Furthermore, $R \oplus S$ is never an integral domain, even when both R and S are. We prove these results below.

Theorem 3.5.1. *There always exist non-zero elements of $R \oplus S$ which have no inverses, so that $R \oplus S$ is never a field.*

If (x, y) has an inverse it must be (x^{-1}, y^{-1}) where x^{-1} and y^{-1} are inverses in R and S. (For if (ξ, η) is the inverse we have $(x\xi, y\eta) = (e, f)$.)

Now consider the element $(0, y)$ where $y \neq 0$. Then this is non-zero, and its inverse, if it exists, must be $(0^{-1}, y^{-1})$. But 0 has no inverse, and so neither has $(0, y)$. We see in fact that if R and S are fields, the elements of $R \oplus S$ which have no inverses are those of either of the forms $(0, y)$ or $(x, 0)$.

Theorem 3.5.2. *The Cancellation Law never holds in $R \oplus S$, so that $R \oplus S$ is never an integral domain.* (*We ignore the trivial case.*)

We wish to find elements a, b and c such that $ab = ac$ but neither $a = 0$ nor $b = c$.

Let $a = (0, y)$ for some non-zero y in S, and let

$$b = (x_1, 0), \quad c = (x_2, 0) \quad \text{where} \quad x_1 \neq x_2.$$

Then $b \neq c$ and $a \neq 0$, but $ab = ac = (0, 0)$.

We recall that if we form the direct product of the group of reals under addition with itself we obtain the group of two-dimensional vectors. We may expect that we could form the direct sum of the *ring* of reals with itself and so obtain a good multiplication structure for vectors. We may indeed obtain a *ring* by this means, but the above theorems show us that, although the reals form a field, the vectors thus obtained do not, neither do they obey the Cancellation Law, and so are of very limited use.

We may extend the idea of direct sum to 3 or more rings in an obvious way, by combining ordered n-tuples, and it is immediate that

$$R \oplus S \cong S \oplus R \quad \text{and} \quad (R \oplus S) \oplus T \cong R \oplus (S \oplus T) \cong R \oplus S \oplus T.$$

Example. Let R_n denote the ring of residues modulo n. Then $R_2 \oplus R_3 \cong R_6$.

The 6 elements of $R_2 \oplus R_3$ are (0, 0), (0, 1), (0, 2), (1, 0), (1, 1), (1, 2). Make these correspond to the elements of R_6 as follows:

$$(0, 0) \leftrightarrow 0, \quad (1, 1) \leftrightarrow 1, \quad (0, 2) \leftrightarrow 2, \quad (1, 0) \leftrightarrow 3,$$
$$(0, 1) \leftrightarrow 4 \quad \text{and} \quad (1, 2) \leftrightarrow 5.$$

Then it is easily seen that this gives an isomorphism (the additive structure is clearly preserved, and it follows that the multiplicative structure is also, since any element, in either ring, may be expressed as $r.1$, when r is an integer).

Rings of given order

To any Abelian group there exists a zero-ring (defined by $xy = 0$ for all x and y). We will ignore such trivial rings in what follows.

To identify all rings of a certain order n, we notice that the additive structure must be an Abelian group of order n, and we start by finding all such groups (if possible). For each group there may or may not exist a corresponding ring, and in some cases there may be more than one such ring.

There is at least one ring of any given order, viz. the ring of residues modulo that order.

Theorem 3.5.3. *The only ring with unity whose additive structure is the cyclic group C_n is the ring of residues R_n.*

By theorem 3.4.5, if the unity e had additive order m, then the additive order of any element must be a factor of m. But there exist elements (the generators) of order n, and hence e has order n and so is a generator. We may take it, without loss of generality, to be the residue 1, so that the general residue r may be written as the sum of r 1's. Hence by the Distributive Law we see that the ring product of r and s must be the residue rs, and so the ring is merely R_n.

It follows that there is just one ring with unity of any prime order, and also of order 6 (since C_6 is the only *Abelian* group of order 6).

The investigation of rings with other Abelian groups as additive structure, and hence the identification of all rings of a given non-prime order is tedious, even if we restrict ourselves to rings with unity, and will not be attempted here.

3.6. Polynomial rings

Sets of polynomials form one of the most important and interesting types of ring. Two polynomials can clearly be added, subtracted and multiplied, but not always divided exactly, and hence we expect sets of polynomials to be rings, rather than fields or merely groups and, provided the coefficients are themselves elements of a ring, this is so.

Many of the most important properties of polynomials, those connected with factorisation, depend on the Cancellation Law of multiplication, i.e. on the polynomials forming an integral domain, and are dealt with in chapter 5. Here we will give the general abstract definition of polynomials over any ring and a few fundamental properties.

Much of this work has been already dealt with in volume 1 (chapter 7) and, although it is not necessary for the reader to have read that chapter, we shall deal fairly briefly, and in an abstract way, with the basic definitions and algebra. The reader who has volume 1 to hand is advised to refer to it for elaboration where necessary.

Polynomials over a ring in 1 *indeterminate*

Suppose we have any ring R, not necessarily commutative nor having a unity. Then we define a polynomial P as the ordered set $(a_0, a_1, \ldots)$ where all the a_i's are elements of R and every a_i after a certain one (say after a_n) is zero. The elements a_i are called the *coefficients* of P and n is the *degree* of P and is denoted by $d(P)$. We define the sum and product of two polynomials P and Q as follows.

If

$$P = (a_i) \quad \text{and} \quad Q = (b_i)$$

then $$P+Q = (a_i+b_i) \quad \text{and} \quad PQ = (c_i),$$
where $$c_i = \sum_{r+s=i} a_r b_s,$$
the sums and products of the coefficients being, of course, as defined in R.

Note. For this definition to be meaningful R must possess both addition and multiplication (and must be a ring if the set of polynomials is to be a ring), although multiplication need not be commutative.

It requires only a little thought to see that the set of all polynomials with these definitions of sum and product forms a ring. The zero polynomial is $(0, 0, \ldots)$ and the negative of (a_i) is $(-a_i)$ where 0 and $-a_i$ are the zero and negative in R. This ring is called the *ring of polynomials over* R and denoted by $R[x]$. Strictly speaking we should say that $R[x]$ is the ring of polynomials over R in one indeterminate.

The above abstract definition is of course based on our elementary ideas of polynomials. To obtain the more usual everyday notation we note that
$$(a_i) = \sum_n (0, 0, \ldots, a_n, 0, 0, \ldots)$$
and we then write the polynomial with zero as every coefficient except the $(n+1)$th and a_n as the $(n+1)$th in the form $a_n x^n$, where x^n is merely a symbol and at present has no other meaning. We then have $(a_i) = a_0+a_1x+a_2x^2+\ldots$. The symbol x is called an *indeterminate*, and must not be thought of as being anything other than a symbol. Its importance lies in the fact that we may substitute particular elements of R in its place, but we must take care not to assume that x itself 'stands for' an element of R, or indeed for any particular element or object.

The form of $R[x]$ *for special types of ring* R

R is finite. The ring of polynomials $R[x]$ will still be infinite.

R is commutative. $R[x]$ will also be commutative, as can be seen at once from the definition of product, which is now symmetrical. The converse is also true: if $R[x]$ is commutative so is R, as can be seen by considering products of polynomials which have every coefficient except the first zero.

R has a unity. If R has a unity 1, then $R[x]$ has a unity $(1, 0, 0, \ldots)$, as can easily be verified.

In this, common, case we may obtain the familiar form of writing polynomials more systematically. Denote the polynomial $(0, 1, 0, \ldots)$ where all other coefficients are 0, by x. Then the product $x.x$ (i.e. the polynomial product) is $(0, 0, 1, 0, \ldots)$ by the rule for multiplying polynomials, $x.x.x$ is $(0, 0, 0, 1, 0, \ldots)$ and so on, so that (a_i) may be written in the form

$$a_0.(1, 0, \ldots)+a_1.(0, 1, 0, \ldots)+\ldots$$

or $a_0+a_1x+a_2x^2+\ldots$, where a_0 is short for $a_0.1$. x is still an indeterminate, but is identified with the particular polynomial $1.x$, or more precisely with $0+1.x+0.x^2+0.x^3+\ldots$.

R satisfies the Cancellation Law. Then $R[x]$ also satisfies this law. For suppose $PQ = PR$, so that $P(Q-R) = 0$. Let $P = (a_i)$ and $Q-R = (b_i)$, and suppose $d(P) = m$, $d(Q-R) = n$, so that a_m and b_n are not zero, but $a_i = 0$ for $i > m$ and $b_j = 0$ for $j > n$. Then the $(m+n+1)$th coefficient of $P(Q-R)$ is $\sum_{r+s=m+n} a_r b_s = a_m b_n$ and is zero since $P(Q-R) = 0$. Hence, since R satisfies the Cancellation Law, either a_m or $b_n = 0$, which is contrary to hypothesis unless either $P = 0$ or $Q-R = 0$, i.e. unless $P = 0$ or $Q = R$. Similarly if $QP = RP$.

Conversely if $R[x]$ satisfies the Cancellation Law so must R, as may be proved by considering the polynomials with every coefficient zero except the first.

Note. In this case $d(PQ) = d(P)+d(Q)$, as may be proved by a method similar to that used in the above proof. [See theorem 5.6.1.]

The division algorithm. As we have indicated, factorisation theory is not applicable to polynomials over any ring, and a general statement of the fundamental theorem known as the 'Division Algorithm' requires R to be a field. But a modified form of the algorithm holds generally, and leads to the remainder theorem and factor theorem.

Theorem 3.6.1. *The division algorithm.*

If the coefficient ring R has a unity 1, *and if A and B are two polynomials in* $R[x]$ *with B having leading* (*i.e. highest non-zero*) *coefficient* 1, *then there exist unique polynomials* Q *and* S *such that* $A = QB+S$ *and such that* $d(S) < d(B)$. (S *may be* 0.)

To prove that Q and S exist we use induction, and assume the result true for all polynomials of degree less than A, i.e. if A' has degree less than $d(A)$, then Q' and S' exist, with

$$d(S') < d(B), \quad \text{such that} \quad A' = Q'B+S'.$$

If $d(A) < d(B)$ the result is trivial, with $Q = 0$. Suppose then that $d(A) \geqslant d(B)$. Let $A = a_n x^n + \ldots$ and $B = x^m + \ldots$ (remembering that B must have leading coefficient 1) and consider the polynomial $A' = A - a_n x^{n-m} B$. Then A' has degree at most $(n-1)$, i.e. less than $d(A)$, and so by the inductive hypothesis we can find Q' and S' with $d(S') < d(B)$ such that $A' = Q'B+S'$. Hence $A = QB+S$, where $Q = Q'+a_n x^{n-m}$ and $S = S'$, so that $d(S) < d(B)$ as required. The result is true for

$$d(A) < d(B)$$

and so the induction can start. Hence the result is true generally.

To prove uniqueness, suppose that

$A = QB+S$ and $A = Q'B+S'$ with $d(S)$ and $d(S')$ both $< d(B)$.

Then $(Q-Q')\,B = S-S'$, where $d(S-S') < d(B)$. But if $d(Q-Q') = k$ and $d(B) = m$, so that $Q-Q' = c_k x^k + \ldots$ and $B = x^m + \ldots$, $(Q-Q')\,B = c_k x^{k+m} + \ldots$ and so has degree $k+m$, and this is $\geqslant m$ unless $Q-Q' = 0$. Hence, since

$$d(S-S') < m$$

we must have $Q-Q'$, and so $S-S'$, $= 0$ and the uniqueness follows.

If R is a field the algorithm holds for any non-zero B (see volume 1, theorem 7.4.2), but for general rings R (with unity) we must impose the above restriction on the leading coefficient of B. Note that we have not assumed R commutative, although in most practical cases this will be the case.

Euclid's algorithm, to find an H.C.F. for two polynomials, depends on repeated use of the division algorithm with the divisor not necessarily having leading coefficient unity, and so cannot be applied in the general case.

Polynomial functions. Much of the value of the study of polynomials arises from the fact that we may replace the indeterminate x in $P(x) = a_n x^n + \dots + a_0$ by an element of the coefficient ring R, and obtain another element of R by adding and subtracting according to the usual rules in R. Thus if we replace x by the element r we obtain $P(r) = a_n r^n + \dots$, and the mapping $r \leftrightarrow P(r)$ is a mapping of R into itself, called a *polynomial function*. The sum of two polynomials gives rise, when considered as a function, to the sum of the corresponding functions, and so does the product, *provided R is commutative* (more precisely, provided all the coefficients commute with all elements of R, since we obtain terms of the form $a_\lambda r^\lambda b_\mu r^\mu$ and wish to put them in the form $a_\lambda b_\mu r^{\lambda+\mu}$). It is convenient therefore to restrict ourselves to commutative coefficient rings and also, since we wish to use the division algorithm, to rings with unity.

Theorem 3.6.2. *The remainder theorem.*

If $A(x)$ is a polynomial over the coefficient ring R, where R is commutative and has a unity, then there exists a (*unique*) *polynomial $Q(x)$ such that $A = (x-c)\,Q + A(c)$, where c is a given element of R.*

Apply the division algorithm with $B = x-c$ (note that this has leading coefficient unity). Then there exist Q and S such that $A = (x-c)\,Q+S$ and $d(S) < d(x-c)$, i.e. $d(S) = 0$ and so S is a constant (i.e. an element of R). Thus

$$A(x) = (x-c)\,Q(x)+S,$$

and this is true when we substitute any element of R for x. Substitute c for x. Then $A(c) = 0.Q(c)+S$, so that $S = A(c)$ and the result follows.

Corollary. *The factor theorem.*

$(x-c)$ is a factor of $A(x)$ if and only if $A(c) = 0$. We say in this case that c is a *zero* of $A(x)$.

Polynomials over a ring in more than 1 *indeterminate*

If R is a ring then so is $R[x]$ and so we may form polynomials over $R[x]$ and obtain a further ring, which we will write as $R[x_1, x_2]$. It consists merely of the set of formal sums $\Sigma a_{ij} x_1^i x_2^j$, where $a_{ij} = 0$ for all i and j greater than some m, i.e. of the set of finite polynomials in two indeterminates x_1 and x_2.

We may similarly define polynomials in n indeterminates $x_1, x_2, \ldots, x_n$, the best way of so doing being by induction by

$$R[x_1, x_2, \ldots, x_n] = R[x_1, x_2, \ldots, x_{n-1}]\,[x_n].$$

We obtain a ring at each stage.

By what we previously showed, if R is commutative, has a unity or satisfies the Cancellation Law, then the ring of polynomials in n indeterminates over R is commutative, has a unity or satisfies the Cancellation Law.

3.7. Subrings

When considering any algebraic structure we are always interested in subsets of the elements which themselves form an algebraic structure of the same type, under the same rules of combination. In the case of rings such substructures are called *subrings*. Obvious examples are the ring of even integers as a subring of the ring of integers, and the latter as a subring of the ring of reals. For a finite example we note that the residues 0, 2 and 4 (modulo 6) may be added, subtracted and multiplied within themselves and thus form a subring of the ring of residues modulo 6.

For a subset S of a ring R to be a subring it is necessary that the sum and product of any two elements of S must be in S. We also require the zero of R to be in S and also the negative of any element of S. Then the two Assocative Laws, the Commutative Law of addition and the Distributive Law are automatically satisfied in S, since they are in R. Thus we have the following definition.

Definition of a subring

A subset S of a ring R is a subring of R if:

(*a*) Given any pair (s_1, s_2) of elements of S the sum s_1+s_2 (as defined in R) is in S.

(*b*) Given any ordered pair (s_1, s_2) of elements of S the product $s_1 s_2$ (as defined in R) is in S.

(*c*) The zero of R is in S.

(*d*) Given any element s in S the negative $-s$ (as defined in R) is in S.

In other words, S must be a subgroup of the additive group of R, and the product of any two elements of S must be in S.

Note that even if R has a unity, S need not contain this. For example, the subring of even integers of the ring of integers does not contain the unity 1.

If R is commutative then S must be also, but the converse need not hold. If R satisfies the Cancellation Law then so must S, but again the converse need not hold.

Examples of subrings

Trivial subrings. The subset consisting of the zero alone is a subring, since it forms a subgroup under addition and the product of 0 with itself is 0 and is in the subset. Also the whole ring R is a subring. These two are called the *trivial* subrings of R—all other subrings are called *non-trivial* or *proper*.

Subrings generated by subsets of elements. Suppose we have a subset T of elements of a ring R, and consider all elements of R which may be obtained by adding or multiplying a finite number of elements of T, or 0, or the negatives of elements of T, some elements possibly being repeated. The set obtained is clearly a subring, and is known as the subring generated by the elements of T. T of course need not itself be a subring.

In many cases the subring obtained will be the trivial subring R. Thus in the ring of integers the subring generated by 1 is the whole ring, while in the ring of Gaussian integers the subring generated by the two elements $1+i$ and $2+3i$ (to take an example) is the whole ring.

Examples where the subring is non-trivial exist in the ring of Gaussian integers where the subring generated by the element 1 (i.e. $1+0i$) is the ring of integers. Similarly in the ring of integers the subring generated by the integer 3 is the subring of multiples of 3.

In a group the powers of a single element always form a subgroup (that generated by the element). In a ring, however, neither the additive powers or multiplicative ones, nor indeed both together, form a subring. The subring generated by a single element is usually considerably larger than the set of powers. As an example consider the ring of rationals, and the subring generated by the element $\frac{1}{2}$. This is easily seen to include all fractions of the form $a/2^n$ for all $n \geqslant 0$.

Subrings of polynomial rings. An important subring of $R[x]$ is the set of all polynomials of the form $(a, 0, 0, \ldots)$ where all coefficients are zero except the first. This is clearly isomorphic to R with the element given corresponding to the element a, and will sometimes be identified with the coefficient ring R itself.

If R is commutative, so that the idea of a polynomial function is useful, then the set of polynomials in $R[x]$ which have given zeros forms a subring of $R[x]$: for if $P(c) = Q(c) = 0$, then both $P(c)+Q(c)$ and $P(c)\,Q(c) = 0$. In particular the polynomials with a zero at 0, i.e. all those with first coefficient zero, form a subring.

Similarly all those functions in a given ring of functions which have given zeros form a subring.

Direct sums

The set of all elements $(x, 0)$ in $R \oplus S$, where 0 is the zero of S, forms a subring isomorphic to R, and the set of elements $(0, y)$ forms a subring isomorphic to S.

Subrings of the rings of residues

Theorem 3.7.1. *Corresponding to each factor m of n there is a subring of the ring of residues modulo n, consisting of all those residues which are multiples of m.*

Let $n = m.r$. Then the sum $\lambda m + \mu m = (\lambda+\mu)\, m$ is a member of the subset, as is the product $(\lambda m)\,(\mu m) = (\lambda \mu m)\, m$. Also the zero is in the subset and so is the negative of λm, viz. $(n-\lambda)\, m$. Hence the subset is a subring, of order r.

Note that although the additive group of this subring is the cyclic group of order r, the subring is not in general isomorphic to the ring of residues modulo r, since it will not in general have a unity.

Theorem 3.7.2. *If S and T are subrings, so is $S \cap T$.*

If r_1 and r_2 are in $S \cap T$ they are in both S and T, and so r_1+r_2 and $r_1 r_2$ are also in both S and T, since these are subrings, and hence are in $S \cap T$. Similarly for 0 and $-r_1$.

Note that $S \cup T$ need not be a subring. For example the even integers and the set of multiples of 3 are both subrings of the ring of integers. Their intersection is the set of multiples of 6 and is also a subring, but their union is clearly not.

The additive structure of a subring

Since a subring is a subgroup of the additive group structure of the ring, all the results for subgroups apply. (But of course a subgroup need not be a subring.)

In particular Lagrange's theorem holds (theorem 1.9.3).

Theorem 3.7.3. (*Immediate deduction from Lagrange's theorem.*) *The order of any subring of a finite ring is a factor of the order of the ring.*

Note that the additive structure of any ring is commutative: there is no distinction between left and right cosets.

By theorem 1.9.6 and theorem 3.5.3 we obtain at once

Theorem 3.7.4. *The only ring with unity of order p, where p is prime, is the ring of residues R_p.*

Since much more of the importance of ring theory lies in rings of infinite order than is the case with groups, Lagrange's theorem is not so vital here as it was in group theory and, of course, it tells us nothing of the product structure of any ring that we might be discussing.

3.8. Ring-homomorphisms

When considering any type of structure we are interested in studying mappings which preserve this structure, and we stated in chapter 2 that such mappings are known as *homomorphisms*. In chapter 2 we studied group-homomorphisms: that is, mappings which preserved the group structure. Mappings which preserve the ring structure are likewise known as *ring-homomorphisms*, and these are the mappings about which we are interested when considering the mapping of a ring into another ring. When it is clear that we are considering rings we may refer simply to a *homomorphism* of one ring into another. The nomenclature and theory are very similar to that for groups, and we will deal fairly briefly with the subject.

Definition of ring-homomorphism

If R and S are rings, then $\theta: R \to S$ is a ring-homomorphism if for any two elements r, r' in R, *both*

$$(r+r')\,\theta = r\theta + r'\theta \quad \textit{and} \quad (rr')\,\theta = (r\theta)\,(r'\theta).$$

$r+r'$ and rr' are the ring sum and product as defined in R, whereas $r\theta + r'\theta$ and $(r\theta)\,(r'\theta)$ are those as defined in S. In words, the conditions are that the images of the sum and product of two elements of R are the same as the sum and product respectively of the images of the elements.

For the mapping to be a homomorphism we must have both sum and product structures preserved. A mapping θ may be a group-homomorphism between the group structures of the rings R and S without being a homomorphism of the rings: thus the mapping of the ring of integers into itself defined by $a\theta = 2a$ preserves the additive structure but not the multiplicative and hence is not a homomorphism. (Similarly $a \to a^2$ preserves product but not sum.)

Because a ring-homomorphism is a homomorphism of the additive group structure we may apply theorems 2.2.1 and 2.2.2 and obtain

Theorem 3.8.1. *If $\theta: R \to S$ is a homomorphism, the zero of R maps into the zero of S and negatives in R map into negatives in S.*

If R has a unity 1, then S need not have and, even if S also has a unity, the image of 1 need not be this. A trivial example is given by the mapping which maps the whole of R into the zero of S, which is clearly a homomorphism, and exists for arbitrary S (including those without unity). A less trivial example is given by $R =$ the ring of residues modulo 2 and $S =$ the ring of subsets of a given set T under the laws given on p. 68. The mapping θ such that $0\theta = \varnothing$ and $1\theta =$ any proper subset A of T is a homomorphism, since the 'sum' of A and A is $\varnothing$ and $AA = A$, $A\varnothing = \varnothing A = \varnothing$. But the unity of S is the improper subset T, and not A.

R may be commutative and S non-commutative, and vice versa. Thus $a \to a + 0i + 0j + 0k$ is a homomorphism of the commutative ring of reals into the non-commutative ring of quaternions, while the trivial homomorphism of the ring of quaternions into the zero of the ring of reals is an example of the contrary case.

As for groups, the set of elements of R which map into the zero of S is called the *kernel* of θ, and the set of images of all elements of R is called the *image* of θ. We have already proved that these are subgroups of R and S respectively, but in fact they are also *subrings*.

Theorem 3.8.2. *The kernel of a homomorphism θ is a subring of R.*

The kernel is a subgroup of R by theorem 2.3.1. If r and r' are any two elements in the kernel, $r\theta = r'\theta = 0$, the zero in S, and so $(rr')\theta = (r\theta)(r'\theta) = 0 \,.\, 0 = 0$, showing that rr' is in the kernel, and hence that the kernel is a subring.

Theorem 3.8.3. *The image of a homomorphism θ is a subring of S.*

The image $R\theta$ is a subgroup by theorem 2.3.3. Let any two elements of $R\theta$ be $r\theta$ and $r'\theta$. Then $(r\theta)(r'\theta) = (rr')\theta$ and hence is in $R\theta$, showing that $R\theta$ is a subring.

Types of ring-homomorphism

We use the same names as for group-homomorphisms. Thus a ring-homomorphism $\theta: R \to S$ which is *onto* S, i.e. such that $R\theta = S$, is called an *epimorphism*. If θ is 1–1, such that

$$r\theta = r'\theta \Rightarrow r = r',$$

θ is a *monomorphism*, and we see at once from theorem 2.4.1 that the condition for this is that the kernel consists of the zero of R alone.

If θ is both an epimorphism and a monomorphism it is known as an *isomorphism*: each element of S is the image of one and only one element of R, and in this case we have an inverse mapping θ^{-1} which is an isomorphism of S into R. (That θ^{-1} is a ring-homomorphism may be proved exactly as in theorem 2.4.2, considering both the product structure and the sum structure.) An isomorphism of R into S is a 1–1 correspondence which preserves the ring structure, and if such a correspondence exists we say that R and S are *isomorphic* rings and write $R \cong S$ exactly as we did for groups. From the abstract point of view they are identical. Note that if R and S are isomorphic as rings then their additive structures are isomorphic as groups, but the converse is not true. For example, the residues modulo 4 and the even residues modulo 8 have isomorphic additive structures, but are not isomorphic as rings (the first has a unity, and the second has not).

A homomorphism $\theta: R \rightarrow R$ is called a *ring-endomorphism* and an isomorphism of R onto itself is a *ring-automorphism.*

Examples of ring-homomorphisms

The identity homomorphism. The homomorphism of R into itself which sends each element into itself is an automorphism of R and is denoted by I_R.

The trivial homomorphism. We have already seen that the mapping which sends every element of R into the zero of S is a homomorphism: the kernel is the whole ring R and the image is the trivial subring consisting of the zero alone.

Homomorphisms from the ring of integers into rings of residues. For any positive integer n, we may map an integer into the residue class modulo n containing the integer. We obtain a homomorphism from the ring of integers into the ring of residues modulo n. This is onto but not 1–1.

Projections and injections

The mapping of $R \oplus S$ into R which sends (r, s) into $(r, 0)$ is an epimorphism of $R \oplus S$ onto R, known as a *projection.* We may similarly project $R \oplus S$ onto S.

If U is a subring of V then the mapping which sends u into itself (considered as an element of V) is a monomorphism of U into V, known as an *injection.* In particular we have an injection of R (or S) into $R \oplus S$.

Homomorphisms of polynomial rings

Let R be a commutative ring, and let b be any fixed element in R. Consider the mapping of the ring of polynomials $R[x]$

into R which is defined by any polynomial $P(x)$ being mapped into its value $P(b)$. This mapping is a homomorphism, since sum and product are at once seen to be preserved (remember that R must be commutative). Such a homomorphism of $R[x]$ into R exists for each b in R.

Theorem 3.8.4. *If R has a unity every homomorphism of $R[x]$ into R is of this form for some b.*

If $\theta: R[x] \to R$ let $x\theta = b$, an element of R. (Note that, since R has a unity, x is certainly a polynomial of $R[x]$.) Then since θ is a homomorphism, we see that any polynomial

$$a_0 + a_1 x + a_2 x^2 + \dots$$

maps into $a_0 + a_1 b + a_2 b^2 + \dots$, and so θ must be of the required form.

All these homomorphisms are onto, since any element r is the image of the polynomial $r + 0x + 0x^2 + \dots$, but none are 1–1, since both x and $b + 0x + 0x^2 + \dots$ map into b. The kernel is the subring of polynomials which have a zero at b.

Combination of ring-homomorphisms

The product of two ring-homomorphisms is defined in exactly the same way as for groups: thus if $\theta: R \to S$ and $\phi: S \to T$ are two homomorphisms, and if $r\theta = s$ and $s\phi = t$, then $\theta\phi$ is a mapping from R into T defined by $r\theta\phi = t$: that this is a homomorphism is proved precisely as in theorem 2.6.1 (proving for product as well as sum) and will not be repeated here.

If θ is an isomorphism from R into S, then θ^{-1} exists and is an isomorphism from S into R, and if $\theta: R \to S$ and $\phi: S \to T$ are isomorphisms then $(\theta\phi)^{-1} = \phi^{-1}\theta^{-1}$. Further $\theta\theta^{-1} = I_R$ and $\theta^{-1}\theta = I_S$, where I_R is the identity isomorphism in R.

The set of all automorphisms of R into itself forms a group, and this will be a subgroup of the group of automorphisms of the additive structure of R.

3.9. The ring of endomorphisms of an Abelian group

Suppose G is an Abelian group (for which we will use the additive notation) and consider all endomorphisms of G, i.e. all group-homomorphisms of G into itself. Then we can combine two of these in two ways. First we have the product of any two such defined in the normal way, and this is of course an endomorphism. Secondly we have the sum defined as in §2.6 (by $g(\theta+\phi) = g\theta+g\phi$) and by theorem 2.6.2 this is also an endomorphism, since G is Abelian. It turns out that under these two compositions the set of endomorphisms for a given G forms a ring, and we now prove this.

Theorem 3.9.1. *The endomorphisms of an Abelian group G into itself form a ring.*

The sum and product are always endomorphisms and are unique. We prove the laws one by one.

A1 $g(\theta+\phi) = g\theta+g\phi = g\phi+g\theta$ (since G is Abelian) $= g(\phi+\theta)$. Hence $+$ is commutative.

A2 The associativity follows from the associativity in G.

A3 The trivial endomorphism which maps all G into the neutral element is a zero for the set of endomorphisms, as can at once be seen.

A4 The negative of θ is given by ϕ, where ϕ is defined by $g\phi = -(g\theta)$, and this is an endomorphism since

$$(g+h)\,\phi = -(g+h)\,\theta = -g\theta-h\theta,$$

since G is Abelian

$$= g\phi+h\phi.$$

M2 Product is associative by the ordinary law of associativity for general mappings.

D1 $g(\theta(\phi+\psi)) = g\theta(\phi+\psi) = g\theta\phi+g\theta\psi = g(\theta\phi+\theta\psi)$.

Also, for the other law,

$$g((\theta+\phi)\,\psi) = (g(\theta+\phi))\,\psi = (g\theta+g\phi)\,\psi = g\theta\psi+g\phi\psi,$$

since ψ is a homomorphism

$$= g(\theta\psi+\phi\psi).$$

The ring of endomorphisms of G always possesses a unity, viz. the identity automorphism. It may or may not be commutative. Since inverses exist only for the *automorphisms* the ring will not usually be a field.

Worked exercises

1. In a ring prove that the Cancellation Law of multiplication is true if and only if the ring has no zero divisors (i.e. if and only if $xy = 0 \Rightarrow x = 0$ or $y = 0$).

Only if. Given the Cancellation Law we have to prove that

$$xy = 0 \Rightarrow x = 0 \quad \text{or} \quad y = 0.$$

If $xy = 0$ we see that $xy = x0$ and so either $x = 0$ or $y = 0$ by the Cancellation Law.

If. We have to prove the Cancellation Law, that

$$xy = xz \Rightarrow x = 0 \quad \text{or} \quad y = z.$$

If $xy = xz$ we have $xy - xz = 0$, i.e. $x(y-z) = 0$ and so since we are given that there are no zero divisors either $x = 0$ or $y - z = 0$, i.e. either $x = 0$ or $y = z$.

Similarly $\qquad yx = zx \Rightarrow x = 0 \quad \text{or} \quad y = z.$

2. Give addition and multiplication tables for the ring $R_2 \oplus R_2$. The 4 elements will be denoted as follows: (0, 0) by 0, (1, 1) by 1, (1, 0) by a and (0, 1) by b, and we see that the tables are as in figures 6 and 7.

	0	1	a	b
0	0	1	a	b
1	1	0	b	a
a	a	b	0	1
b	b	a	1	0

Fig. 6

	0	1	a	b
0	0	0	0	0
1	0	1	a	b
a	0	a	a	0
b	0	b	0	b

Fig. 7

3. Give a multiplication table for, and identify the subring of multiples of 3 in the ring of residues modulo 12.

The elements are 0, 3, 6 and 9, and the multiplication table is given in figure 8.

	0	3	6	9
0	0	0	0	0
3	0	9	6	3
6	0	6	0	6
9	0	3	6	9

Fig. 8

We see that 9 is a unity and, bearing in mind that the additive structure is the cyclic group C_4, we use theorem 3.5.3 and deduce that the subring is the ring of residues R_4. We identify 0, 3, 6, 9 with the residues (modulo 4) 0, 3, 2, 1 respectively.

4. Does the mapping of the ring of polynomials over the rationals into the ring of reals of the form $a+b\sqrt{2}$, where a and b are rational, defined by $f(x) \rightarrow f(\sqrt{2})$ give a ring-homomorphism? If it does, give the kernel and image.

The mapping is given by

$$a_0 + a_1x + \ldots + a_nx^n \rightarrow (a + 2a_2 + 2^2a_4 + \ldots) + (a_1 + 2a_3 + \ldots)\sqrt{2}.$$

This is clearly a homomorphism, since the algebraic processes of finding either the sum or product on either side correspond. The kernel is the set of all polynomials which have x^2-2 as a factor, and the image is the whole ring $\{a+b\sqrt{2}\}$, since any such $a+b\sqrt{2}$ is the image of the polynomial $a+bx$. The homomorphsim is of course an epimorphism.

5. What is the ring of endomorphisms of the cyclic group C_n?

There is a unique endomorphism in which the element 1 maps into any given r, since this ensures that s must map into rs and such a mapping certainly gives a group-homomorphism. Denote this endomorphism by $\tilde{r}$.

Consider $\tilde{r}+\tilde{s}$. This maps 1 into $1\tilde{r}+1\tilde{s} = r+s$, and so since it is an endomorphism, must be $\widetilde{(r+s)}$.

Consider $\tilde{r}.\tilde{s}$. $1\tilde{r}.\tilde{s} = r\tilde{s} = rs$ and so $\tilde{r}.\tilde{s} = \widetilde{rs}$. Thus the ring of endomorphisms is merely the ring of residues R_n, with $\tilde{r}$ corresponding to the residue r.

Exercises 3A

In these exercises, and in exercises 3B, we will denote the ring of residues modulo n by R_n.

Are the sets in **1–8**, under the usual rules of addition and multiplication, rings? If not state which laws are not satisfied.

1. All complex numbers which lie on either the real or imaginary axis (i.e. are of the form $a+bi$ where at least one of a and b is zero).

2. All real numbers of the form $r/3^s$, where r is an integer and s is a positive or zero integer.

3. All real numbers of the form $a+b\sqrt{5}$ where both a and b are of the form in **2** (i.e. $r/3^s$).

4. As in **3** for all reals of the form $a+b\sqrt{3}$.

5. The residues 1, 3, 5, 7, 9 modulo 10.

6. All polynomials (with real coefficients) which have the first two coefficients zero.

7. All polynomials (with real coefficients) which have the second coefficient zero.

8. All complex numbers of the form $a+bi$ where a is an integer and b is an even integer.

9. Which of the sets in **1–8** which are rings have a unity, which are commutative and which have inverses for all non-zero elements? Are there any which do not have inverses but yet obey the Cancellation Law of multiplication?

10. Show that if a zero-ring (i.e. one for which $xy = 0$ for all x and has a unity then it is the trivial ring consisting of the one element 0.

11. If a ring has its zero as unity prove that it is the trivial ring consisting of this one element.

12. What is the characteristic of R_n?

What are the characteristics of the rings in **13–15**?

13. The ring of subsets of a given set S given on p. 68.

14. $R_3 \oplus R_3$.

15. $R_4 \oplus R_6$.

16. Prove that if an element in a ring has an inverse then this inverse is unique.

17. Prove that if an element x in a finite ring with unity has an inverse then it is a root of unity, i.e. $\exists$ a positive integer n such that $x^n = 1$ but $x^r \neq 1$ for $r < n$.

18. Prove that if an element x in a ring with unity is a root of unity then it has an inverse.

19. State which elements of R_{12} are nth roots of unity and give n.

Give addition and multiplication tables for the rings in **20–22.**

20. R_2.

21. R_5.

22. $R_2 \oplus R_4$.

23. If S is a set with two elements, identify the ring of subsets of S as defined on p. 68.

Find Q and S such that $A = QB+S$ and $d(S) < d(B)$ for the pairs of polynomials in **24** and **25**, working over the given coefficient rings R.

24. $A \equiv 2x^3+5x, \quad B \equiv x+2, \quad R = R_6$.

25. $A \equiv x^5, \quad B \equiv x^2+2x+1, \quad R = R_4$.

26. Find all factors of the form $x-c$ of the polynomial x^2-1, where the coefficient ring is R_8.

27. Repeat **26** for the polynomial x^4 and the coefficient ring R_{16}.

28. Prove that the subring of the ring of Gaussian integers generated by $i(=0+1i)$ is the whole ring. What is the subring generated by $2i$?

29. In the ring of the reals prove that the subring generated by $\frac{1}{2}$ and $\frac{1}{3}$ is the same as that generated by $\frac{1}{6}$.

30. Give an example in which R is not commutative but contains a commutative non-trivial subring S.

Let $R_n(m)$ be the subring of multiples of m in the ring of residues modulo n, where m is a factor of n (see theorem 3.7.1). Give multiplication tables and identify the subrings in **31–36**, giving the unities where applicable.

31. $R_4(2)$.

32. $R_6(2)$.

33. $R_6(3)$.

34. $R_8(4)$.

35. $R_{10}(2)$.

36. $R_{10}(5)$.

37. Give the multiplication table for $R_8(2)$.

38. Prove that theorem 3.5.3 does not necessarily hold if we include rings without unity.

Which of **39–43** give ring-homomorphisms? For those which do, state the type (e.g. epimorphism, etc.) and give the kernel and image.

39. The complex numbers into the reals by $a+bi \to a$.

40. $R \oplus S \to R \oplus R$ by $(a, b) \to (a, a)$.

41. $R_n \to R_{2n}$ by $x \to 2x$.

42. $R_{2n} \to R_n$ by $x \to x$ $(0 \leqslant x \leqslant n-1)$ and $x \to x-n$ $(n \leqslant x \leqslant 2n-1)$.

43. Polynomials over the integers into the ring of reals of the form $a+b\surd 2$, where a and b are integers, by $f(x) \to f(2\surd 2)$.

Identify the rings of endomorphisms of the Abelian groups in **44–45**.

44. The finite cyclic group.

45. $C_2 \times C_3$.

46. Show that the ring of endomorphisms of $C_2 \times C_2$ has sixteen elements, all except 0 having additive order 2. Why is it not isomorphic to

$$R_2 \oplus R_2 \oplus R_2 \oplus R_2\,?$$

47. What is the group of automorphisms of the ring of integers?

48. What is the group of automorphisms of the ring $R_2 \oplus R_2$?

49. If $\theta: R \to S$ is a homomorphism and R is commutative prove that the image $R\theta$ is commutative (although S need not be).

50. If R in **49** has a unity prove that the image $R\theta$ has a unity (although S need not have).

51. If $\theta: R \to S$ is a homomorphism and A is a subring of R prove that $A\theta$ is a subring of S.

52. If $\theta: R \to S$ is a homomorphism and B is a subring of S, prove that the set of elements of R whose images are in B forms a subring of R.

Exercises 3B

1. Prove that the characteristic of $R_m \oplus R_n$ is the L.C.M. of m and n. Does the same result hold for the direct sum of more than two rings of residues?

2. If S is a finite set of n elements prove that the ring of mappings from S into a ring R is isomorphic to $R \oplus R \oplus \ldots \oplus R$, where there are n terms.

3. Let the set S consist of n elements. Show that the ring of subsets of S as given on p. 68 is isomorphic to $R_2 \oplus R_2 \oplus \ldots \oplus R_2$, where there are n terms. (Cf. volume 1, exercises 12B, no. 14.)

4. Let R be the set of 4 elements $0, 1, x, 1+x$ and let sum and product be defined in the normal way with the conditions that

$$1+1 = 0 \quad \text{and} \quad 1+x+x^2 = 0.$$

Prove that $x+x = 0$ and $x^3 = 1$ and show that R is a ring. Give addition and multiplication tables for R and identify the additive structure of R.

Deduce that theorem 3.5.3 does not necessarily apply to groups other than C_n.

5. Prove that, if p and q are co-prime, $R_p \oplus R_q \cong R_{pq}$.

6. In the ring of reals prove that, if m and n are integers: (i) the subring generated by m and n is the subring generated by the H.C.F. of m and n; (ii) the subring generated by $1/m$ and $1/n$ is the subring generated by $1/l$, where l is the L.C.M. of m and n.

7. In R_n prove that the subring generated by the residue r is the set of all residues which are multiples of the H.C.F. of n and r.

8. Show by considering the subrings of multiples of 2 and the subring of multiples of 4 in the ring of residues modulo 12 that (i) a ring without unity may have a subring which has a unity; (ii) a ring in which the Cancellation Law of multiplication doesn't hold may have a subring in which it does.

9. Find counter-examples as in **8** by considering $R \oplus S$ and its subring R for suitably chosen R and S.

10. There are four mappings of the ring R_2 into itself. Express each as a polynomial function.

11. Express the 27 mappings of R_3 into itself as polynomial functions of degree $\leqslant 2$.

12. Give the addition and multiplication tables for $R_2 \oplus R_2 \oplus R_2$. Which elements have inverses?

13. If R is a ring with unity and has finite characteristic m, prove that any ring of functions whose image space is R also has characteristic m.

14. Investigate the validity of the result and proof of theorem 3.6.1 when $A = QB+S$ is replaced by $A = BQ+S$ (assuming R is non-commutative).

15. If S is any subset of a ring R, not necessarily a subring, we denote the set of all elements of R that commute with every element of S by $C(S)$. Prove:

(i) $C(S)$ is always a subring;
(ii) $S_1 \supseteq S_2 \Rightarrow C(S_1) \subseteq C(S_2)$;
(iii) $C(C(S)) \supseteq S$;
(iv) $C(C(C(S))) \subseteq C(S)$;
(v) $C(C(C(S))) = C(S)$.

16. If a is any fixed element of a ring R, and x is a variable element of R, prove that the mapping $x \to xa$ is an endomorphism of the additive group of R into itself.

Denoting this endomorphism by $\tilde{a}$, prove that the mapping $a \to \tilde{a}$ is a ring-homomorphism of R into the ring of endomorphisms of the additive group of R.

If R has a unity prove that the kernel of the homomorphism $a \to \tilde{a}$ consists of the element 0 only, and deduce that in this case $R \cong$ a subring of the ring of endomorphisms of the additive group of R.

4

FIELDS

4.1. Definition of a field

We have seen that a group is a set which has a structure admitting the addition or subtraction of any two elements, or alternatively their multiplication and division, but not all four rules. If we use the addition notation and introduce a product we obtain a ring, which still will not necessarily admit of division. If now we extend our requirements to include division, so that all four rules are present, we obtain a structure called a *field.*

A field then is a ring with additional properties. For division to be possible we must have a unity and also an inverse to each element (except 0). Commutativity of addition is essential for the Distributive Law to be useful (as is the case for general rings) but commutativity of multiplication need not be insisted upon. We find however that nearly all fields of any practical interest possess this property and that if we assume it some of our theory is made simpler, so we will postulate it as one of our field axioms. (Nearly all writers follow the same convention, but not all, and the reader must be careful to check this when reading any new book or paper.) A structure that obeys all the field axioms except for commutativity of multiplication will be called a *skew field* (sometimes also called a *division ring*, or even a *sfield*) and in §4.9 we will indicate ways in which these differ from fields. The most important skew field is the set of quaternions.

Thus in a field all four basic rules are present and elements are combined according to the fundamental laws of algebra. It follows that the algebra of fields is similar to ordinary elementary algebra, except that we do not have any properties of order. Because they possess so much structure fields have plenty of properties, largely those of combination of their elements. On the other hand there are far fewer examples of fields than of rings or groups: the large number of axioms restricts us tremendously and the classification of fields is a much more straightforward matter than that of rings or groups. Many of the properties of groups, which of course hold for fields, become trivial owing to the restricted types (for example there is no real theory of field homomorphisms, as will be seen in §4.5) and the theory is much more akin to that in elementary algebra (dealing largely with elements) than to the abstract synthetic theory so dominant for groups and to a lesser extent for rings. We will see that our usual topics of substructures and homomorphisms may be dealt with quickly, and towards the end of the chapter we will give some powerful classification theorems, completing this process in chapter 8.

The importance of fields lies of course in their similarity to the ordinary sets of numbers used in elementary work. The sets of rationals, reals and complex numbers are all fields (but not the set of integers), as are the sets of residues modulo a prime number. In many branches of mathematics we use such sets as a background. Obvious examples are in vectors and co-ordinate geometry, where our co-ordinates are usually taken to be either real or complex numbers. We may however take elements of any field as co-ordinates and obtain vector and geometrical spaces with the usual properties but differing in important ways from the familiar ones (thus we obtain finite spaces by taking the field of residues with a prime modulus as a base).

It is convenient to state all the laws and to give an abstract definition of a field. Most of the laws are the same as those for a ring in chapter 3, and all were previously stated in chapter 10 of volume 1.

Abstract definition of a field

A set F of elements forms a field if to any two elements x and y of F taken in a particular order there are associated two elements of F, called the sum and product of x and y and denoted by $x+y$ and xy respectively, both being unique, and such that the following laws are satisfied:

A1 *For any two elements x and y, $x+y = y+x$.*

A2 *For any three elements* x, y *and* z, $(x+y)+z = x+(y+z)$.
A3 *There is an element called the zero and denoted by* 0 *which has the property that* $x+0 = 0+x = x$ *for all elements* x.
A4 *Corresponding to each element* x *there is an element* $-x$ *called the negative of* x *which has the property that*

$$x+-x = -x+x = 0.$$

M1 *For any two elements* x *and* y, $xy = yx$.
M2 *For any three elements* x, y *and* z, $(xy)z = x(yz)$.
M3 *There is an element called the unity and denoted by* 1 *which has the property that* $x1 = 1x = x$ *for all elements* x.
M4 *Corresponding to each element* x *except* 0 *there is an element* x^{-1} *called the inverse (or reciprocal) of* x *which has the property that* $xx^{-1} = x^{-1}x = 1$.
D1 *For any three elements* x, y *and* z

$$(x+y)z = xz+yz, \quad x(y+z) = xy+xz.$$

X1 $1 \neq 0$.

Since both sum and product are commutative we do not strictly need both equations in A3, A4, M3, M4 and D1, but have included both to keep in line with previous work.

Note that 0 is excluded in M4, i.e. has no inverse. This is easily seen to be necessary, since it is an immediate theorem, quoted in the next section, that $x0 = 0$ for any x (for any a, $xa+x0 = x(a+0) = xa = xa+0$ and so, adding $-xa$ to each side, $x0 = 0$) and so no element 0^{-1} can exist so that $0^{-1}0 = 1$.

Axiom X1 postulates that 1 and 0 are distinct, so that a field has at least two elements. If $1 = 0$ it follows that $x1 = 0$ and so $x = 0$ for all x, and so the set has just one element: such a trivial case is excluded by X1. There *is* of course a field with 2 elements (the field of residues modulo 2).

4.2. Elementary properties

It is immediate from the definition that a field is an Abelian group under addition, since all the group axioms are included in the field axioms. It is almost immediate that the elements of a field excluding 0 form an Abelian group under multiplication (the unity and inverses cannot be 0, and the product of two non-zero elements cannot be 0, by theorem 4.2.8). Thus, as we have seen, we obtain groups of the rationals, reals and complex numbers under addition and also, excluding 0 in each case, under multiplication: we also have groups of residues under addition and groups of residues excluding 0, modulo a prime number, under multiplication.

We see that a field is always a commutative ring with a unity, and so all the ring and group results apply but, as said previously, they often become trivial when applied to the restricted case of fields.

Because a field is a ring, all the theorems 3.2.1–3.2.12 apply, and we also have the following theorems about the unity and inverses, which were proved in chapter 10 of volume 1.

Theorem 4.2.1. *The uniqueness of the unity.*

There cannot be two different elements each having the unity property.

Theorem 4.2.2. *The uniqueness of the inverse.*

For any x there cannot be two different elements each having the property possessed by x^{-1}.

Theorem 4.2.3. *The inverse of* 1 *is* 1.

Theorem 4.2.4. *The inverse of x^{-1} is x.*

Theorem 4.2.5. *The Cancellation Law of multiplication.*

$$xy = xz \Rightarrow y = z \quad \text{unless} \quad x = 0.$$

Our proof depends on multiplication of both sides by x^{-1}, which is not possible if $x = 0$. In the case when $x = 0$, *xy always* equals xz ($= 0$) and the result is not true. Thus *we cannot divide by* 0.

Theorem 4.2.6. *The equation $ax = b$ has the unique solution $x = a^{-1}b$, unless $a = 0$, in which case there is no solution unless also $b = 0$, when every x is a solution.*

This theorem shows us that every first degree equation has a unique solution (unless the coefficient of x is 0). Note that the solution $a^{-1}b$ may also be written ba^{-1} (since multiplication is assumed commutative) and is obtained by dividing b by a. It is often written $\frac{b}{a}$ (or b/a) and this causes no confusion (it does in the non-commutative case).

Theorem 4.2.7. $(xy)^{-1} = x^{-1}y^{-1}$.

A theorem which is easily proved but often used (e.g. in the solution of a quadratic equation by factors) is the following. (See Worked exercises 1 of chapter 3 for its equivalence to the Cancellation Law in a ring.)

Theorem 4.2.8. *The absence of zero divisors.*

$xy = 0 \Rightarrow x = 0$ *or* $y = 0$, *i.e. the product of two non-zero elements cannot be zero.*

Suppose $x \neq 0$. Then it has an inverse x^{-1}, so that

$$x^{-1}xy = x^{-1}0 = 0.$$

But $x^{-1}xy = 1y = y$ and so $y = 0$.

(A proof from the Cancellation Law may be given:

$$xy = 0 = x0$$

and so either $x = 0$ or $y = 0$.)

Corollary. $xyz = 0 \Rightarrow$ *at least one of* x, y *or* z *is zero.*

For either $x = 0$ or $yz = 0$ by the theorem, and in the latter case either $y = 0$ or $z = 0$.

We may similarly extend the result to the product of n elements.

Additive powers of an element

The work that was done for rings on additive powers is of course valid for fields and, since we always have a unity, we may identify any positive integer n with the sum of n 1's, and then nx is either the sum of n x's or the field product of n and x, and these are the same by theorem 3.4.3. We may also identify $-n$ with the sum of n -1's (where -1 is of course the negative of 1), so that $-nx$ is both the product of $-n$ and x and the negative of nx. $0x$ is of course 0.

Theorem 4.2.9. *The Index Laws for addition.*

If m and n are integers, positive, negative or zero, then

(i) $(m+n)x = mx+nx$;

(ii) $(m-n)x = mx-nx$;

(iii) $(mn)x = m(nx)$.

These may be proved in the same way as was done for rings

and groups, but note that for fields they are direct consequences of the Distributive and Associative Laws, if we identify n with $n1$.

Multiplicative powers of an element

As for rings, we write x^2 for xx, x^3 for xxx, x^1 for x, etc. We may always write x^0 for 1, since this exists, and may use negative powers, so that x^{-n} means either the inverse of x^n or $(x^{-1})^n$, and these are the same, since

$$(x^n)^{-1} = (xx...x)^{-1} = x^{-1}x^{-1} \dots x^{-1} = (x^{-1})^n.$$

Theorem 4.2.10. *The Index Laws for multiplication.*

If m and n are integers, positive, negative or zero, then

(i) $x^m x^n = x^{m+n}$;

(ii) $x^m/x^n = x^{m-n}$;

(iii) $(x^m)^n = x^{mn}$.

These are proved as for groups.

Note. The above are valid for all x other than 0: there is no meaning to be given to 0^{-n} (since 0 has no inverse) and (i) and (iii) are valid for $x = 0$ only when m and n are positive or zero (if we allow 0^0 to be defined as 1).

The question of order of elements of a field is dealt with in §4.6.

The ordinary rules of elementary algebra for manipulation of fractions are valid for fields, and some will be given in the exercises at the end of the chapter. We give one below to show how such results may be established. We recollect that a/b means ab^{-1} (or $b^{-1}a$) and that a and b are any elements of the field (not of course necessarily 'integers') except that $b \neq 0$.

Example. Show that $(a/b)+(c/d) = (ad+bc)/(bd)$ where b and d are not 0, and so $bd \neq 0$ by theorem 4.2.8.

$$\begin{aligned}(a/b)+(c/d) &= ab^{-1}+cd^{-1}\\ &= adb^{-1}d^{-1}+bcb^{-1}d^{-1}, \text{ since product is commutative,}\\ &= (ad+bc)\, b^{-1}d^{-1}, \text{ by the Distributive Law,}\\ &= (ad+bc)\,(bd)^{-1}\\ &= (ad+bc)/(bd).\end{aligned}$$

4.3. Examples of fields

The large number of field axioms impose heavy restrictions on a structure, and it is to be expected that there are comparatively few examples of fields, particularly when we consider the rich possibilities for groups and, to a lesser extent, rings. We will find that the examples given below are mostly familiar ones, falling into three types: the real or complex numbers and subsets of these, rational functions, and fields of residues or extensions of these. The restriction to these three types is no accident: we will see in §8.6 that all commutative countable fields are included in this, admittedly rough, classification.

In the examples it is usually fairly easy to prove that a given one is a field, and we will omit detailed verification of the laws: usually the most difficult to verify is M4 and we will sometimes show that inverses exist by mentioning them explicitly. Since most of the examples are already familiar, from elementary algebra or as groups or rings, we will be able to deal with them quite quickly.

Elementary number fields

The rationals, the reals and the complex numbers all form fields. The integers, of course, do not.

Subsets of the real numbers

Consider all real numbers of the form $a+b\sqrt{2}$, where a and b are rational. These form a field, all the axioms being easily verifiable except possibly the existence of inverses. This is shown by rationalising the denominator of $1/(a+b\sqrt{2})$ so that it becomes

$$\frac{a}{a^2-2b^2}-\frac{b}{a^2-2b^2}\sqrt{2},$$

which is of the desired form and hence shows that $a+b\sqrt{2}$ has an inverse, in the set. (Note that the inverse exists unless

$$a^2-2b^2 = 0,$$

which is so only if $a = b = 0$, since a and b are rational, giving inverses to every element except 0.)

In a similar way it may be shown that all real numbers of the form $a+b\sqrt{n}$, where a and b are rational and n is any rational number not a perfect square, form a field, the inverse of $a+b\sqrt{n}$ being

$$\frac{a}{a^2-nb^2}-\frac{b}{a^2-nb^2}\sqrt{n}.$$

Note that if n were a perfect square we would still obtain a field, but it would be merely the field of rationals.

Another field is obtained by taking the set of all real numbers of the form $a+b\sqrt[3]{3}+c\sqrt[3]{9}$, where a, b and c are rational. Since $(\sqrt[3]{3})^2 = \sqrt[3]{9}$ and $(\sqrt[3]{3})^3 = 3$, the product of two such numbers is in the set, as are the sum and the negative. It is more difficult to show that any number of this form has an inverse of the same form, and we will not do this here. (See §8.4, where this type of field is discussed.)

Subsets of the complex numbers

A field in many respects similar to those above is obtained as follows.

Let ω be a complex cube root of unity, so that $\omega^2+\omega+1 = 0$. Consider all complex numbers of the form $a+b\omega$, a and b both being rational. The sum of two such numbers is of the same form, as is the negative of any and, by using the fact that $\omega^2 = -\omega-1$, the product of any two is in the set. Again it is the inverse that causes most trouble, but in this case we see that

$$\frac{1}{a+b\omega} = \frac{(a+b)(a+b\omega^2)}{a^3+b^3} = \frac{(a+b)(a-b-b\omega)}{a^3+b^3}$$

and so, since

$$\frac{a^2-b^2}{a^3+b^3} = \frac{a-b}{a^2-ab+b^2} \quad \text{and} \quad \frac{-b(a+b)}{a^3+b^3} = \frac{-b}{a^2-ab+b^2}$$

are rational and exist (unless $a = b = 0$), the inverse is in the set and the set is a field (0 and 1 are of course in it).

Residues with a prime modulus

The set of residues modulo n always forms a ring, but is a field only if n is prime, for the residue r has an inverse if and only if r is prime to n, and this is true for all r except 0 only when n is prime.

Thus corresponding to any prime p the set of residues modulo p is a field of order p. These are the basic, and most important, finite fields and are particularly useful in giving finite examples of structures that behave in most important respects like the ordinary sets of real or complex numbers. We used some of them in this way to illustrate the fundamental laws of algebra in chapter 10 of volume 1. The finite field of residues modulo p is often called the 'arithmetic modulo p'.

An extension of the field of residues modulo 2

Working with coefficients in the field of residues modulo 2, consider polynomials in x, where we take x as a root of the equation $x^2+x+1 = 0$. (This is analogous to the subset $\{a+b\omega\}$ of the field of complex numbers discussed above, but in the present case we should not think of x as being a cube root of unity in the usual sense but merely as a symbol which happens to satisfy the given equation.) Then

$$x^2 = -x^2 \text{ (we are working modulo 2) } = x+1,$$

$$x^3 = x(x+1) = x^2+x = x+1+x = 1,$$

and so on, so that we need only consider polynomials of degree 0 or 1. Thus we have the four elements $0, 1, x, x+1$. Combined according to the usual rules, working modulo 2 and remembering that $x^2 = x+1$, it is obvious that they satisfy all the field axioms except possibly the existence of inverses. (The negatives are the elements themselves.) We may quickly verify that inverses do exist, namely that $1^{-1} = 1$, $x^{-1} = x+1$, $(x+1)^{-1} = x$. Thus the four elements form a finite field of order 4. The addition and multiplication tables are given in figure 9, from which it is seen that all non-zero elements have additive order 2, the additive group being the Vierergruppe, and the multiplicative group of non-zero elements is cyclic, with x and $x+1$ as generators.

	0	1	x	$x+1$
0	0	1	x	$x+1$
1	1	0	$x+1$	x
x	x	$x+1$	0	1
$x+1$	$x+1$	x	1	0

	0	1	x	$x+1$
0	0	0	0	0
1	0	1	x	$x+1$
x	0	x	$x+1$	1
$x+1$	0	$x+1$	1	x

Fig. 9

The above example is an example of the so-called 'Galois fields' which are further discussed in §4.8, and later in §8.5.

Fields of rational functions

In chapter 3 we showed that the set $R[x]$ of polynomials with coefficients in a ring R is a ring, and has a unity if R has. It is however never a field, even when R is.

Theorem 4.3.1. *If R is a ring with unity, the element x of $R[x]$ has no inverse and so $R[x]$ is not a field.*

Suppose x has an inverse $\sum_{r=0}^{n} a_r x^r$.

Then $1 = x\sum_0^n a_r x^r = \sum_0^n a_r x^{r+1}$ and so, equating coefficients of x^0, we have $1 = 0$, which is impossible.

To obtain a field in this connection we must consider the so-called *rational functions* over the coefficients set R. A rational function is, briefly, the ratio $P(x)/Q(x)$ of two polynomials, where P/Q and S/T are taken as the same rational function if and only if $PT = QS$, and neither denominator (Q or T) may be the zero polynomial. For this definition to work we must assume the coefficient set R to be commutative and also to satisfy the Cancellation Law of multiplication (i.e. R must be an integral domain). The necessity of commutativity is fairly obvious (we wish to write P/Q rather than PQ^{-1} or $Q^{-1}P$) and that of the Cancellation Law may be seen by considering the example of residues modulo 6, say, where it is not satisfied. In this case we would wish the rational function $1/2$ to be the same as $1.3/2.3$, but this latter is $3/0$, which is inadmissible.

The full details of the definition are discussed in §8.2, where the general concept of the field of fractions of any integral domain is introduced: the present case is that in which the domain is a set of polynomials and is discussed intuitively, merely in order to provide examples of fields.

It is fairly obvious that the rational functions (with coefficients in any integral domain) form a field.

If $Q \neq 0$ and $T \neq 0$, $P/Q \pm S/T = (PT \pm QS)/QT$ and is an element of the set,

$$(P/Q).(S/T) = (PS)/(QT) \quad \text{and} \quad (P/Q)^{-1} = Q/P \text{ if } P \neq 0,$$

the zero is $0/P$ for any non-zero P and the zero polynomial 0, the unity being P/P for any non-zero P. The fact that P/Q may be replaced by any equivalent rational function $(AP)/(AQ)$ without these sums or products being affected is also straightforward, though somewhat tedious, to prove, and this again is done thoroughly in §8.5.

We obtain various fields, all infinite, by this means, though all are not necessarily different. Thus the coefficient sets of the rationals and the integers lead to the same field of rational functions. The finite fields lead to infinite fields of rational functions (consider the field of residues modulo 2) and even domains without unity lead to fields (thus the rational function with coefficients in the set of even integers, which has no unity, form a field, whose unity is $2/2$ (or $4/4$, x/x, etc.)).

Direct sums of fields

By theorem 3.5.1 the direct sum of two fields is never a field, although it is of course always a ring.

4.4. Subfields

As with groups and rings, we define a subfield in an obvious way as follows.

Definition of subfield

A subset D of the elements of a field F forms a subfield of F if

(i) for any pair (d_1, d_2) of elements in D, d_1+d_2 and $d_1 d_2$ are both in D;

(ii) for any non-zero d in D, $-d$ and d^{-1} are in D;

(iii) 0 and 1, the zero and unity of F, are both in D.

The whole field F forms a trivial subfield, but there is no trivial subfield consisting of just one element, since any field must contain at least 2 (viz. 0 and 1). Of course, 0 and 1 together will not in general give a subfield, since any such must contain $1+1$, $(1+1)^{-1}$, etc.

There are some obvious non-trivial examples of subfields. Thus the rationals form a subfield of the reals, and the reals of the complex numbers. The set $\{a+b\sqrt{2}\}$, a and b rational, is a subfield of the reals, and the set $\{a+b\omega\}$ is a subfield of the complex numbers.

The theory of subfields is not on the whole very important, particularly when compared with that of subgroups, which is fundamental in the study of group structure. This arises partly from the fact that subgroups are connected intimately with group-homomorphisms (the kernel and the image of any homomorphism are subgroups) and, as we will see in the next section, there are no non-trivial field-homomorphisms. Another reason for the lack of importance of subfields is due to the types of field that exist being restricted—the most important subfield of any field, called its *prime field*, must be one of two types and the field may then be built up from this in standard ways. We return to the subject in §4.7.

4.5. Field homomorphisms

A mapping $\theta: F \to E$ is a *field-homomorphism* (or merely *homomorphism* if it is understood that we are dealing with fields) if it preserves the sum and product structures, i.e. if for any two elements f_1 and f_2 in F, $(f_1+f_2)\,\theta = f_1\theta+f_2\theta$ and $(f_1 f_2)\,\theta = (f_1\theta)\,(f_2\theta)$. In other words, θ must be a group-homomorphism for both the additive and multiplicative groups of F. It follows from theorems 2.2.1 and 2.2.2 that the zero and unity in F map into the zero and unity in E, and that negatives and inverses are preserved.

The following theorem shows us that all field-homomorphisms are essentially trivial (a monomorphism is in effect merely an isomorphism between F and a subfield of E and tells us little about their structure), and hence there is no important theory connected with them.

Theorem 4.5.1. *If $\theta: F \to E$ is a field-homomorphism, then either θ is a monomorphism (with zero kernel) or θ maps the whole of F onto the zero of E (having zero image). (In the latter case the image has just one element and so is not a subfield.)*

If the kernel consists merely of the zero 0 of F, θ is a monomorphism (for $f_1\theta = f_2\theta \Rightarrow (f_1-f_2)\,\theta = 0$, the zero in E, and so $f_1-f_2 = 0$, showing that $f_1 = f_2$).

Now suppose that the kernel contains a non-zero element x of F. Then $x\theta = 0$.

For any y in F, $y = yx^{-1}x$ and so

$$\begin{aligned} y\theta &= (y\theta)(x^{-1}\theta)(x\theta), \\ &= (y\theta)(x^{-1}\theta)\, 0 \\ &= 0. \end{aligned}$$

Hence the kernel is the whole of F and so θ maps the whole of F onto the zero of E.

Note. The essential point about the above proof is that x has an inverse, a property that does not necessarily apply for rings.

The automorphisms of a field

All non-trivial homomorphisms of a field F into itself will be automorphisms. As in the case of groups and rings, the product of any two of these will itself be an automorphism, as will be the inverse of any, and we always have the identity automorphism. Hence the set of automorphisms of F will form a group.

4.6. The characteristic of a field

In this section we start the process of classifying the possible fields that may exist, by discovering stringent conditions on their additive structure. Later, and in the next section, we see that all fields contain as a subfield one of a very limited class of fields, and in §4.8 we classify completely the finite fields, reserving a further consideration of the infinite case until chapter 8.

Since the elements of a field form a group under addition, every element has an additive order, and by Lagrange's theorem this is a factor of the order of the field, in the finite case. Furthermore, since the elements except 0 form a group under multiplication, every element except 0 is a root of unity, which was not the case for rings. If a finite field has order n, the multiplicative order of any element must be a factor of $n-1$, by Lagrange's theorem, since the multiplicative group contains $n-1$ elements, viz. all in the field except 0. The multiplicative group is in fact cyclic (see theorem 4.8.6).

With regard to the additive order we bear in mind that a field is also a ring, and apply theorem 3.4.5 to see that if the unity e

has order m, then the order of any element is a factor of m. In the special case of fields we may however go further than this.

As an example take the field of residues modulo a prime p. Then the order of the unity is p, and so is that of any other non-zero element, since p is prime (this is not of course the case for the ring of residues modulo a composite number). In the field of rationals no element except 0 has a finite order, as is the case in the fields of reals, complex numbers, and the fields $\{a+b\sqrt{2}\}$, $\{a+b\omega\}$ which were considered earlier. In the extension field of residues modulo 2, consisting of the four elements $0, 1, x, x+1$, every non-zero element has order 2. Thus in all cases all non-zero elements have the same order, finite or infinite, and we now prove that this is true for any field whatsoever.

Before we give the proof we recollect two things. First, the symbol mx, where m is a positive integer and x an element of the field, is strictly speaking ambiguous, meaning either the sum of m x's or the field product of m (meaning me, where e is the unity) with x. As for rings with unity this ambiguity causes no trouble since these two expressions are always the same, and we will use the symbol mx freely. Secondly we recollect the definition of additive order of x as the least positive multiple mx of x that is zero.

Theorem 4.6.1. *In any field F, all non-zero elements have the same additive order. This is called the* characteristic *of F and is said to be zero if all non-zero elements have infinite order.*

Suppose that any given non-zero element y has finite order m. Then if x is any other non-zero element,

$$(mx)y = m(xy) = x(my) = 0.$$

But $y \neq 0$ and so $mx = 0$. (This is a step that could not be performed in the case of rings, which may have zero divisors.)

Also, suppose $rx = 0$ for some $r < m$.

Then $x(ry) = (rx)y = 0$ as before and since $x \neq 0$, $ry = 0$, which is impossible since $r < m$ and m is the order of y. Hence x also has order m.

Thus if *any* non-zero element has finite order m, every other such element has the same order. Thus either they all have the same finite order or all have infinite order.

Note. The characteristic is of course also the order of the unity.

It is not possible for a field to have *any* finite characteristic: this must be a prime. This need not be true for the characteristic of a ring, but is in line with the fact that residues form a field only when the modulus is a prime. The proof is fairly simple.

Theorem 4.6.2. *If F has a finite characteristic, this must be a prime.*

Suppose F has as characteristic a composite number hk. If a is any non-zero element of F, a^2 is non-zero and so has order hk, so that $hka^2 = 0$, i.e. $(ha)(ka) = 0$ and so either ha or ka is zero, which is impossible since both h and k are less than hk and a must have order hk.

Thus any field has characteristic either 0 or a prime p, and in either case the field must contain a certain subfield, as we see in the next two theorems.

Theorem 4.6.3. *If a field F has characteristic p, it contains a subfield isomorphic to the field of residues modulo p.*

Consider the unity 1 and as usual denote $1+1$ by 2, $1+1+1$ by 3, etc. Then 1 has order p, and so $p = 0$ (i.e. the sum of p 1's is the zero element).

Now consider the subset of elements $0, 1, 2, \ldots, (p-1)$. The sum and difference of any two of these is in the set, as are 0 and 1. Furthermore the product of any two is in the set, since the product is the ordinary product of residues modulo p, by the Distributive Law. (For example,

$$2.3 = (1+1)(1+1+1) = 1+1+1+1+1+1 = 6.)$$

Since sum and product behave as they do for residues modulo p the subset is isomorphic to the set of residues modulo p under residue sum and product and so, since p is prime, inverses exist in the subset and the subset is isomorphic to the field of residues modulo p.

Theorem 4.6.4. *If a field F has characteristic 0, it contains a subfield isomorphic to the field of rational numbers.*

The unity 1 has infinite order and so the elements $0, 1, 2, \ldots$ are all distinct. Consider the subset of all elements m/n, where

m and n are co-prime, n is non-zero and m/n means $m.n^{-1}$. It is easily seen that this subset contains the sum, difference, product and quotient of any two of its elements (except for dividing by 0, of course) and also contains 0 and 1, and so is a subfield, and is trivially isomorphic to the field of rationals. (Note that m stands for the element that is the sum of m 1's, and it is easily seen that km/kn is the same element as m/n.)

4.7. Prime fields

A field is called a *prime* field if it possesses no proper subfields: i.e. no subfields except that consisting of the whole field. (Remember that the set consisting of 0 alone is not a subfield, since a field must contain at least 2 elements, a zero and a unity.)

Theorem 4.7.1. *The only prime fields are the fields of residues modulo p and the field of rationals.*

We first show that these fields are prime. Any subfield must contain 1 and hence also 2, 3, ... Hence for the residue fields any subfield must be the whole field, and for the field of rationals any subfield must contain all n and hence also all n^{-1} and so all m/n, and hence must again be the whole field.

To show that these are the only prime fields we use theorems 4.6.3 and 4.6.4 to show that any field F has one of these as a subfield and so, if F is prime, one of these must be the whole field F (strictly speaking must be isomorphic to F).

Note that we have proved that every field contains a prime subfield, and that if F has characteristic p its prime subfield is isomorphic to the field of residues modulo p, whereas if F has characteristic zero its prime subfield is isomorphic to the field of rationals.

We may consider the prime subfield of a field in a slightly different way as follows.

Theorem 4.7.2. *The prime subfield of a field F is the intersection of all subfields of F.*

The intersection *is* a subfield, as is easily proved (if x and y are in it they are in all subfields of F and hence $x+y$ is in all subfields and so in their intersection, and similarly for the other operations). Also the intersection has no proper subfields, since any such would be subfields of F and hence must contain the intersection, which is impossible. Hence the intersection is a prime field. It is of course the smallest subfield of F and is contained in any subfield.

Note. F cannot contain two distinct prime subfields, since their intersection would also be a subfield and hence must be identical with both.

4.8. Finite fields

Finite (commutative) fields are known as *Galois fields*, named after Evariste Galois (1811–32) and studied by him in connection with his work on the roots of polynomial equations. (Galois was the first man to give a 'simple' proof of the insolubility of the general quintic.) The study of Galois fields is intimately connected with the advanced theory of roots of polynomials, and is far too difficult to be included in any detail here. It is possible however to classify such fields completely, and we will give the main theorems, which are interesting and rather surprising. The proofs of many of them raise too many deeper issues to be given, and will be omitted. The interested reader will find them in both *A Survey of Modern Algebra*, by G. Birkhoff and S. MacLane, and *Modern Algebra*, by B. L. van der Waerden: the former is more straightforward, the latter more complete, and in neither is the work easy.

The finite fields that we have already met are the fields of residues and the extension of the field of residues modulo 2 given on p. 100. In fact all such fields are similar to these.

We first prove an obvious result.

Theorem 4.8.1. *Any finite field has characteristic p, for p a finite prime (i.e. it cannot have characteristic* 0).

For by theorem 4.6.4 a field with characteristic 0 contains an infinite subfield and so must itself be infinite.

Theorem 4.8.2. *If F is a finite field with characteristic p, then the number of elements in F is a power p^n of p.*

We do not prove this properly here, but a rough outline is as follows:

By theorem 4.6.3 F contains a subfield isomorphic to the field of residues modulo p—call this subfield J_p. It can be shown that we can find a set of elements $f_1, f_2, \ldots, f_n$ of F such that

(*a*) $j_1 f_1 + j_2 f_2 + \ldots + j_n f_n = 0 \Rightarrow j_i = 0$ for all i, where $j_i \in J_p$ for all i,

(*b*) every element of F can be expressed in the form

$$j_1 f_1 + j_2 f_2 + \ldots + j_n f_n \quad \text{for some} \quad j_1, j_2, \ldots, j_n \in J_p.$$

It then follows that all such linear combinations of the f_i's are distinct elements of F, and so F contains exactly p^n elements, since there are p possibilities for each of the n j_i's. (F is in fact a vector space over J_p: see §9.16.)

The two following theorems are fundamental, but are too difficult for us to prove here.

Theorem 4.8.3. *If F and F′ are two finite fields with the same number of elements then they are isomorphic.*

Theorem 4.8.4. *For any prime p and positive integer n there does exist a finite field with p^n elements.*

As a consequence of these theorems we see that, corresponding to each prime p and positive integer n there is one and only one field with p^n elements, and that these are the only finite fields that exist. The field of order p^n is called the *Galois field of order* p^n, and is denoted by $GF[p^n]$.

Examples of Galois fields

$GF[p]$ is of course merely the field of residues modulo p.

$GF[2^2]$ is our extension field on p. 100, its addition and multiplication tables being shown in figure 9.

Unfortunately, even the proofs of theorem 4.8.4 given in the standard works do not give us a straightforward method of actually constructing the general Galois field. In §8.5 we give a method that can be applied in simple cases, though it becomes tedious for large n and p. We give below, for interest and without giving reasons for our methods, constructions for $GF[2^3]$ and $GF[3^2]$.

Construction of $GF[2^3]$

With coefficients in the field of residues modulo 2, consider all polynomials in x of degree $\leqslant 2$, and define addition and multiplication in the usual way, replacing x^3 by $x+1$. We obtain the required field, with its 2^3 elements being all the polynomials of the given form, and give its addition and multiplication tables in figure 10 for interest. Note that we have not here proved that we *do* obtain a field but the reader can easily verify the existence of inverses (the only non-obvious axiom) from the tables.

$+$	0	1	x	$x+1$	x^2	x^2+1	x^2+x	x^2+x+1
0	0	1	x	$x+1$	x^2	x^2+1	x^2+x	x^2+x+1
1	1	0	$x+1$	x	x^2+1	x^2	x^2+x+1	x^2+x
x	x	$x+1$	0	1	x^2+x	x^2+x+1	x^2	x^2+1
$x+1$	$x+1$	x	1	0	x^2+x+1	x^2+x	x^2+1	x^2
x^2	x^2	x^2+1	x^2+x	x^2+x+1	0	1	x	$x+1$
x^2+1	x^2+1	x^2	x^2+x+1	x^2+x	1	0	$x+1$	x
x^2+x	x^2+x	x^2+x+1	x^2	x^2+1	x	$x+1$	0	1
x^2+x+1	x^2+x+1	x^2+x	x^2+1	x^2	$x+1$	x	1	0

$\times$	0	1	x	$x+1$	x^2	x^2+1	x^2+x	x^2+x+1
0	0	0	0	0	0	0	0	0
1	0	1	x	$x+1$	x^2	x^2+1	x^2+x	x^2+x+1
x	0	x	x^2	x^2+x	$x+1$	1	x^2+x+1	x^2+1
$x+1$	0	$x+1$	x^2+x	x^2+1	x^2+x+1	x^2	1	x
x^2	0	x^2	$x+1$	x^2+x+1	x^2+x	x	x^2+1	1
x^2+1	0	x^2+1	1	x^2	x	x^2+x+1	$x+1$	x^2+x
x^2+x	0	x^2+x	x^2+x+1	1	x^2+1	$x+1$	x	x^2
x^2+x+1	0	x^2+x+1	x^2+1	x	1	x^2+x	x^2	$x+1$

Fig. 10

Construction of $GF[3^2]$

Working now with coefficients in the field of residues modulo 3, consider the 9 polynomials in x of degree 0 or 1. Under the ordinary addition and multiplication, and writing x^2 as 2 whenever it occurs, these form the $GF[3^2]$. The addition and multiplication tables are given in figure 11, and we see that inverses exist for all non-zero elements.

The additive structure of Galois fields

The additive group of $GF[p]$ is obviously C_p, the cyclic group of order p. If we look at the addition tables of $GF[2^2]$, $GF[2^3]$

+	0	1	2	x	$x+1$	$x+2$	$2x$	$2x+1$	$2x+2$
0	0	1	2	x	$x+1$	$x+2$	$2x$	$2x+1$	$2x+2$
1	1	2	0	$x+1$	$x+2$	x	$2x+1$	$2x+2$	$2x$
2	2	0	1	$x+2$	x	$x+1$	$2x+2$	$2x$	$2x+1$
x	x	$x+1$	$x+2$	$2x$	$2x+1$	$2x+2$	0	1	2
$x+1$	$x+1$	$x+2$	x	$2x+1$	$2x+2$	$2x$	1	2	0
$x+2$	$x+2$	x	$x+1$	$2x+2$	$2x$	$2x+1$	2	0	1
$2x$	$2x$	$2x+1$	$2x+2$	0	1	2	x	$x+1$	$x+2$
$2x+1$	$2x+1$	$2x+2$	$2x$	1	2	0	$x+1$	$x+2$	x
$2x+2$	$2x+2$	$2x$	$2x+1$	2	0	1	$x+2$	x	$x+1$

×	0	1	2	x	$x+1$	$x+2$	$2x$	$2x+1$	$2x+2$
0	0	0	0	0	0	0	0	0	0
1	0	1	2	x	$x+1$	$x+2$	$2x$	$2x+1$	$2x+2$
2	0	2	1	$2x$	$2x+2$	$2x+1$	x	$x+2$	$x+1$
x	0	x	$2x$	2	$x+2$	$2x+2$	1	$x+1$	$2x+1$
$x+1$	0	$x+1$	$2x+2$	$x+2$	$2x$	1	$2x+1$	2	x
$x+2$	0	$x+2$	$2x+1$	$2x+2$	1	x	$x+1$	$2x$	2
$2x$	0	$2x$	x	1	$2x+1$	$x+1$	2	$2x+2$	$x+2$
$2x+1$	0	$2x+1$	$x+2$	$x+1$	2	$2x$	$2x+2$	x	1
$2x+2$	0	$2x+2$	$x+1$	$2x+1$	x	2	$x+2$	1	$2x$

Fig. 11

and $GF[3^2]$ we see that the first is that of the Vierergruppe $C_2 \times C_2$ (generators 1 and x), the second is $C_2 \times C_2 \times C_2$ (generators 1, x, x^2) and the third is $C_3 \times C_3$ (generators 1 and x). This suggests that the additive group of $GF[p^n]$ is merely $C_p \times C_p \times \ldots \times C_p$, with n terms, and we proceed to prove this.

Theorem 4.8.5. *The additive group of $GF[p^n]$ is $C_p \times C_p \times \ldots \times C_p$, where there are n terms.*

In the outline proof of theorem 4.8.2 we saw that the elements of $GF[p^n]$ are all of the form $\sum_{i=1}^{n} j_i f_i$, where $j_i \in J_p$ for all i.

But considering only the additive structure, that of J_p is C_p, and we see that our field is isomorphic to $C_p \times C_p \times \ldots \times C_p$, the isomorphism being given by $\sum_{i=1}^{n} j_i f_i \leftrightarrow (j_1, j_2, \ldots, j_n)$.

The multiplicative structure of Galois fields

The multiplicative group of $GF[p^n]$ is of order $p^n - 1$. For $GF[2^2]$ this equals 3 and so the group must be C_3, with either x or $x+1$ as generator. Similarly $GF[2^3]$ has its multiplicative group of order 7 and so this is C_7, with any non-unity element as generator. In the case of $GF[3^2]$ the order is 8, which gives us three possibilities for the group (since this is Abelian): namely C_8, $C_4 \times C_2$, $C_2 \times C_2 \times C_2$. If we consider the element $x+1$ we see after a little study of the multiplication table that it has order 8, and so the group must be C_8.

In all three examples above the multiplicative group is cyclic, and it is a very surprising fact that this is always the case. The proof of this, though not easy, is fairly straightforward and interesting, and we proceed to give it. The reader may omit it if he finds it difficult, and merely bear in mind the result.

Theorem 4.8.6. *The multiplicative group of $GF[p^n]$ is C_{p^n-1}.*

For convenience write the order $p^n - 1$ as t, and factorise it into prime factors so that

$$t = q_1^{\lambda_1} q_2^{\lambda_2} \dots q_r^{\lambda_r},$$

where the q_i's are distinct primes.

Lemma 1. *There is at least one element c_i of $GF[p^n]$ of multiplicative order $q_i^{\lambda_i}$, for any i, $1 \leqslant i \leqslant r$.*

Denote the field $GF[p^n]$ by F. The coefficients of all polynomials used in this Lemma will be understood to be in F.

By Lagrange's theorem the multiplicative order of any non-zero element of F is a factor of t, and so all non-zero elements are zeros of the polynomial $x^t - 1$. Hence by the factor theorem (theorem 3.6.2, Corollary, which certainly applies when the coefficient ring is a field), $x - f$ is a factor of $x^t - 1$ for any non-zero element f of F. It is an immediate corollary that the product of all such linear factors is a factor of the polynomial (see p. 117 of volume 1 for a specific proof), and so, since there are t such factors and the polynomial is of degree t, with leading coefficient 1, we must have

$$x^t - 1 \equiv \Pi(x - f),$$

where the product is taken over all non-zero elements of F.

Now write $x^{q_i^{\lambda_i}} = z$, so that $x^t - 1 \equiv z^{q_2^{\lambda_2} \dots q_r^{\lambda_r}} - 1$, and this has a factor $z - 1$, so that $x^t - 1$ has a factor $x^{q_i^{\lambda_i}} - 1$.

It will be proved in § 5.6 that factorisation of polynomials with coefficients in a field is unique, and hence it follows that

$x^{q_i^{\lambda_i}}-1$ is a product of linear factors of the form $x-f$, for $f \in F$: i.e. that $x^{q_i^{\lambda_i}}-1 = 0$ has $q_i^{\lambda_i}$ roots in F.

Similarly we may show that $x^{q_i^{\lambda_i-1}}-1$ has $q_i^{\lambda_i-1}$ roots, all in F, and hence there is at least one element c_i of F (there will generally be more than one) which is a solution of $x^{q_i^{\lambda_i}} = 1$ but not of $x^{q_i^{\lambda_i-1}} = 1$. Thus the order of c_i is a factor of $q_i^{\lambda_i}$ but not of $q_i^{\lambda_i-1}$, and hence *is* $q_i^{\lambda_i}$, since q_i is prime.

Lemma 2. *If f and g are elements of F with orders μ and ν, where μ and ν are co-prime, then the order of fg is $\mu\nu$.*

$$(fg)^{\mu\nu} = f^{\mu\nu}g^{\mu\nu} = 1 \quad \text{since} \quad f^{\mu} = g^{\nu} = 1.$$

Suppose now that $(fg)^{\kappa} = 1$, so that $f^{\kappa}g^{\kappa} = 1$. Then

$$f^{\kappa\mu}g^{\kappa\mu} = 1^{\mu} = 1, \quad \text{and} \quad f^{\kappa\mu} = 1,$$

since $f^{\mu} = 1$, so $g^{\kappa\mu} = 1$. Hence since g has order ν, ν is a factor of $\kappa\mu$, and being co-prime to μ must be a factor of κ.

Similarly μ is a factor of κ and so κ is a multiple of $\mu\nu$, proving the lemma.

Proof of the theorem

Consider the element $c_1 c_2 \ldots c_r$, where the c_i's are as in lemma 1.

Since c_1 has order $q_1^{\lambda_1}$ and c_2 has order $q_2^{\lambda_2}$, and since $q_1^{\lambda_1}$ and $q_2^{\lambda_2}$ are co-prime, it follows by lemma 2 that $c_1 c_2$ has order $q_1^{\lambda_1}q_2^{\lambda_2}$.

Similarly since $q_1^{\lambda_1}q_2^{\lambda_2}$ and $q_3^{\lambda_3}$ are co-prime $c_1 c_2 c_3$ has order $q_1^{\lambda_1}q_2^{\lambda_2}q_3^{\lambda_3}$.

Continue for r steps. Then $c_1 c_2 \ldots c_r$ has order $q_1^{\lambda_1}q_2^{\lambda_2} \ldots q_r^{\lambda_r}$, i.e. t. Thus it is a generator of the multiplicative group of F and so this group must be cyclic, being $C_t = C_{p^n-1}$.

Corollary. *If p is prime the group of residues modulo p under multiplication is cyclic.*

For it is the multiplicative group of $GF[p]$.

Note. The group of residues prime to a composite n under multiplication may or may not be cyclic: if $n = 6$ the residues 1 and 5 form the cyclic group C_2, whereas if $n = 8$ the residues 1, 3, 5, 7 form the Vierergruppe, which is not cyclic.

4.9. Skew fields

A skew field is a structure that obeys all the axioms for a field except that of commutativity of multiplication. Some of the work of this chapter is valid in this case but not all.

The only skew field of any importance is the set of quaternions, previously met in §3.1, example 5. Here product is not commutative (for example $ij = k$ but $ji = -k$) but inverses do exist, since it may be verified at once that

$$\frac{a-bi-cj-dk}{a^2+b^2+c^2+d^2}.(a+bi+cj+dk) = 1$$

and $$(a+bi+cj+dk).\frac{a-bi-cj-dk}{a^2+b^2+c^2+d^2} = 1,$$

so that $a+bi+cj+dk$ has inverse

$$\frac{a-bi-cj-dk}{a^2+b^2+c^2+d^2},$$

and this always exists unless $a = b = c = d = 0$, i.e. unless the given element is zero.

We will show briefly what part of our work on fields applies also to skew fields. Theorems 4.2.1 to 4.2.6 apply, but we must be careful in 4.2.6 to give the equation as $ax = b$ and the solution as $x = a^{-1}b$, or alternatively $xa = b$ and $x = ba^{-1}$ respectively. Theorem 4.2.7 becomes $(xy)^{-1} = y^{-1}x^{-1}$ as for groups. If n means the sum of n 1's then n and x always commute (since 1 and x do so) and of course m and n do so as well. The index laws are valid as stated, but many of the ordinary manipulative rules of elementary algebra must be modified: thus it is no longer true that $ab^{-1}+cd^{-1} = (ad+bc)(bd)^{-1}$, and we can of course no longer write a/b without ambiguity.

In our discussion of rational functions we needed the coefficient ring R commutative, and then the field of rational functions over R was commutative, so that skew fields do not enter into this work either as possible coefficient rings or as the resulting fields.

The theories of subfields and field-homomorphisms apply equally well to skew fields, including the fundamental theorem 4.5.1.

All of our theory of characteristics is valid for skew fields, since $(mx)y = x(my)$ as before, $mna^2 = (ma)(na)$ and additive powers of unity and their inverses are still commutative. Similarly the theory of prime fields is also applicable.

Theorems 4.8.1 and 4.8.2 are still true for skew fields, but theorem 4.8.3 is not, so that our simple classification breaks down, although it is not easy to give examples of finite skew fields.

Although much of our elementary work is still valid for skew fields, it is found that in many important applications commutativity of multiplication is needed. Thus both factorisation theory of polynomials and the study of polynomial *functions* require the coefficient set to be commutative, and the use of fields as sets of scalars for vector spaces and as sets of co-ordinates for geometry is much simpler when they are commutative. Hence skew fields are of comparatively little importance.

Worked exercises

1. Prove that for any field, if a and b are any elements but $b \neq 0$,

$$(a/b)+(-a/b) = 0.$$

$$\begin{aligned}(a/b)+(-a/b) &= ab^{-1}+(-a)\,b^{-1}\\ &= (a+-a)\,b^{-1} \quad \text{by law D1,}\\ &= 0.b^{-1} \quad \text{by A4,}\\ &= 0 \quad \text{by theorem 3.2.10, which applies for fields.}\end{aligned}$$

2. In the field of residues modulo 11 find the inverses of 3 and 7 and hence solve the equations $3x = 10$, $10x = 3$.

In such a simple case we may proceed by trial and error to find the inverses, and since $3.4 = 7.8 = 1 \pmod{11}$, we see that $3^{-1} = 4$ and $7^{-1} = 8$.

If $3x = 10$, $x = 10.3^{-1} = 10.4 = 7$.
If $10x = 3$, $x = 3.10^{-1} = (10.3^{-1})^{-1} = 7^{-1} = 8$.

3. In the field of residues modulo 59 find the inverse of 26.

In difficult cases such as this trial and error methods may be tedious, and it is usually best to proceed by the general method of using Euclid's algorithm as given below. For a full discussion of Euclid's algorithm see volume 1.

We express 1 (the H.C.F. of 59 and 26) in terms of 59 and 26 as follows.

$$59 = 2.26+7,$$
$$26 = 3.7+5,$$
$$7 = 1.5+2,$$
$$5 = 2.2+1.$$

Hence
$$\begin{aligned} 1 &= 5-2.2. \\ &= 5-2(7-1.5) = 3.5-2.7, \\ &= 3(26-3.7)-2.7 = 3.26-11.7, \\ &= 3.26-11(59-2.26) = 25.26-11.59. \end{aligned}$$

Thus modulo 59 we have $1 = 25.26$ and so $26^{-1} = 25$.

4. Identify the group of automorphisms of the field $\{a+b\omega\}$, where

$$\omega^2+\omega+1 = 0$$

and a, b are rational.

If θ is any automorphism of the field we know that $1\theta = 1$ and so $n\theta = n$ for any integer n, and hence $m^{-1}\theta = m^{-1}$ and so $(n/m)\,\theta = n/m$, i.e. θ leaves all elements $a+0\omega$ invariant. Hence θ is completely determined by the image $\omega\theta$.

Let $\omega\theta = \omega'$. Then $(\omega^2+\omega+1)\,\theta = \omega'^2+\omega'+1 = 0$ since the image of 0 is 0.

Thus ω' is a root of $x^2+x+1 = 0$ and so either $\omega' = \omega$ or $\omega' = -1-\omega$, since the sum of the two roots is -1. Hence there are just 2 automorphisms, viz. the identity I and the automorphism X given by

$$(a+b\omega)\,X = a+b(-1-\omega).$$

Hence the group must be C_2. (It is easily verified that $X^2 = I$ but this is strictly unnecessary, since the only group with 2 elements is C_2.)

Exercises 4A

In these exercises, and in exercises 4B, we denote the field of residues modulo p by J_p.

Are the sets in **1**–**6**, under the usual rules of addition and multiplication, fields? If not state which laws are not satisfied.

1. The set of irrational real numbers.

2. All real or complex numbers that are the root of a quadratic equation with real coefficients.

3. All complex numbers of the form $a+bi$ where a and b are rational.

4. All real numbers of the form $a+b\sqrt{2}+c\sqrt{3}$ where a, b, c are rational.

5. All real numbers of the form $a+b\sqrt{2}+c\sqrt{3}+d\sqrt{6}$ where a, b, c, d are rational.

6. All complex numbers of the form $(a+bi)/(c+di)$ where a, b, c, d are integers and c, d are not both zero.

7. Prove that a zero-ring (i.e. a ring in which $xy = 0$ for any pair of elements x and y) cannot be a field.

8. A set F is an Abelian group under addition, the non-zero elements of F form an Abelian group under multiplication and the Distributive Law is true in F. Prove that (i) $x0 = 0x$ for all $x \in F$; (ii) $(xy)0 = x(y0)$ for all x and $y \in F$ (zero or non-zero); (iii) F is a field.

Prove the results **9–14** for any field, stating at each step which law is being used. (a/b means ab^{-1} and $b, d \neq 0$.)

9. $a/b = c/d \Leftrightarrow ad = bc$.

10. $(a/b)(c/d) = ac/bd$.

11. $a+b/d = (ad+b)/d$.

12. $(a/b)/(c/d) = ad/bc$.

13. $(-a)/(-b) = a/b$.

14. $(-a)/b = a/(-b) = -(a/b)$.

15. In J_5 let $a = 2$, $b = 3$, $c = 4$ and $d = 1$. Evaluate both sides of the equations in **10–14** and verify that the equations are satisfied in this particular case.

16. In J_7 solve the equations $3x = 6$, $3x = 5$, $5x = 3$, $6x = 3$.

17. List the 4 polynomials of the second degree with coefficients in J_2 and use the factor theorem to factorise them where possible. Are there any which do not factorise?

18. Solve the simultaneous equations, working in J_5,

$$3x-2y = 2, \quad x+4y = 3.$$

19. The set of non-zero elements of a field F is denoted by $F-\{0\}$. Prove that $(F_1-\{0\})\times(F_2-\{0\})$ is a group under a multiplication defined by $(f_1, f_2).(f_1', f_2') = (f_1 f_1', f_2 f_2')$. Identify this group when $F_1 = F_2 = J_3$.

20. Prove that if D and E are subfields of F, $D \cap E$ is a subfield. Give an example to show that $D \cup E$ need not be a subfield.

21. If $\theta: F \to E$ is a field homomorphism, investigate under what conditions the kernel and image of θ are subfields of F and E respectively.

22. If $\theta: F \to F$ is an automorphism of a field F, prove that $x\theta = x \; \forall \; x \in$ the prime subfield of F. Prove further that if x satisfies a polynomial equation with coefficients in the prime subfield of F then $x\theta$ is a root of the same equation.

Identify the groups of automorphisms of the fields in **23–26**, using the result of **22**.

23. J_p.

24. The field of complex numbers $a+bi$, a and b rational.

25. The field $\{a+b\sqrt[3]{3}+c\sqrt[3]{9}\}$, a, b, c rational.

26. The field $\{a+b\sqrt{2}+(c+d\sqrt{2})\,i\}$, a, b, c, d, rational.

27. Give an example to show that all elements (except 0 and 1) of a field *need not* have the same multiplicative order.

28. Give an example to show that all elements (except 0 and 1) of a field *may* have the same multiplicative order.

29. What is the characteristic of the field of rational functions with coefficients in J_p? What is the prime subfield of this field?

30. Give the multiplicative inverses of the non-zero elements of $GF[2^2]$, $GF[2^3]$, $GF[3^2]$.

31. Show that the only proper subfield of the field $\{a+b\sqrt{2}\}$, a and b rational, is the prime subfield, i.e. the rationals.

32. Prove that the sum of all the elements of a Galois field is zero, and the product of all the non-zero elements is -1. (Hint. Use the identity $x^t-1 \equiv \Pi(x-f)$ derived in the proof of theorem 4.8.6.)

33. From **32** deduce Wilson's theorem, that $(p-1)! \equiv -1 \pmod p$ if p is prime.

Exercises 4B

1. How many possible linear equations (i.e. of the form $ax = b$ where $a \neq 0$) are there in J_p? In the set of solutions of all these, do all elements appear the same number of times?

2. Use Euclid's algorithm to find integers s and t such that $1 = 139s+21t$ and hence find the inverse of 21 in J_{139}.

3. Prove that, working in J_p for $p > 2$, the equation $x^2 = a$ where $a \neq 0$ has either 0 or 2 solutions and deduce that there are $\frac{1}{2}(p-1)$ distinct values of a for which it has two solutions. Do these results hold for $p = 2$?

4. Derive the general solution of the simultaneous equations

$$a_1x+b_1y = c_1, \quad a_2x+b_2y = c_2$$

working in any field, and state when this solution is not valid.

5. If R is a subring of a field F show that the set $\{ab^{-1}\}$, where $a, b \in R$ and $b \neq 0$, is a *subfield* $\tilde{R}$ of F. Show further that any subfield which contains R also contains $\tilde{R}$. Deduce that $\tilde{R}$ is the intersection of all subfields containing R.

Identify $\tilde{R}$ in the following cases:

(i) F is the field of reals, R is the ring of integers.

(ii) F is the field of rational functions over the integers, R is the ring of polynomials over the integers.

6. Prove that the field of rational functions over the integers (i.e. with the integers as its coefficient set) is isomorphic to that over the rational numbers. Are these isomorphic to the field of rational functions over the domain of even integers?

7. Find the inverse of $a+b\sqrt[3]{3}+c\sqrt[3]{9}$ in the field of numbers of this form with a, b, c rational by the following two methods:

(i) letting the inverse be $\alpha+\beta\sqrt[3]{3}+\gamma\sqrt[3]{9}$ obtain three linear equations in α, β, γ and solve by determinants.

(ii) use Euclid's algorithm to find polynomials S and T such that $1 \equiv S(cx^2+bx+a)+T(x^3-3)$, giving S as the inverse when we put $x = \sqrt[3]{3}$.

8. By considering the group-automorphisms of their additive and multiplicative groups, or otherwise, show that the groups of automorphisms of $GF[2^3]$ and $GF[3^2]$ are C_3 and C_2 respectively.

9. Let F be a field with finite characteristic p and let a, b be any two elements of F. Assuming the binomial theorem, that

$$(a+b)^n = \sum_{r=0}^{n} \frac{n!}{(n-r)!\,r!} a^{n-r}b^r,$$

prove the following sequence of results:

(i) $\dfrac{p!}{(p-r)!\,r!}$ is a multiple of p for $1 \leqslant r \leqslant p-1$;

(ii) $(a+b)^p = a^p+b^p$;

(iii) $(a-b)^p = a^p - b^p$;

(iv) $\theta: a \to a^p$ is a homomorphism of F into itself;

(v) if F is finite, $\theta: a \to a^p$ is an automorphism of F.

10. In **9**, if $F = J_p$, show that θ is the identity automorphism of F, while if $F = GF(p^n)$ with $n > 1$, θ is *not* the identity automorphism.

11. Prove that in $GF[2^n]$ there is no element of multiplicative order 2, and that in $GF[p^n]$ for $p > 2$ the only element of multiplicative order 2 is -1.

12. Prove by using Lagrange's theorem for the additive and multiplicative groups, or otherwise, that any subfield of $GF[p^n]$ must have order p^{n_1}, where n_1 is a factor of n.

5

INTEGRAL DOMAINS

5.1. Definition and elementary properties

In the two previous chapters we have discussed rings and fields, the former being structures which admit of addition, subtraction and multiplication, while the latter possess a division process also. We have seen that rings possess comparatively little elementary theory, whereas elementary field theory is of course similar to elementary algebra.

There are some rings, however, which are not fields yet possess many properties that are not shared by all rings. Thus the rings of integers and polynomials over any field possess a great deal of important theory (that concerned with primes and factorisation) that is not applicable to rings in general. There are comparatively few such rings (the integers and polynomials form the main examples) but they are so important that it is worthwhile giving them a distinct name and unifying the theory, rather than dealing with each example separately. (We saw this process at work in volume 1, where much of the polynomial theory in chapter 7 was an exact replica of that for integers in chapter 4.) We are thus led to the idea of an *integral domain*.

We will see that the theory of factorisation in a ring depends basically on the Cancellation Law of multiplication, that $xy = xz \Rightarrow y = z$ or $x = 0$. This is not possessed by all rings (for example the residues modulo a composite number and the rings of functions discussed in chapter 3 do not have the property) but it is a property of the integers and polynomials over a field. It is slightly more convenient to put this in another form: that $xy = 0 \Rightarrow x = 0$ or $y = 0$—it is easily shown (see below) that this is equivalent to the Cancellation Law. This property is often known as the 'absence of zero divisors' since it states that zero cannot be expressed as the product of non-zero factors. As an example where this does not hold consider the residues modulo 6, where $2.3 = 0$.

Thus we will define an integral domain as a ring with no zero divisors. We also insist on it being commutative and having a unity. These conditions are both necessary for most of the theory and are possessed by all important examples and so we choose to adopt them at the outset, although some authors use the term 'integral domain' to include structures without unity. We then have the following definition.

Definition of integral domain

An integral domain is a commutative ring with a unity that has the additional property that $xy = 0 \Rightarrow x = 0$ or $y = 0$.

Note. We will often shorten the term 'integral domain' in this chapter to 'domain'. This is not recommended in general work, since the term 'domain' has many meanings in mathematics, some precise and others more vague, but it will cause no confusion here.

Theorem 5.1.1. *The absence of zero divisors is equivalent to the Cancellation Law.*

I.e. if $xy = 0 \Rightarrow x = 0$ *or* $y = 0$ *then*

$$xy = xz \Rightarrow y = z \quad \text{or} \quad x = 0,$$

and conversely.

Suppose first that there are no zero divisors. Then if

$$xy = xz,\ xy - xz = 0 \quad \text{and so} \quad x(y-z) = 0.$$

Hence either $x = 0$ or $y - z = 0$ by hypothesis, i.e. $x = 0$ or $y = z$.

Conversely let the Cancellation Law be true and suppose that $xy = 0$. Then since $x0 = 0$, $xy = x0$ and so by the Cancellation Law $x = 0$ or $y = 0$.

Theorem 5.1.2. *In an integral domain* $xyz = 0 \Rightarrow$ *at least one of* x, y *or* $z = 0$ *and similarly for the product of more than three elements.*

This is immediate. For $xyz = 0 \Rightarrow x = 0$ or $yz = 0$ and the latter condition implies that y or z is zero.

In theorem 4.2.5 we proved the Cancellation Law of multiplication to hold in any field and so a field is always an integral domain. For an integral domain to be a field it is only necessary that inverses should exist. That this is not necessarily true can be seen by considering the domain of integers. It is no accident however that the important examples of integral domains (integers and polynomials) are all of infinite order (even for polynomials over a finite field). In the finite case inverses always exist.

Theorem 5.1.3. *A finite integral domain is always a field.*

It is sufficient to prove that every non-zero element x has an inverse.

Consider all products xy, where x is fixed and y ranges over all elements of the domain.

Since $x \neq 0$, $xy = xz \Rightarrow y = z$ and so all products xy are different. But the total number of elements is finite, say n, and so as y takes all n values xy also takes n *different* values. Hence xy takes *all* possible values, and so $xy = 1$ for some y, and this y is the inverse of x. ($yx = 1$ also since multiplication is commutative.)

We see the necessity for finiteness. In the domain of integers consider $x = 2$. Then $2y$ takes distinct values for distinct values of y, but this set, the set of even integers, does not include 1.

Theorem 5.1.3 is important, for the case where our integral domain is a field not only enables it to possess the field properties, but means that it does *not* possess most interesting integral domain properties. For fields the *domain* theory is largely trivial: there are no primes, and factorisation theory is irrelevant, since any non-zero element is a factor of any other. It is only in the intermediate case between a ring and a field that the work of this chapter is important: the main examples being, as stated before, the integers and polynomials.

Note that a finite ring that is not an integral domain need not be a field (e.g. the residues modulo a composite integer).

Because an integral domain is a commutative ring with unity, all properties of such rings hold. We also have multiplicative orders of elements (see exercise 5B, no. 1) but the concept is not important in this case: since only in limited cases is this finite. Nor does Lagrange's theorem have any meaning since integral domains have infinite order (if they are not fields).

We recollect from theorem 3.5.2 that the direct sum of two integral domains is never an integral domain (although it *is* a ring, commutative and with unity).

Whenever we have considered new types of structure the two basic topics considered have always been those of substructures and homomorphisms. In the case of integral domains neither is important.

Any subring of an integral domain is itself an integral domain *provided* it contains the unity (the even integers provide a counter-example when this is not the case), for it is inevitably commutative and cannot have zero divisors, since the domain containing it has none. Thus the study of sub-domains is merely that of subrings (restricted to those containing 1).

Domain-homomorphisms are ring-homomorphisms of one domain into another (not merely into a ring), and there is no new *process* involved. If D and D' are integral domains, any ring-homomorphism $\theta: D \to D'$ may be thought of as a domain-homomorphism since it preserves all the structure of the domain D.

5.2. Examples

Any field is an integral domain, but considered *as a domain* is of little importance since the domain properties become trivial.

The integers form the most important example. They contain no sub-domain, since any such must contain 1 and hence contains $1+1 = 2$, $2+1 = 3$, etc. and similarly negatives. The theory will be applied to the integers in §5.5.

We saw in §3.6 that the ring of polynomials $R[x]$ is commutative, has a unity or satisfies the Cancellation Law if and only if the coefficient ring R has the same property. Hence $R[x]$ is an integral domain if and only if R is an integral domain. Even if R is a field $R[x]$ cannot be a field, for the element x cannot have an inverse. Polynomials are an extremely important example and are dealt with in §5.6.

The ring of Gaussian integers (complex numbers of the form $a+bi$, where a and b are integers) is commutative, contains a unity and has no zero divisors (since it is a subring of the field of complex numbers) and is therefore an integral domain. It will be considered briefly in §5.8.

The only other type of integral domain that we will consider is that given by the set of numbers (real or complex) of the form $a+b\sqrt{n}$, where n is a positive or negative integer, not being a perfect square (in which case we merely obtain the domain of integers) or minus a perfect square (when we obtain the Gaussian integers), and a and b are integers. This is clearly a ring, has a unity and is commutative, and so to prove it an integral domain it suffices to show that it has no zero divisors. This is proved below.

Theorem 5.2.1. $(a_1+b_1\sqrt{n})(a_2+b_2\sqrt{n}) = 0 \Rightarrow$ *either*

$$a_1 = b_1 = 0 \quad or \quad a_2 = b_2 = 0,$$

if a_1, a_2, b_1, b_2 *are integers and* n *is an integer, not* $\pm$ *a perfect square.*

The zero product implies that

$$a_1a_2+b_1b_2n = 0 \quad \text{and} \quad a_1b_2+a_2b_1 = 0.$$

Multiplying by b_2, a_2 and subtracting, we obtain

$$b_1(b_2^2n-a_2^2) = 0$$

and so either $b_1 = 0$ or $b_2^2n-a_2^2 = 0$.

In the latter case, since n is not $\pm$ a square, $a_2 = b_2 = 0$, while in the first case $b_1 = 0$ and so $a_1a_2 = a_1b_2 = 0$. Thus either $a_2 = b_2 = 0$ again, or $a_1 = 0$, which gives, in conjunction with $b_1 = 0$, $a_1 = b_1 = 0$.

We shall refer to the above domain as $I[\sqrt{n}]$, being obtained by 'adjoining' $\sqrt{n}$ to the integers.

5.3. Factors, units and prime elements

Most of the importance of integral domains is in their factorisation properties, leading to the theory of prime elements and of factorisation into primes. In a field any non-zero element is a factor of any other and the theory is trivial, while since the Cancellation Law is essential to the theory it does not apply to general rings.

We are interested in the possibility of dividing an element

a by an element b, that is in the existence of a solution of $ax = b$. Note that we must not mention or use inverses since we are not dealing with fields—everything must be stated in product language. Note also that our domains are commutative and so we may write either ax or xa.

If we allow zero elements into our discussion we need to make tiresome exceptions to our results, and the applicability of these results to zero is always trivial. *Hence we understand that zero elements are excluded in this work except if otherwise stated.*

Many of the terms used are familiar from number theory (which is the study of the domain of integers) or polynomial theory: our work is in fact a generalisation of that which is common to these two, and which arises because the structures concerned are integral domains. The reader is advised to think of examples from the integers and polynomials to illustrate the work: such examples are usually easy to find and discuss and we will not introduce many here. The application of the theory to them is shown later in the chapter, and much of the work was covered for these special cases in volume 1.

Factors

If $a = xb$ has a solution we say that b is a *factor*, or *divisor*, of a, and that a is *divisible* by b, or is a *multiple* of b. We write this as b/a. (As stated above, it is understood that a and b are non-zero.) Note that *if* a solution exists then it must be unique by the Cancellation Law.

Units

The study of factorisation and primes is complicated by the presence of divisors of unity. Thus for integers the integer 6 may be factorised as 2.3 or $(-2).(-3)$, or as $1.2.3$, or $(-1).2.(-3)$, or in many other ways, whereas the only essentially distinct factorisation is 2.3. The integers 1 and -1 are divisors of 1 and may be introduced *ad lib.* either by themselves or to multiply other factors. The same difficulty occurs with polynomials, where constant multipliers may be introduced without changing the essential substance of a factorisation:

thus $x^2-1 = (x+1)(x-1) = (2x+2)(\frac{1}{2}x-\frac{1}{2})$. For both the integers and polynomials over the rationals factorisation into primes is essentially unique, but only when possible multiplication by divisors of unity is ignored. The same trouble occurs in general and we therefore make the following definition.

In an integral domain a factor of the unity 1 is called a *unit*. That is, u is a unit if there exists v such that $uv = 1$. (Note that v is also a unit.) Take care to distinguish between *unit* and *unity*. Since $1.1 = 1$, 1 is always a unit, and in general there will be others. The units are precisely those elements that possess multiplicative inverses.

Theorem 5.3.1. *If u is a unit then u is a factor of any element a.*

For if $uv = 1$, $a = a.1 = (av)u$.

Thereom 5.3.2. *The units form a group under the domain multiplication.*

If u and u' are units so that there exist v and v' with

$$uv = u'v' = 1, \quad \text{then} \quad (uv)(u'v') = 1,$$

i.e. $(uu')(vv') = 1$ and so uu' is a unit.

If u is a unit it has an inverse v which is also a unit.

1 is a unit.

Hence the units form a group.

Associates

In the examples of factorisation that we gave for integers and polynomials, we obtained different factors by multiplying the original ones by units. Such factors are called *associates* of the given ones. More precisely, a and b are *associates* if $b = ua$ where u is a unit. (This means of course that $vb = vua = a$ and so the definition is a reciprocal one.) We see that if a and b are associates then $a|b$ and $b|a$. The converse is also true.

Theorem 5.3.3. *If $a|b$ and $b|a$ then a and b are associates.*

For $b = xa$ and $a = yb$ by hypothesis.

Hence $b = xyb$ and so by the Cancellation Law, since we assume $b \neq 0$, $xy = 1$.

Hence x is a unit and so a and b are associates.

Note that we must have *both $a|b$ and $b|a$*: the theorem is obviously not true if only one condition is satisfied. Note also that we here use the cancellation property: it is in this important theorem that the theory first breaks down for general rings.

We would expect factorisation properties to be unchanged when elements are replaced by associates. We prove this precisely in the next theorem.

Theorem 5.3.4. *If $b|a$ then $ub|a$, where u is any unit.*

There exists u' such that $1 = uu'$.

We have $a = cb$ for some c.

Hence $a = cu'ub = (cu')ub$ and the result follows.

We now show that the relation of being associates divides the elements of an integral domain into mutually exclusive subsets, by proving that the relation is an equivalence relation. (See theorem 1.1.1.)

Theorem 5.3.5. *The relation xRy where x and y are associates is an equivalence relation.*

Reflexive. xRx since $x = 1x$ and 1 is a unit.

Symmetric. $xRy \Rightarrow y = ux \Rightarrow vy = vux = x \Rightarrow yRx$, where v is such that $uv = 1$.

Transitive. Let xRy and yRz, so that $y = ux$ and $z = u'y$ where u and u' are units. Then $z = u'ux$ and $u'u$ is a unit since the units form a multiplicative group. Hence xRz.

Irreducible elements

By theorem 5.3.1 any element has any unit as a factor. It also has any associate as a factor. If an element has no *proper* factors, that is no factors except units and associates, then it is called *irreducible*, provided it is not itself a unit. (Compare with prime integers, where p is prime if its only factors are, strictly, ± 1 or $\pm p$.)

Theorem 5.3.6. *If p is irreducible, any associate up is also irreducible.*

Suppose up has a factor q, so that $q|up$. Then clearly $q|vup$, and so p $(= vup)$ has a factor q. Hence q is either a unit or an associate of p, and hence an associate of up by transitivity of the associate relation.

Fields

Every non-zero element of a field has an inverse and hence is a unit. Thus any two elements are associates and there are no irreducible elements. In fact any non-zero element is a factor of any other, since $ax = b$ always has a solution.

Thus the theory for fields is trivial.

5.4. Prime factorisation

There are two main questions that interest us in the factorisation theory of integers and polynomials, and therefore of any integral domain. These are: first, is it always possible to express an element as a product of irreducible elements, and secondly, if so is this factorisation unique?

We continue to except zero elements in our discussion. We also note that a unit may be expressed in many ways as the product of units, but never as a product of non-units, and therefore (since units are not classed as primes) units are also excepted.

Factorisation is certainly not unique if we admit the replacement of elements by associates. For suppose an element $x = p_1 p_2 \dots p_n$ where $p_1, p_2, \dots, p_n$ are irreducible. Then if $u_1, u_2, \dots, u_{n-1}$ are units, where $u_i v_i = 1$, x is also equal to $(u_1 p_1)(u_2 p_2) \dots (u_{n-1} p_{n-1})(v_1 v_2 \dots v_{n-1} p_n)$, and the new factors $u_i p_i$ are also irreducible by theorem 5.3.6. Thus in any factorisation we may always replace elements by their associates, although in general we may then need a unit factor in front (i.e. $v_1 v_2 \dots v_n$ if we replace p_n also by $u_n p_n$—$v_1 v_1 \dots v_n$ *is* a unit since it is the product of units).

Hence any question of unique factorisation is always taken of necessity modulo units (i.e. unique only to within associates) and of course also admitting possible changes in the order of the factors. For the rest of the chapter we will find it convenient to use the notation $x \sim y$ to mean 'x and y are associates'. This is not an ideal notation, since the symbol '$\sim$' is an overworked one in mathematics, and we will not use it except in this chapter. Its use here will save us from tedious repetition of the phrase that it replaces.

With the above amendments, our criteria for the possibility and uniqueness of prime factorisations are embodied in the following definition, for domains where these conditions are satisfied.

Definition. *An integral domain is called a* unique factorisation domain, *or* Gaussian, *or merely a* U.F.D., *if*

(i) *any non-zero element that is not a unit can be expressed as the product of a finite number of irreducible elements, with possibly a unit factor also;*

(ii) *such factorisation is unique modulo units, i.e. if*

$$x = up_1p_2 \dots p_m = u'p_1'p_2' \dots p_n'$$

then $m = n$ *and the* p_i*'s and* p_j'*'s may be paired so that* $p_i \sim p_j'$ *for each pair.*

The importance of a domain being a U.F.D. is that, in this case, it is a trivial matter to decide whether any given element b is a factor of another element a, at least provided we can actually *find* the irreducible factors. (This latter process is often difficult, if not impossible in general cases, as for example in the factorisation of the general quintic.) For the condition that b is a factor of a is that all irreducible factors of b are associates of irreducible factors of a, as is easily proved by letting $a = bc$ and expressing b and c in terms of their unique irreducible factors.

Many important domains are U.F.D.S, as for example are the domain of integers, and that of polynomials over any field. That it is not a universal property is shown by the example below. (For an example where factorisation into irreducible elements is not even always possible see exercises 5B, no. 5.)

Example of a domain that is not a U.F.D.

Consider the domain $I[\sqrt{-5}]$ consisting of all complex numbers of the form $a+b\sqrt{-5}$, where a and b are integers. That this is an integral domain is shown by theorem 5.2.1. We will show that the element 9 has two distinct factorisations into irreducible elements, and so unique factorisation does not hold. Consider the two factorisations $9 = 3.3 = (2+\sqrt{-5})(2-\sqrt{-5})$. We first investigate which elements of the domain are units.

Let u be a unit, with $uv = 1$. Then $|u|^2|v|^2 = |1|^2 = 1$. But if x is any element $a+b\sqrt{-5}$, $|x|^2 = a^2+5b^2$ and is a positive integer, since a and b are integers. Hence $|u|^2 = 1$, and so if $u = \alpha+\beta\sqrt{-5}$, $\alpha^2+5\beta^2 = 1$ and this means that $\beta = 0$, $\alpha = \pm 1$, so that the only units are ± 1.

Hence, the factors $2+\sqrt{-5}$, $2-\sqrt{-5}$ are certainly not associates of 3.

It remains to show that all these factors are irreducible. Consider the factor 3 and suppose $3 = xy$. Then $|x|^2|y|^2 = 9$ and so $|x|^2 = 1, 3$ or 9 (since both $|x|^2$ and $|y|^2$ are positive integers). Now if $x = \xi+\eta\sqrt{-5}$ and $|x|^2 = 3$ we have $\xi^2+5\eta^2 = 3$ and this is impossible for integers ξ and η. Hence either $|x|^2 = 1$ or $|x|^2 = 9$, in which case $|y|^2 = 1$. But if $|x|^2 = 1$ we have $\xi^2+5\eta^2 = 1$, $\eta = 0$ and $\xi = \pm 1$ and x is a unit, and y is an associate. Similarly if $|y|^2 = 1$ y is a unit and x an associate. Hence 3 has no proper factors and so is irreducible. That $2+\sqrt{-5}$ and $2-\sqrt{-5}$ are irreducible follows in exactly the same way, since if either $= xy$ we still have $|x|^2|y|^2 = |2\pm\sqrt{-5}|^2 = 9$.

The two conditions for a domain to be a U.F.D. (possibility and uniqueness of prime factorisation) may be put in other forms; that these follow from the conditions is trivial to prove, but that they lead to the original conditions is by no means so obvious. A domain can often be shown to be a U.F.D. by proving that it satisfies these alternative criteria. They are as follows.

Criterion 1. There exists no infinite sequence of elements $\{a_i\}$ such that a_{i+1} is a proper factor of a_i for all i.

Criterion 2. If p is irreducible and p/ab then either p/a or p/b.

We prove first that these hold in any U.F.D.

Theorem 5.4.1. *In any* U.F.D. *criteria* 1 *and* 2 *are true.*

1 Suppose that y is a proper factor of x so that $x = yz$. Let the unique factorisations of x, y and z be as follows, omitting units and using the '$\sim$' notation for associates:

$$x \sim p_1p_2\dots p_n, \quad y \sim q_1q_2\dots q_m, \quad z \sim r_1r_2\dots r_l.$$

Then
$$p_1p_2\dots p_n \sim q_1q_2\dots q_m r_1r_2\dots r_l.$$

Hence by the uniqueness property, $q_1 \sim p_{i_1}$ for some i_1 and so, by the Cancellation Law, $p_1\dots \overset{\uparrow}{i_1} \dots p_n \sim q_2\dots q_m r_1\dots r_l$. Thus

$q_2 \sim p_{i_2}$ for some i_2 and, proceeding similarly, each $q_r \sim p_{i_r}$ for some i_r where the i_r's are all different.

Hence $m \leqslant n$. But if $m = n$ we would have finally that $1 \sim r_1 r_2 \ldots r_l$ and so z is a unit, contrary to the hypothesis that y is a *proper* factor of x. Thus $m < n$. Now suppose there exists a sequence $\{a_i\}$, where a_{i+1} is a proper factor of a_i for each i, and let the factorisation of a_i contain n_i proper factors. Then $n_1 > n_2 > n_3 > \ldots$ and so the sequence must terminate after at most n_1 terms, so that criterion 1 is satisfied.

2 Let $a \sim q_1 q_2 \ldots q_m$ and $b \sim r_1 r_2 \ldots r_l$ be the prime factorisations of a and b. Then $ab \sim q_1 \ldots q_m r_1 \ldots r_l$ and so, if p is an irreducible factor of ab, p must be an associate of either q_i or r_j for some i or j. Hence p is a factor of either a or b.

The proof of criterion 1 given above seems complicated, but the difficulty lies in the notation and not in the ideas, which are elementary. This is characteristic of much of the present work: care is needed with the notation to make the arguments 'water-tight', but the proofs are usually much simpler as regards the principles involved than they appear.

We will now show that criterion 1 implies possibility of prime factorisation and criterion 2 implies uniqueness, so that if both are satisfied in an integral domain the domain is a U.F.D.

Theorem 5.4.2. *If criterion* 1 *is satisfied in an integral domain then any non-zero, non-unit element can be expressed as the product of irreducible elements.*

We show first that any such element x possesses an irreducible factor.

Either x is irreducible, in which case the result is trivial, or it has a proper factor x_1.

If x_1 is irreducible it is an irreducible factor of x, whereas if not it has a proper factor x_2, and x_2 is a factor of x since if $x = x_1 y$ and $x_1 = x_2 z$, $x = x_2(zy)$. Similarly if x_2 is not irreducible it has a proper factor x_3 which is a factor of x, and we proceed in this way obtaining a sequence $x, x_1, x_2, \ldots$, where each x_{i+1} is a proper factor of x_i. But by criterion 1 this sequence must terminate, and hence at some stage we obtain an irreducible factor of x.

We now show that x can be expressed as the product of irreducible elements.

By the above, x possesses an irreducible factor p_1, so that either $x \sim p_1$ or $x = p_1 x_1'$, for example. Then x_1' has an irreducible factor p_2 with $x_1' \sim p_2$ or $x_1' = p_2 x_2'$ and so $x = p_1 p_2 x_2'$. Continuing in this way we obtain a sequence $x, x_1', x_2', \ldots$ where each is a proper factor of the preceding. By the criterion this sequence must terminate, say at x_n', so that x_n' must be irreducible. Then $x = p_1 p_2 \ldots p_n x_n'$ and is expressed as the product of irreducible elements.

Theorem 5.4.3. *If criterion* 2 *is satisfied in an integral domain then prime factorisation is unique modulo units.*

Let $x \sim p_1 p_2 \ldots p_m \sim p_1' p_2' \ldots p_n'$.

By the trivial extension of the criterion to more than two elements ($p/a_1 a_2 \ldots a_n \Rightarrow pa_i$ for some i), p_1'/p_{i_1} for some i_1 and so, since p_{i_1} is itself irreducible and p_1' is not a unit, $p_{i_1} \sim p_1'$. Hence by the Cancellation Law $p_1 \ldots \overset{\uparrow}{i_1} \ldots p_m \sim p_2' \ldots p_n'$.

As before $p_2' \sim p_{i_2}$ for some i_2 (not equal to i_1), and so on. Hence each p_j' may be paired off as an associate of some p_{i_j}, where the i_j's are all different, and so we must have $m = n$, and the factorisation is unique.

Corollary. *If a domain satisfies criteria* 1 *and* 2 *it is a* U.F.D.

Highest common factor

Let a and b be two non-units in a U.F.D. and let their prime factorisations be

$$a \sim p_1 p_2 \ldots p_m q_1 q_2 \ldots q_n,$$

$$b \sim p_1 p_2 \ldots p_m r_1 r_2 \ldots r_l,$$

where no r_j is an associate of any q_i, and some irreducible factors may of course be repeated. (Thus the p_k's are all the irreducible factors common to both a and b, to within associates of course.) The element $h = p_1 p_2 \ldots p_m$ has the properties that (i) it is a factor of both a and b and (ii) any common factor of both a and b must be a factor (proper or improper) of h. h is called the *highest common factor*, or H.C.F., of a and b. It is defined only to within associates (i.e. we may multiply $p_1 p_2 \ldots p_m$ by any unit and obtain another H.C.F.).

We may define H.C.F. in any integral domain by the above properties, as follows.

Definition. *An element h is an* H.C.F. *of the two elements a and b (where a and b are non-zero and non-units) if*

(i) h/a *and* h/b; (ii) *if* c/a *and* c/b *then* c/h.

The above definition says nothing about the existence or otherwise of an H.C.F., but it is easily proved that there cannot be two distinct H.C.F.S.

Theorem 5.4.4. *If a and b have two* H.C.F.*s h and h' then $h \sim h'$.*

By the definition we see that h'/h and h/h' and so theorem 5.3.3 gives the result.

If h is an H.C.F. it follows from theorem 5.3.4 that any associate of h is also an H.C.F. Thus the H.C.F. is only defined to within associates, and when we talk about 'the H.C.F.' we must bear this in mind. If a and b have an H.C.F. it is denoted by (a, b): this symbol standing for any member of the class of associates.

We have seen that an H.C.F. always exists in a U.F.D. It is a remarkable fact that, if we consider only domains which satisfy criterion 1, for any two elements to have an H.C.F. the domain *must be* a U.F.D. In other words, the existence of an H.C.F. implies the validity of criterion 2, and hence ensures that the domain is a U.F.D.

Theorem 5.4.5. *In any integral domain, if any two non-unit (and non-zero) elements have an* H.C.F. *(unique by theorem* 5.4.4), *the domain satisfies criterion* 2. *If it also satisfies criterion* 1 *it is a* U.F.D.

We first consider the H.C.F. of three elements, taken in two different orders.

Let $k \sim (a, (b, c))$. Then k/a and $k/(b, c)$ so that k/b and k/c also.

Now suppose d/a, d/b and d/c. Then $d(b, c)$ and so d/k.

Similarly if $k' \sim ((a, b), c)$ we also have k'/a, k'/b and k'/c and if d/a, d/b and d/c then d/k'.

It follows as in theorem 5.4.4 that $k \sim k'$. (This is called the H.C.F. of a, b and c and written (a, b, c).)

We have proved that

$$(a, (b, c)) \sim ((a, b), c). \tag{1}$$

We now show that

$$\lambda(a, b) \sim (\lambda a, \lambda b). \tag{2}$$

Let $h = (a, b)$ and $h' = (\lambda a, \lambda b)$. Then h/a and h/b and so $\lambda h/\lambda a$ and $\lambda h/\lambda b$. Thus $\lambda h/h'$. Let $h' = \lambda hx$.

We also have $h'/\lambda a$ and $h'/\lambda b$: suppose $\lambda a = h'y$ and $\lambda b = h'z$. Thus $\lambda a = \lambda hxy$ and $\lambda b = \lambda hxz$. Hence by the Cancellation Law $a = hxy$ and $b = hxz$, so that hx/a and hx/b. Hence hx/h and so $h = hxu$ say, so that $1 = xu$ and x is a unit. Thus $h' \sim \lambda h$.

Now suppose p/ab where p is irreducible. Then $(p, ab) \sim p$. Consider (p, a). Since p is irreducible $(p, a) \sim 1$ or p, and, if $(p, a) \sim p$ we must have p/a. Hence if p is not a factor of a then $(p, a) \sim 1$, and similarly if p is not a factor of b. Since we have shown that $(p, ab) \sim p$, we can prove the theorem provided we can show that $(p, a) \sim 1$ and $(p, b) \sim 1$ imply that $(p, ab) \sim 1$. We proceed to do this.

If $(p, a) \sim 1$, $(pb, ab) \sim b$ by (2) above.

Hence $1 \sim (p, b) \sim (p, (pb, ab))$ by the line above,

$\sim ((p, pb), ab)$ by (1),

$\sim (p, ab)$ since $(p, pb) \sim p$.

To show that an H.C.F. exists in the case of the integers or polynomials over a field the most satisfactory method (in that it does not postulate unique factorisation and thus avoids a circular argument) is by the so-called 'Euclid's algorithm', which is a consequence of the 'Division algorithm'. (See volume 1 for details.) This process may be generalised. Thus if we postulate the existence of a division algorithm we may show that H.C.F.s exist and hence, if we also postulate conditions for criterion 1 to be satisfied we may prove that the domain is a U.F.D. Such a domain is called a *Euclidean domain.* For the Euclidean algorithm to work we need a remainder that is 'descending' in a certain sense, so that the process terminates: in the case of integers the modulus decreases, while for polynomials we consider the degree. In the general case we define a function $\delta(x)$ over the elements x.

Definition. *An integral domain D is called a* Euclidean domain *if there exists a function $\delta(x)$ defined for all elements x of D such that*

(1) *$\delta(x)$ is a non-negative integer, with $\delta(x) = 0 \Leftrightarrow x = 0$.*

(2) *$\delta(xy) = \delta(x)\,\delta(y)$.*

(3) *If a and b are any elements of D with $b \neq 0$, then there exist elements q and r such that $a = bq+r$ where $\delta(r) < \delta(b)$.*

(3) is the division algorithm, while (2) ensures that criterion 1 is satisfied.

In the domain of the integers, $\delta(x) = |x|$, and in the domain of polynomials over a field, $\delta(P) = 2^{d(P)}$ (so that condition (2) is satisfied).

Theorem 5.4.6. *If D is a Euclidean domain then (a) criterion 1 is satisfied in D, (b) any two non-unit and non-zero elements of D have an* H.C.F. *It follows that D is a* U.F.D.

(*a*) We show first that $\delta(u) = 1 \Leftrightarrow u$ is a unit. If u is a unit a, $1 = uv$ and so $\delta(1) = \delta(u)\,\delta(v)$. But since $\delta(1x) = \delta(1)\,\delta(x)$ we have $\delta(1) = 1$, and so since $\delta(u)$ and $\delta(v)$ are positive integers, $\delta(u) = 1$.

Conversely suppose $\delta(u) = 1$. Then by (3), with $a = 1$ and $b = u$, we have $1 = uq+r$ where $\delta(r) < \delta(u)$ and so $\delta(r) = 0$. Hence $r = 0$ by (1) and $1 = uq$, showing that u is a unit.

Now suppose that a_{i+1} is a proper factor of a_i. Then $a_i = a_{i+1}y$ for some non-unit y and so by (2), $\delta(a_i) = \delta(a_{i+1})\,\delta(y)$ with $\delta(y) > 1$. Thus $\delta(a_{i+1}) < \delta(a_i)$ and so we cannot have an *infinite* sequence $\{a_i\}$ with a_{i+1} a proper factor of a_i.

(*b*) *Euclid's algorithm*

Apply (3) repeatedly as below.

$$a = bq_1+r_1,$$

$$b = r_1q_2+r_2,$$

$$r_1 = r_2q_3+r_3,$$

$$\dots\dots\dots\dots\dots\dots$$

$$r_{n-2} = r_{n-1}q_n+r_n,$$

$$r_{n-1} = r_nq_{n+1}.$$

(Since $\delta(b) > \delta(r_1) > \delta(r_2) > \ldots$ the process must terminate after a finite number of steps, the remainder in the last line having its 'δ' as 0 and so being itself 0.)

Working upwards we see that r_n is a factor of r_{n-1}, and so of r_{n-2}, ... and finally of b and a. Working downwards, if c/a and c/b, c/r_1, and so c/r_2 and so on. Thus c/r_n. Hence r_n is an H.C.F. of a and b.

Summary of this section

The reader will probably have found the present section difficult. What we have done is to start with unique factorisation domains and to explore successively different criteria that are either equivalent to the definition or which lead to it. Thus we have obtained several different conditions for a domain to possess unique factorisation into irreducible elements: some equivalent to the original in that all U.F.D.s possess them, and some weaker. We have gone backwards, so to speak, through the usual processes adopted when discussing particular domains such as the integers.

The work is not so pointless as it may seem. Although many integral domains possess all the properties given in the various sets of criteria, others do not, and to prove a domain a U.F.D. it is often best to proceed via some of the alternative criteria: for example the Gaussian integers will be shown in §5.8 to form a Euclidean domain, and hence a U.F.D.

As a brief summary we have proved:

(*a*) A U.F.D. satisfies criteria 1 and 2 and conversely any domain satisfying both these is a U.F.D.

(*b*) The existence of an H.C.F. is an alternative to criterion 2. Thus an H.C.F. existing plus criterion 1 implies that a domain is a U.F.D.: conversely an H.C.F. always exists in a U.F.D.

(*c*) A Euclidean domain is a U.F.D., but in this case the converse is not true. (The domain of polynomials over the integers is a U.F.D. but the division algorithm does not apply.)

5.5. The integers

The factorisation theory of the domain of integers was considered in detail in volume 1, chapter 4.

The only units are $+1$ and -1 and, apart from 0, classes of associates each consist of 2 integers, $\pm n$.

The integers form a Euclidean domain, with $\delta(n) = |n|$; the division algorithm was proved in volume 1. Thus they are a U.F.D. (The step from the proof of existence of an H.C.F. to criterion 2, corresponding to theorem 5.4.5, was there proved by a simpler method which is applicable to any Euclidean domain, for which see exercises 5B, no. 6.)

5.6. Domains of polynomials

The ring $R[x]$ of polynomials in an indeterminate x over the ring R is an integral domain if and only if R is an integral domain. In chapter 3 we proved the division algorithm for $R[x]$ where R is any ring with unity, *provided the divisor B had leading coefficient* 1. If R is a field the division algorithm holds generally, but this is not the case otherwise.

Theorem 5.6.1. *If P and Q are any non-zero polynomials in* $R[x]$, *where R is an integral domain, then* $d(PQ) = d(P)+d(Q)$.

Let $d(P) = n$ and $d(Q) = m$, where

$$P = a_n x^n + \ldots,$$

$$Q = b_m x^m + \ldots \quad \text{with} \quad a_n \neq 0 \quad \text{and} \quad b_m \neq 0.$$

Then $PQ \equiv a_n b_m x^{n+m} + \ldots$ (where $a_n b_m x^{n+m}$ is the leading term) and $a_n b_m \neq 0$ by the absence of zero divisors in R. The result follows.

Units in polynomial domains

Theorem 5.6.2. *The units in* $R[x]$ *correspond to the units in* R: *i.e. P is a unit if and only if it has degree* 0 *and is* $(u, 0, \ldots)$ *where u is a unit in R.*

$(u, 0, \ldots)$ is certainly a unit in $R[x]$, since if $uv = 1$ in R then $(u, 0, \ldots)(v, 0, \ldots) = (1, 0, \ldots)$ in $R[x]$. Conversely suppose U is a unit so that $UV = 1$, the unit polynomial. Then by theorem 5.6.1, $d(U)+d(V) = d(1) = 0$ and so $d(U) = d(V) = 0$. Hence $U = (u, 0, \ldots)$ and $V = (v, 0, \ldots)$ for some u and v, and $1 = UV = (uv, 0, \ldots)$, so that $uv = 1$ in R and u is a unit in R.

It follows at once that P and Q are associates if and only if $Q = uP$ for a unit u in R.

If R is a field all its non-zero elements are units, and so all polynomials in $R[x]$ of degree 0 (this does not of course include the zero polynomial) are units, and $P \sim cP$ for any non-zero c in R.

Irreducible elements of polynomial domains

If R is a field any polynomial of degree 1 is irreducible, for if $P = QR$ then by theorem 5.6.1 one of Q and R must have degree 0 and so be a unit. But there may be irreducible polynomials of higher degree. (For example x^2+1 over the reals and x^3-2 over the rationals.) By the Fundamental Theorem of Algebra there are no irreducible polynomials of degree greater than 1 if R is the field of complex numbers, and if R is the field of reals there are none of degree greater than 2. If R is not a field there may be polynomials of degree 1 which are *not* irreducible, for example if R is the integers $2x+2 = 2(x+1)$ and neither 2 nor $x+1$ are units. The whole question of irreducibility for polynomials over various domains is a difficult one and we studied it in more detail in volume 1, to which the reader is referred for more information.

Polynomials over a field

If R is a field, define $\delta(P)$, for any P in $R[x]$, to be $2^{d(P)}$ if $P \neq 0$, and 0 if $P = 0$. Then certainly $\delta(P)$ is a non-negative integer and is zero if and only if $P = 0$.

That $\delta(PQ) = \delta(P)\,\delta(Q)$ follows from theorem 5.6.1 if P and Q are non-zero, and is immediate if one or both is 0. (Note that we could define $\delta(P)$ equally well to be $k^{d(P)}$ for any positive integer $k > 1$.)

That the division algorithm holds in $R[x]$ for general B is proved as in theorem 3.6.1, the proof in this case depending on the possibility of dividing coefficients, and thus on the fact that R is a field. We will not repeat the proof here, since the modifications are fairly obvious. (For a detailed proof in the general case see volume 1, p. 113.) We note that the proof gives us the fact that $d(S) < d(B)$, which of course implies that $\delta(S) < \delta(B)$.

Thus the three conditions for a Euclidean domain are satisfied, and so $R[x]$, for R a field, is a U.F.D.

Polynomials over an integral domain

If R is an integral domain but not a field, then the division algorithm is not true in general. As a simple example, we cannot divide x^2 by $2x+1$ *over the integers* to obtain a remainder of degree 0, since $2x$ is not a factor of x^2. Thus $R[x]$ is not Euclidean.

Thus we cannot prove that $R[x]$ is a U.F.D. by the usual chain of reasoning, and the question of whether or not it *is* a U.F.D. arises. It is fairly obvious that if R is itself not a U.F.D. then $R[x]$ cannot be, for the polynomials of degree 0 behave like elements of R. (Thus the domain of poynomials over $I[\sqrt{-5}]$ is not a U.F.D., for the *polynomial* 9 possesses two distinct prime factorisations.)

The important question therefore is: 'If R is a U.F.D. then is $R[x]$ necessarily a U.F.D.?' This is in fact the case, and is a deep and important theorem in domain theory; so important that we will state it formally.

Theorem 5.6.3. *If R is a* U.F.D. *then $R[x]$ is a* U.F.D.

The proof of this theorem is difficult, and involves the concept of the 'Field of fractions' of the domain R, to be discussed in chapter 8. We will not give the proof, which may be found in any standard work, for example in Birkhoff and MacLane, *A Survey of Modern Algebra*, chapter III, theorem 16.

An important special case of theorem 5.6.3 is that of the polynomials over the integers. In this and the other cases we do of course have H.C.F.'s always existing, although we cannot find them by the Euclidean algorithm.

5.7. The characteristic of an integral domain

Much of the theory of characteristics of fields as given in §4.6 applies to general integral domains. Theorems 4.6.1 and 4.6.2 are true, the proofs being valid since they depend basically on the absence of zero divisors. Thus in any integral domain all non-zero elements have the same additive order, called the *characteristic* of the domain, and this is either a prime p or 0 (in the case where all non-zero elements have infinite order).

Theorems 4.6.3 and 4.6.4 apply in a modified form: a domain with characteristic p contains a sub-domain isomorphic to the domain (in fact field) of residues modulo p, while a domain with characteristic 0 contains a sub-domain isomorphic to the domain of integers (not rationals as in the case of fields).

It is interesting to note that we may have a domain with characteristic p which is not finite and not a field: an example is the domain of polynomials over the field of residues modulo p.

5.8. Gaussian integers

We have already shown that the ring of complex numbers $a+bi$, where a and b are integers, is an integral domain, called the domain of *Gaussian integers*. It is a good example of a domain that satisfies all the factorisation properties of this chapter, but for which their application to particular cases is not always easy to effect. Work with the Guassian integers is interesting and shows the power of our general methods. We start by considering the units.

Theorem 5.8.1. *The units of the domain of Gaussian integers are* $+1$, -1, $+i$, $-i$.

That these are units is immediate, since

$$1 = 1.1 = (-1)(-1) = i(-i).$$

To prove that they are the only units, let $u = a+bi$ be a unit with $uv = 1$. Then $|u|^2|v|^2 = |1|^2 = 1$ and so $|u|^2 = 1$, i.e. $a^2+b^2 = 1$. But a and b are integers, and so we must have either $a = 0, b = \pm 1$ or $b = 0, a \pm 1$, and the result follows.

Corollary. *The associates of* $x = a+bi$ *are* $-x$, ix *and* $-ix$.

The question of whether or not a given Gaussian integer is irreducible is, as might be expected, a difficult one. All we can do is to give examples of two apparently similar integers, one of which is irreducible and one not.

Example 1. The Gaussian integer 5 is not irreducible. For $5 = (1+2i)(1-2i)$.

Example 2. The Gaussian integer 3 is irreducible. Let $3 = xy$. Then $|x|^2|y|^2 = 9$ and so either $|x|^2 = 1$, and x is a unit, $|x|^2 = 9$ and y is a unit, or $|x|^2 = 3$. In this latter case we have, if $x = a+bi$, $a^2+b^2 = 3$, which is impossible. Hence either x or y is a unit and 3 is irreducible.

We now prove the main theorem of the section, that the Gaussian integers form a Euclidean domain, and hence that H.C.F.'s exist and they are a U.F.D. The only difficult (and interesting) part of the proof is in establishing the division algorithm.

Theorem 5.8.2. *The Gaussian integers are a Euclidean domain.*

If $x = a+bi$, define $\delta(x) = |x|^2 = a^2+b^2$. Then certainly $\delta(x)$ is a non-negative integer and $= 0$ if and only if $x = 0$.

We also have at once that $\delta(xy) = \delta(x)\,\delta(y)$.

It remains to prove the division algorithm.

Let α and β be any two Gaussian integers with $\beta \neq 0$.

Considering $\alpha:\beta$ as a complex number, let

$$\alpha:\beta = \rho = p_1+p_2 i,$$

where p_1 and p_2 will not in general be integers.

$\exists$ *integers* q_1 and q_2 such that $|p_1-q_1| \leqslant \frac{1}{2}$, $|p_2-q_2| \leqslant \frac{1}{2}$. If $\sigma = q_1+q_2 i$, consider $|\rho-\sigma|$.

$$\begin{aligned} |\rho-\sigma| &= |p_1-q_1+(p_2-q_2)\,i| \\ &= \sqrt{\{(p_1-q_1)^2+(p_2-q_2)^2\}} \\ &\leqslant \sqrt{(\tfrac{1}{4}+\tfrac{1}{4})} \\ &< 1. \end{aligned}$$

Now let $\alpha-\beta\sigma = \omega$ (so that $\alpha = \beta\sigma+\omega$). Then since α, β, σ are Gaussian integers, so is ω. Also

$$\left|\frac{\omega}{\beta}\right| = \left|\frac{\alpha}{\beta}-\sigma\right| = |\rho-\sigma| < 1.$$

Therefore $|\omega| < |\beta|$, i.e. $\delta(\omega) < \delta(\beta)$.

Worked exercises

1. If r/s is a rational zero, in its lowest terms, of the polynomial

$$a_n x^n + a_{n-1} x^{n-1} + \ldots + a_0$$

where all a_i are integers, then r/a_0 and s/a_n.

We have that
$$0 = a_n \left(\frac{r}{s}\right)^n + a_{n-1} \left(\frac{r}{s}\right)^{n-1} + \ldots + a_0$$

$$\text{i.e.} \quad 0 = a_n r^n + a_{n-1} r^{n-1} s + \ldots + a_0 s^n,$$

multiplying by s^n.

Thus $a_n r^n = -s(a_{n-1} r^{n-1} + \ldots + a_0 s^{n-1})$ and so s is a factor of $a_n r^n$.

Let $a_n r^n = st$, and decompose s into prime factors, a typical factor being $p_i^{\nu_i}$.

Since s and r are co-prime (r/s is in its lowest terms), p_i is not a factor of r, and so not of r^n, and hence $p_i^{\nu_i}$ must be a factor of a_n. Hence s/a_n. Similarly $a_0 s^n = -r(a_n r^{n-1} + \ldots + a_1 s^{n-1})$ and r/a_0.

2. Find all irreducible cubic polynomials over the field of residues modulo 2.

The domain of such polynomials is a U.F.D., the only unit being 1. If a cubic has a non-unit, non-associate factor this must have degree 1 or 2, and in the latter case the other factor has degree 1. Hence if we consider all products of a linear factor with a quadratic, we will obtain all reducible cubics (including those that have 3 linear factors since our quadratic need not be irreducible).

Now the only linear polynomials are x and $x+1$, and the quadratics are x^2, x^2+x, x^2+1, x^2+x+1. Forming the 8 products we obtain the following cubics:

$$x^3, \quad x^3+x^2, \quad x^3+x, \quad x^3+x^2+x, \quad x^3+x^2, \quad x^3+x, \quad x^3+x^2+x+1,$$
$$x^3+1.$$

Two of these cubics are repeated, and there are precisely two of the eight possible different cubics that do not appear, viz. x^3+x^2+1 and x^3+x+1, and hence these are the only 2 irreducible cubics.

3. Find Gaussian integers q and r with $|r| < |b|$ such that $a = bq+r$, when $a = 14-3i$ and $b = 4+7i$. Find the H.C.F. of a and b and represent it in the form $sa+tb$.

As a complex number
$$\frac{a}{b} = \frac{(14-3i)(4-7i)}{65}$$
$$= \frac{35-110i}{65}$$
$$= \tfrac{7}{13} - \tfrac{22}{13}i.$$

We choose q so that both real and imaginary parts of q lie within $\frac{1}{2}$ of those of a/b. Thus $q = 1-2i$.

Then $a-bq = 14-3i-(4+7i)(1-2i) = -4-2i$. Hence $r = -4-2i$.

Now we use Euclid's algorithm to find the H.C.F. To save space we give the results only for each line: the detailed working is as above.

$$\begin{aligned} 14-3i &= (4+7i)(1-2i)+(-4-2i), \\ 4+7i &= (-4-2i)(-1-i)+(2+i), \\ -4-2i &= (2+i)(-2). \end{aligned}$$

Hence the H.C.F. is $2+i$ (or, multiplying by units, $-2-i$, $i(2+i) = -1+2i$, or $1-2i$).

To find s and t we use the algorithm above and obtain

$$\begin{aligned} h = 2+i &= b-(-1-i)(-4-2i) \\ &= b-(-1-i)(a-(1-2i)b) \\ &= b+(1+i)a+(-3+i)b \\ &= (1+i)a+(-2+i)b. \end{aligned}$$

4. (A difficult but instructive and exciting example.) Show that any prime $p = 4n+1$ may be expressed as the sum of 2 squares of integers.

Working over the field of residues modulo p consider the polynomial $x^{4n}-1 \equiv (x^{2n}-1)(x^{2n}+1)$.

Now if x is any non-zero element of the field, $x^{4n} = 1$ by Lagrange's theorem since the $4n$ non-zero elements form a multiplicative group, and hence the order of x is a factor of $4n$. Hence $x^{4n}-1$ has $4n$ zeros, and so by the factor theorem, and since the domain of polynomials is a U.F.D., has $4n$ linear factors. Hence $x^{2n}+1$ has $2n$ linear factors, and thus certainly has at least 1 zero, m say. I.e. there exists m such that

$$m^{2n}+1 = 0 \pmod p.$$

Hence $m^{2n}+1 = kp$ for some integers m and k.

Now put $m^n = t$. Then $t^2+1 = kp$. Now consider the domain of Gaussian integers, which contains all real integers as a sub-domain, so that the above equation, $t^2+1 = kp$, is true in the domain of Gaussian integers. But in this domain $t^2+1 = (t+i)(t-i)$.

Now p is a factor of t^2+1 by the above work, and it is not a proper factor of either $t+i$ or $t-i$, since it is a real integer and does not divide the imaginary part of these ($p(a+bi) = pa+pbi$ and $pb \neq \pm 1$). Hence p is not a prime, since it contradicts criterion 2.

Let p have a proper factor $c+di$. Then since it is real it also has $c-di$ as a factor, and so c^2+d^2 is a factor of p^2, and this is true in the domain of integers, since the other factor must of course also be an integer. Hence c^2+d^2 is either 1, p or p^2 since p is prime.

If $c^2+d^2 = 1$, $c+di = \pm 1$ or $\pm i$ and is a unit. If

$$c^2+d^2 = p^2, \quad \frac{p}{c+di}\cdot\frac{p}{c-di} = 1$$

and so $p/(c+di)$ is a unit and $c+di$ is an associate of p. Hence as $c+di$ is a *proper* factor of p we must have $c^2+d^2 = p$, and the result is proved since c and d are integers, as $c+di$ is a Gaussian integer.

Exercises 5A

There are many routine examples on the use of Euclid's algorithm and on factorisation of polynomials in volume 1, chapters 4 and 7. The reader who does not possess that volume can easily construct such practice examples for himself: we will not repeat them here.

1. If a/b and b/c prove that a/c.

2. If a is a proper factor of b and b/c prove that a is a proper factor of c.

3. Prove that any factor of a unit is itself a unit.

4. Prove that any associate of a unit is itself a unit.

5. Prove that any 2 units are associates.

6. Prove that if $a^2 = a$ in an integral domain then a is either the unity or zero.

7. Prove that if a/b and a/c then $a/(b+c)$.

8. If $a \sim a'$ and $b \sim b'$ show that $ab \sim a'b'$.

9. If $ax \sim by$ and $a \sim b$, $a \neq 0$, prove that $x \sim y$.

10. If p/a or p/b whenever p/ab prove that p is necessarily irreducible.

11. If $(a, c) = 1$ and c/ab prove that c/b.

12. If $(a, c) = 1$ and a/m and c/m then ac/m.

13. Prove that $I[\sqrt{(nk^2)}]$ is a proper sub-domain of $I[\sqrt{n}]$, where k is any integer whose modulus is greater than 1 and n is not a perfect square.

14. Prove that $\pm 1 \pm \sqrt{2}$, $\pm 3 \pm 2\sqrt{2}$ are units in $I[\sqrt{2}]$. If $a^2 - 2b^2 = \pm 1$ prove that $a + b\sqrt{2}$ is a unit.

15. By considering moduli prove that the only units in $I[\sqrt{(-n)}]$, where n is a positive integer > 1, are ± 1.

16. Find a unit, other than ± 1, in $I[\sqrt{3}]$.

17. Let functions $f(x)$, $g(x)$ be defined by

$$f(x) = 0, \quad 0 \leqslant x \leqslant \tfrac{1}{2}, \qquad f(x) = x - \tfrac{1}{2}, \quad \tfrac{1}{2} \leqslant x \leqslant 1$$
$$g(x) = \tfrac{1}{2} - x, \quad 0 \leqslant x \leqslant \tfrac{1}{2}, \qquad g(x) = 0, \quad \tfrac{1}{2} \leqslant x \leqslant 1.$$

Show that $fg = 0$ in $[0, 1]$ and deduce that the ring of continuous real-valued functions defined in $[0, 1]$ is not an integral domain.

18. If D is an integral domain of characteristic p show that $\theta: x \to x^p$ is a ring-endomorphism of D. (See exercises 4B, no. 9.)

19. Identify θ in **18** when D is the field of residues modulo p. (See exercises 4B, no. 10.)

In **20–23** use the factor theorem to find all factors of the form $x-c$ of the polynomials given, working over the field of residues modulo p.

20. $x^2+1, p = 2.$

21. $x^3-x, p = 3,$

22. $x^2+2, p = 5.$

23. $2x^3-1, p = 7.$

24. Prove that the factors found in **20–23** include *all* linear factors, to within associates.

25. Find all irreducible quadratic polynomials over the field of residues modulo 3.

26. Prove that an integral domain that is not a field must possess a subset of its elements that can be put in 1–1 correspondence with the set of all elements. (*Hint.* Consider the proof of theorem 5.1.3, and consider $\{xy\}$ for some fixed x which has no inverse.)

27. If $\theta: D \to D'$ is a ring-homomorphism between the integral domains D and D', and if $D\theta \neq \{0\}$, prove that $1\theta = 1'$, where $1, 1'$ are the unities in D and D' respectively.

28. Give an example to show that if $\theta: D \to R$ is a ring-homomorphism from the integral domain D into the ring R, then $D\theta$ need not be an integral domain.

29. By considering the residues 2 and 4 in the ring of residues modulo 6, prove that theorem 5.3.3 is not necessarily true in a ring that is not an integral domain.

30. In $I[\sqrt{-3}]$ express 4 in two ways as the product of irreducible elements and hence show that $I[\sqrt{-3}]$ is not a U.F.D.

31. In $I[\sqrt{-3}]$ prove that 5 is irreducible, but 4 is not.

32. An H.C.F. of n elements $a_1, \ldots, a_n$ is defined to be an element h, where (i) h/a_i for all i; (ii) if c/a_i for all i then c/h.

Prove that the H.C.F. is unique modulo units, if it exists.

33. Prove that if $a_n x^n + \ldots + a_0$ is irreducible then so is $a_0 x^n + \ldots + a_n$.

34. In $I[\sqrt{2}]$ prove that if $a+b\sqrt{2}$ is a unit then $a^2-2b^2 = \pm 1$. Prove that $\pm(1 \pm \sqrt{2})^n$ is a unit for all positive integers n.

In **35–37** find Gaussian integers q and r with $|r| < |b|$ such that $a = bq+r$.

35. $a = 2,\ b = 1+i.$

36. $a = 8-i,\ b = 2-i.$

37. $a = -10+11i,\ b = 7+4i.$

38. Express $3+i$ as the product of irreducible Gaussian integers.

39. Find the H.C.F. of a and b in **38** and express it in the form $sa+tb$.

40. Give an example to show that the q and r in the division algorithm for Gaussian integers are not necessarily unique.

Exercises 5B

1. Prove that in any integral domain (i) $x^n = 0$ for a positive integer $n \Rightarrow x = 0$; (ii) $x^m = x^n$ for positive integers m and n with $n > m$ and $x \neq 0 \Rightarrow x^{n-m} = 1$.

Deduce that, if $x^n = 1$ for some positive integer n and $x^m \neq 1$ for $1 \leqslant m < n$, then the elements $1, x, x^2, \ldots, x^{n-1}$ are distinct. (n is the *multiplicative order* of x: if no such n exists we say that the multiplicative order is infinite.)

2. Prove that if x has a finite multiplicative order then it is a unit. Give an example to show that the converse is not true.

3. By finding the factorisation into irreducible polynomial, or otherwise, write down *all* factors (to within associates) of $x^4+3x^3+4x^2+x+4$, working with coefficients in the field of residues modulo 5.

4. By considering moduli prove that criterion 1 is satisfied in $I[\sqrt{(-n)}]$ for any positive integer n and deduce that prime factorisation is always *possible* (though not unique). Find a prime factorisation in $I[\sqrt{-2}]$ of

$$\text{(i) } 1+\sqrt{-2}, \quad \text{(ii) } 2+\sqrt{-2}, \quad \text{(iii) } 5+\sqrt{-2}.$$

5. Let S be the set of all expressions of the form $a_1x^{\alpha_1}+a_2x^{\alpha_2}+\ldots+a_nx^{\alpha_n}$, where the a_i's are rational numbers and the α_i's are non-negative rational numbers. Define addition and multiplication in the obvious way (using $x^\alpha x^\beta = x^{\alpha+\beta}$ and $a_1x^\alpha+a_2x^\alpha = (a_1+a_2)\,x^\alpha$). Show that S is an integral domain. Show that the element x of S is not a unit, but that it does not possess a factorisation into irreducible elements, so that S is a domain that does not even possess prime factorisation.

6. Prove that in a Euclidean domain if $h = (a, b)$ then there exist elements s and t such that $h = sa+tb$. *Deduce* criterion 2.

7. Show that the result of **6** is not necessarily true for a U.F.D. that is not Euclidean by considering the domain of polynomials over the integers with $a = 2$, $b = x$. (Note that an H.C.F. exists in such a domain.)

8. The domain of polynomials $R[x, y]$ in two indeterminates over the field of rationals is a U.F.D. by theorem 5.6.3 applied to the Euclidean domain $R[x]$.

Prove that the result of **6** is not true for this domain by taking $a = x$, $b = y^2+x$ and deduce that $R[x, y]$ is not Euclidean.

9. A *least common multiple*, or L.C.M., of the two elements a and b is defined as an elment l such that

(i) a/l and b/l; (ii) if a/c and b/c then l/c.

Prove that, if it exists, the L.C.M. is unique modulo units. Prove also that in a U.F.D. an L.C.M. always exists, and that if h is an H.C.F. of a and b, then $hl \sim ab$.

10. Prove that a polynomial of degree n over an integral domain has at most n distinct zeros. Deduce

(i) if $a_n x^n + \ldots + a_0$ has more than n distinct zeros then

$$a_n = a_{n-1} = \ldots = a_0 = 0;$$

(ii) if

$$P \equiv a_n x^n + \ldots + a_0,$$

$$Q \equiv b_n x^n + \ldots + b_0$$

and $P(y) = Q(y)$ for more than n elements y, then $a_i = b_i$ for all i and $P(y) = Q(y)$ for all y.

11. If R is an integral domain show that $\theta: R[x] \to R[x]$ defined by $P(x)\,\theta = P(-x)$ (e.g. $(x^3+x^2)\,\theta = -x^3+x^2$) is a ring-*automorphism*.

12. Show that $\theta: R[x] \to R[x]$ defined by $P(x)\,\theta = P(ax)$ for some fixed $a \neq 0$ is a ring-*automorphism* if and only if a is a unit of R. Deduce that if R is a field, θ is a ring-automorphism for all non-zero a.

13. Working over a field, prove that there is at most one polynomial of degree n which takes specified values b_i at the $n+1$ elements a_i,

$$i = 1, 2, \ldots, (n+1).$$

Prove that there *does* exist such a polynomial by showing that the polynomial $P(x)$ satisfies the conditions, where

$$P(x) \equiv \sum_{i=1}^{n+1} \frac{(x-a_1)\ldots(x-a_{i-1})\,(x-a_{i+1})\ldots(x-a_{n+1})}{(a_i-a_1)\ldots(a_i-a_{i-1})\,(a_i-a_{i+1})\ldots(a_i-a_{n+1})}\,b_i.$$

(The formula for $P(x)$ is known as *Lagrange's Interpolation Formula.*)

6

INVARIANT SUBGROUPS

6.1. Description of the problem

The advanced theory of groups depends on the idea of an *invariant* (or *normal* or *self-conjugate*) *subgroup*. Beyond the elementary work on subgroups leading up to Lagrange's theorem, and homomorphisms, it is not possible to progress far with the study of group theory without introducing and using this concept. Its importance cannot be over-emphasised.

Although so important, the idea of an invariant subgroup is not easy to grasp for a student meeting it for the first time. It is easy to give a definition, but the significance of this and its far-reaching implications are difficult to appreciate before more experience is gained. There are in fact several approaches to the problem.

In this chapter we will develop the essential theory of invariant subgroups. In §6.2 and §6.3 we give what seems to be the most illuminating definition, and in §6.4 give an alternative and important approach. In §6.7 and §6.8 the important connection between invariant subgroups and homomorphisms is explained; in §6.9 the concept is linked with the idea of direct products.

In the remainder of the present section we try to explain some of the ideas that lie behind the definition, attempting to show intuitively some of the importance of the concept. The approach is informal, the rigorous treatment being left until the next sections.

The problem is basically that of 'dividing' one group by another: that is in 'dividing' a group G by a subgroup H to obtain another group, a 'quotient group'. The idea is similar to that of a direct product, where we 'multiply' the groups G and H to form a new 'product group' $G \times H$. Of course the words 'divide' and 'multiply' are used in a loose sense, but the analogy is useful if not over-emphasised. Thus if G and H have finite orders m and n then the order of $G \times H$ is mn.

Let us investigate some examples where we have such a 'division'.

The most obvious case is that of 'dividing' $A \times B$ by B (strictly, by the subgroup of elements of the form (e, b)). We would expect the 'quotient' to be A (or at least isomorphic to A). Suppose we consider the cosets of B. A typical one will contain an element of the form (a, f) where f is the neutral element of B, and will consist of all elements (a, b) for fixed a and variable b. It may be written aB for convenience. Now multiply any element in aB by any element in $a'B$. The product is $(a, b).(a', b') = (aa', bb')$ for some b and b' in B, and is in *$aa'B$ whichever elements we took in B*. Thus we may say that the 'product' of the cosets aB and $a'B$ is the coset $aa'B$, and under this definition it is easily seen that the cosets are the elements of a group which is isomorphic to A. We have thus obtained a 'quotient group' of $A \times B$ divided by B by considering the cosets of B and forming a group with these as elements.

As a special case of the above let both A and B be the group of real numbers under addition. Then $A \times B$ is the group of 2-dimensional vectors, which are best represented for our present purpose by the points of a plane. Then the process described consists of 'dividing' a plane by one of its axes (say the y-axis) and obtaining as a quotient the other axis or, more correctly, the set of lines parallel to $x = 0$ which form a group by considering their points of intersection with the x-axis.

The direct product example is easy to understand but by no means covers all possibilities. As a very interesting and important example let us investigate the group of residues modulo n under addition.

Let G be the group of integers (under addition) and H the subgroup of all multiples of n. The residue r (modulo n) is defined as the equivalence class (under the relation of congruence) which contains all integers congruent to r. The reason why residues form a group is that if *any* member of one class is added to *any* member of another then the sum is always in the same class whichever members were chosen. (See volume 1 for full details, but note that we are concerned only with the sum and difference structure here, not with the product and possible quotient properties.) But the class of all integers congruent to r is precisely the coset rH in the group G (cosets are in fact defined in general as equivalence classes in a similar way to residues), and the addition property of residues mentioned above, that the sum is independent of the particular members chosen, is expressed in the product notation by the fact that the group product of any element in rH and any element in $r'H$ is always in $rr'H$, i.e. that the 'product' of the cosets rH and $r'H$ is $rr'H$. Thus the cosets of H are the elements of the group of residues, which may be thought of as the 'quotient group' of G divided by H. This is the same procedure as was adopted for direct product, but there is an important difference. In the direct product case the quotient group was isomorphic to A, a subgroup of $A \times B$. But the group of residues is *not* isomorphic to any subgroup of the group of integers. (This is easily seen by noting that the residue 1 has order n, and no integer has additive order n for any finite $n > 1$.) In fact, although that lying between 0 and $n-1$ seems superficially to be the most important member of a residue class, and gives its name to that class, it cannot be isolated as the element (g, f) could be: the properties of the group of residue classes are *not* those of these particular members, but are properties of the classes (or cosets) as a whole.

Let us now consider some multiplication tables of finite groups: for convenience we will take groups of small order. We start with a direct product, the Vierergruppe $C_2 \times C_2$.

	e	a	b	ba
e	e	a	b	ba
a	a	e	ba	b
b	b	ba	e	a
ba	ba	b	a	e

Fig. 12

	ϵ	β
ϵ	ϵ	β
β	β	ϵ

Fig. 13

Figure 12 shows the multiplication table of $G = C_2 \times C_2$ partitioning with respect to the cosets of $H = \{e, a\}$. We note that in each square thus formed all the elements belong to the same coset. If we denote the coset H by ϵ and the other coset

$\{b, ba\}$ by β and fill in the partitions by the coset which every element belongs to we obtain the table in figure 13, which is the multiplication table of the group C_2 with elements ϵ and β. This illustrates the 'division' of $C_2 \times C_2$ by the coset $H = C_2$ to form the 'quotient' C_2, being the group of cosets.

	e	a^2	a	a^3
e	e	a^2	a	a^3
a^2	a^2	e	a^3	a
a	a	a^3	a^2	e
a^3	a^3	a	e	a^2

Fig. 14

	ϵ	α
ϵ	ϵ	α
α	α	ϵ

Fig. 15

Figures 14 and 15 show the same process for $G = C_4$ and $H = C_2$. We still obtain a 'quotient' group C_2. There is one difference between this and the previous example. In the first case we could have written the cosets as eH and bH and obtained a group isomorphic to the subgroup $\{e, b\}$ of $C_2 \times C_2$. In the second case we could still have written the cosets as eH and aH, but the group of cosets would *not* have been isomorphic to the subset $\{e, a\}$, since although $(aH)^2 = eH$ it is not true that $a^2 = e$, and in fact the subset $\{e, a\}$ is not even a subgroup. This state of affairs is similar to that which occurred with residues. (That G does contain a subgroup, namely H, isomorphic to the quotient group is beside the point: the important fact is that we cannot select members from the different cosets which themselves form a subgroup.)

We now give a rather more interesting example, where $G = D_6$, the dihedral group of order 6, which is of course non-Abelian. We take first the subgroup $H = \{e, a, a^2\}$. The multiplication table in figure 16, when partitioned into products of cosets, exhibits the same property as before, that all elements in any square formed by the partitioning are in the same coset, and we obtain the table in figure 17, giving us the 'quotient group' C_2. (Note here that although the elements e and b of the cosets form a group $\cong$ the quotient group, yet G is not $H \times C_2$.)

Now let us take the same $G = D_6$, and let H be the subgroup $\{e, b\}$. (We write ab for b^2a and a^2b for ba for convenience.)

	e	a	a^2	b	ba	ba^2
e	e	a	a^2	b	ba	ba^2
a	a	a^2	e	ba^2	b	ba
a^2	a^2	e	a	ba	ba^2	b
b	b	ba	ba^2	e	a	a^2
ba	ba	ba^2	b	a^2	e	a
ba^2	ba^2	b	ba	a	a^2	e

Fig. 16

	ϵ	β
ϵ	ϵ	β
β	β	ϵ

Fig. 17

	e	b	a	ab	a^2	a^2b
e	e	b	a	ab	a^2	a^2b
b	b	e	a^2b	a^2	ab	a
a	a	ab	a^2	a^2b	e	b
ab	ab	a	b	e	a^2b	a^2
a^2	a^2	a^2b	e	b	a	ab
a^2b	a^2b	a^2	ab	a	b	e

Fig. 18

This time, as can be seen in figure 18, the elements in a square do not necessarily belong to the same coset: this may happen of course as in the first column, but in general it will not. In this example we cannot form a group from the cosets in any natural way. (We could form them into a group, namely C_3, by placing an arbitrary multiplicative structure on them, but this would bear no relation to the original group and subgroup and would be of no use nor interest in connection with these: thus if we called the coset $\{a, ab\}$ α and considered α^2 to be $\{a^2, a^2b\}$ then it is true that the *particular elements* a and a have a product in α^2, but this is not true for, say, ab and ab.)

The last example shows us that it is not always possible to form a group from the cosets of an arbitrary subgroup H of G. Thus our 'division' process is not always possible. If the cosets *do* form a group in the way that we have considered then H is said to be an *invariant* subgroup of G, and in the next section we will find the conditions for this to happen.

The reader may wonder why this idea is so important. As usual it is difficult to give specific examples from the advanced theory that would be understood at this stage, but we can understand something of the reasons if we bear in mind that two of the most important subjects for study in group theory are those of subgroups and homomorphisms. When dealing with subgroups an obviously important question is that of the decomposition of the whole group into its cosets: it was used in the proof of Lagrange's theorem. If the cosets themselves form a group then we can apply the results of group theory to them, and this is clearly an illuminating procedure in its own right, besides giving us a method of forming new groups from given ones.

This is not however, the whole story. The second topic, the study of homomorphisms, turns out to be intimately connected with that of invariant subgroups: to every invariant subgroup there corresponds a 'natural' homomorphism, and conversely, so that the study of homomorphisms is in effect the study of invariant subgroups. This point is dealt with later in the chapter.

For specific examples, some will emerge in the present chapter; one important one, the formation of the group of residues from the group of integers, has already been mentioned. An important advanced example, well beyond our scope to prove, is in the theory of the insolubility of the general quintic, where invariant subgroups play a key part in the work.

6.2. Invariant subgroups

We wish to investigate the condition for the cosets of a subgroup H of a group G to form a group. It is clearly necessary for us to be able to multiply any two cosets together, in an obvious way by multiplying typical elements. (We repeat that it is no use giving an *arbitrary* group structure to the cosets: our group of cosets must be linked to the group structure of G.) Thus we require the product of any two elements in two given cosets always to lie in the same coset.

We will at present use left cosets for convenience (although in the present work the distinction between left and right cosets soon disappears) and we will use the Frobenius notation freely.

Thus our condition is that if $x \in gH$ and $y \in g'H$ then xy is always in the same coset. But by taking $x = g$ and $y = g'$

this must be the coset $gg'H$. Thus we require xy to be in $gg'H$ or, in Frobenius notation, $(gH)(g'H) \subseteq gg'H$, for any pair of cosets gH and $g'H$.

Remembering that product of complexes is associative, this means that $gHg'H \subseteq gg'H$ or, pre-multiplying by g^{-1}, that $Hg'H \subseteq g'H$. But in this case, by choosing the particular element e of H (the unity is contained in any subgroup) in the last H on the left-hand side, we must have that $Hg' \subseteq g'H$, i.e. that for any $h \in H$, hg' can be expressed in the form $g'h'$ for some $h' \in H$.

Now if $Hg' \subseteq g'H$ for any g' we must have $Hg \subseteq gH$ for any g (changing g' to g for convenience) and also $Hg^{-1} \subseteq g^{-1}H$. This latter condition gives us $gHg^{-1}g \subseteq gg^{-1}Hg$, or $gH \subseteq Hg$, so that we must in fact have $gH = Hg$, for all $g \in G$.

Thus a necessary condition for the cosets to form a group is that the left and right cosets coincide. It can easily be shown, as below, that this is also sufficient, and thus this condition gives us our definition of invariant subgroup.

Definition. *H is an invariant subgroup of G if it is a subgroup and if the left and right cosets of H in G coincide*; *i.e. if*

$$gH = Hg \; \forall \; g \in G.$$

Invariant subgroups are often called *normal* subgroups, or *self-conjugate* subgroups. The latter term arises naturally from the approach adopted in §6.4, but both it and the term 'normal' are rather over-worked words in mathematics, and we prefer to use 'invariant'.

If the left coset gH *is* a right coset then it must contain g $(= eg)$ and so *must* be the right coset Hg. Hence if every left coset is a right coset (or vice versa) H is invariant.

Note that the definition does *not* mean that $gh = hg \; \forall \; h \in H$. This is the case in Abelian groups (where every subgroup is invariant) and may happen for particular cosets but for non-Abelian groups it will not happen generally. All that is required is that $gh = h'g$ *for some* $h' \in H$. As an example consider $G = D_6$ and $H = \{e, a, a^2\}$ as in figure 16. Then the coset H is of course both left and right, and the other left coset bH is the set

$\{b, ba, ba^2\}$, whereas Hb is $\{b, ab, a^2b\}$, that is $\{b, ba^2, ba\}$, which consists of the same elements *but in a different order.*

Note also that *all* the cosets must coincide. The coset H is always both left and right, and there may be others, but H need not be invariant.

In figure 18 we see why $H = \{e, b\}$ is not invariant in D_6. For $aH = \{a, ab\}$ and $Ha = \{a, ba\}$ or $\{a, a^2b\}$ and they do not coincide.

We now prove the sufficiency of this condition for a product to be defined for cosets.

Theorem 6.2.1. *If H is invariant in G then $(gH)(g'H) = gg'H$ for all $g, g' \in G$.*

$gHg'H = g(Hg')H = g(g'H)H$ since $Hg' = g'H$ as H is invariant,

$= gg'H^2 = gg'H$ since $H^2 = H$ as H is a subgroup.

Note that it would have been sufficient to prove that

$$(gH)(g'H) \subseteq gg'H,$$

but our result is just as easy to establish, and in fact follows from the inclusion (see exercises 6A, no. 12).

The ease with which, by use of the Frobenius notation, we prove this fundamental theorem may obscure the important steps in the proof. In terms of single elements we are trying to show that for any $h, h' \in H$, $ghg'h' = gg'h''$ for some $h'' \in H$. Because H is invariant we know that $hg' = g'k$ for some $k \in H$ and so $ghg'h' = gg'kh'$ and, since H is a subgroup, $kh' = h''$ where h'' *is* in H.

We may put the condition for an invariant subgroup in a slightly different form. If $gH = Hg$ then $H = g^{-1}Hg$ and conversely. We immediately obtain the following theorem.

Theorem 6.2.2. *A subgroup H is an invariant subgroup of G if and only if $g^{-1}Hg = H$ for all $g \in G$.*

A very common way to prove H invariant in practice is to show that for any $h \in H$ and $g \in G$, $g^{-1}hg$ is always in H. That this implies that H is invariant is proved in the next theorem.

Theorem 6.2.3. *If H is a subgroup of G and if for all $h \in H$ and $g \in G$, $g^{-1}hg \in H$, then H is an invariant subgroup of G.*

The condition tells us that $g^{-1}Hg \subseteq H \; \forall \; g \in G$. Hence, replacing g by g^{-1}, $gHg^{-1} \subseteq H$. Thus $g^{-1}gHg^{-1}g \subseteq g^{-1}Hg$, i.e. $H \subseteq g^{-1}Hg$ and so $g^{-1}Hg = H$. The result follows from theorem 6.2.2.

6.3. Quotient groups

We have seen that the necessary and sufficient condition for the product of cosets to be well defined is that H is an invariant subgroup of G. In the examples of §6.1 we saw that under this definition of product the cosets always formed a group. It is easy to prove that this is true generally.

Theorem 6.3.1. *If H is an invariant subgroup of G, the cosets of H in G form a group under the group product given by*

$$(gH)(g'H) = gg'H.$$

The product is well defined and unique, and of course is always a coset.

It is associative, since

$$((gH)(g'H))(g''H) = gg'g''H = (gH)((g'H)(g''H)).$$

(This works because product is associative in G, or simply because product of complexes is associative.)

The coset $H \; (= eH)$ is a unity since

$$(gH)(eH) = geH = gH \quad \text{and} \quad (eH)(gH) = egH = gH.$$

The coset gH has an inverse $g^{-1}H$, since

$$(gH)(g^{-1}H) = gg^{-1}H = eH = H \quad \text{and} \quad (g^{-1}H)(gH) = H$$

similarly.

This group formed by the cosets of H in G is called the *Quotient Group of H in G*, or sometimes the *Factor Group of G relative to H*. It is denoted by G/H. It is vital to realise *the quotient group exists only when H is an invariant subgroup of G.*

Notice that although theorem 6.3.1 is a vital one in showing that such a group exists, its proof was really trivial, once theorem 6.2.1 was established: the invariancy of H is necessary for a product of cosets to be definable at all but once this is shown possible the fact that the cosets form a group follows at once from the group properties of G.

If G and H are finite, with orders n and m, then there are n/m cosets (an integer by Lagrange's theorem) and so the order of G/H is n/m, the index of H in G.

G/H need not be Abelian, although it must be if G is Abelian. For example if $G = A \times B$ and $H = B$, $G/H \cong A$ and is not Abelian if A is not.

At the risk of labouring the point we must emphasise yet again that the choice of g in the coset gH is arbitrary, and the properties of the product of cosets do not depend on its choice. g is *any* element of the coset and, for example, the fact that $(gH)(g'H) = H$ does *not* mean that $gg' = e$. We refer again to the example of the residues modulo n, where the coset rH consists of all integers congruent to r and, although the 'group product' of rH and $(n-r)H$ is H, it is not true that the sum of r and $n-r$, considered as integers between 0 and $n-1$, is 0. The elements of G/H are *cosets*.

The reader should make absolutely sure that he understands the method of forming the quotient group, and should refer again to the examples in §6.1, studying them and seeing the process at work. More examples will be given later.

If H is invariant we have 'divided', in a certain sense, the group G by H. The resulting group is written G/H by analogy with ordinary division, but the analogy must not be pressed too far. That the order of G/H is the ratio of the orders of G and H is as expected (in the finite case). But the symbol $1/H$ is meaningless, and one must never write GH^{-1} instead of G/H, since H^{-1} means 'the set $\{h^{-1}: h \in H\}$' and is in fact H when H is a subgroup.

When dealing with the quotient group G/H we may think in terms of working 'modulo H'. Thus in the residues case two integers are in the same coset if they are 'congruent modulo n', that is if they differ by a multiple of n. In general two elements are in the same coset of H if their quotient is in H (i.e. if they are g and k we must have $k^{-1}g \in H$).

6.4. Conjugacy

Another approach to the idea of invariant subgroups is through the idea of conjugate elements and sub-groups: this work on conjugacy is important and gives a method of finding the invariant subgroups of a given group that is often useful, but it is not so illuminating intuitively as our approach in §6.2.

We first define what is meant by conjugate elements.

Definition. *The elements x and $g^{-1}xg$, where x and g are elements of a group G, are called conjugate elements in G: in other words, x and y are conjugate if and only if there exists an element $g \in G$ such that $y = g^{-1}xg$.*

The important fact about the relation of conjugacy is that it is an equivalence relation. This is fairly easily proved.

Theorem 6.4.1. *The relation of conjugacy between elements of any group G is an equivalence relation.*

Reflexive. Since $x = e^{-1}xe$, x is conjugate to itself.

Symmetric. If $y = g^{-1}xg$ we have $x = gyg^{-1} = (g^{-1})^{-1}\,yg^{-1}$.

Transitive. If $y = g^{-1}xg$ and $z = h^{-1}yh$ we have

$$z = h^{-1}g^{-1}xgh = (gh)^{-1}\,x(gh).$$

It follows that the relation divides the elements of G into mutually exclusive equivalence classes; these are called *conjugacy classes* in G.

If the conjugacy class of x consists of just x itself then x is known as a *self-conjugate* element. The condition for this is that $g^{-1}xg = x \;\forall\, g \in G$, i.e. that $gx = xg$, that x commutes with all elements of G.

In an Abelian group all elements are self-conjugate and the concept is trivial, as is all the work of the present section: it is only in the non-Abelian case that the idea of conjugacy is fruitful. In any group the neutral element is self-conjugate: there may or may not be other such elements.

Example. Consider D_6 with the notation used in §6.1.

e is self-conjugate.

Consider the conjugacy class containing a. If $g = e$, a or a^2, then $g^{-1}ag = a$. If $g = b$, $g^{-1}ag = b^{-1}ab = bab = bba^2 = a^2$.

Similarly if $g = ba$ or ba^2, $g^{-1}ag = a^2$. Hence $\{a, a^2\}$ forms a conjugacy class.

Now consider the class containing b.

$$a^{-1}ba = a^2ba = baa = ba^2,$$

whereas $$(a^2)^{-1}\,ba^2 = aba^2 = ba^2a^2 = ba.$$

Hence $\{b, ba, ba^2\}$ is the third conjugacy class.

Theorem 6.4.2. *Conjugate elements have the same order.*

Suppose x and y are conjugate, with $y = g^{-1}xg$. Then $y^2 = g^{-1}xgg^{-1}xg = g^{-1}x^2g$, $y^3 = g^{-1}x^3g$, and so on. Hence if $x^n = e$, $y^n = g^{-1}eg = e$. Similarly if $y^n = e$ then so does x^n. Hence x and y have the same order, finite or infinite.

We now extend the idea of conjugacy to subgroups. If H is a subgroup we consider the set, or complex, $g^{-1}Hg$. We first prove that this is a subgroup.

Theorem 6.4.3. *If H is a subgroup of G then $g^{-1}Hg$ is a subgroup for any element $g \in G$.*

For $(g^{-1}Hg)^2 = g^{-1}Hgg^{-1}Hg = g^{-1}HHg = g^{-1}Hg$ since $H^2 = H$ as H is a subgroup. Also

$$(g^{-1}Hg)^{-1} = g^{-1}H^{-1}g = g^{-1}Hg \quad \text{since} \quad H^{-1} = H.$$

Hence (theorem 1.8.3) $g^{-1}Hg$ is a subgroup.

We say that H and $g^{-1}Hg$ are *conjugate subgroups* in G. By precisely the same argument as in theorem 6.4.1 we obtain:

Theorem 6.4.4. *The relation of conjugacy between subgroups of G is an equivalence relation.*

We thus obtain *conjugacy classes of subgroups* of G, and notice in passing that all subgroups in the same class have the same order.

An immediate application of theorem 6.2.2 gives us now the fundamental condition, in terms of conjugacy, for H to be invariant.

Theorem 6.4.5. *H is an invariant subgroup of G if the conjugacy class of subgroups of G which contains H consists of just H itself: i.e. if the only subgroup conjugate to H is H.*

We see now the reason for the use of the term 'self-conjugate subgroup' in place of 'invariant subgroup'.

Example. Consider D_6 and all its subgroups.

The trivial subgroups $\{e\}$ and D_6 are clearly invariant. The only subgroup of order 3 is $\{e, a, a^2\}$ and hence all subgroups conjugate to this must be the subgroup itself (the reader is advised to check this for one or two elements g by finding $g^{-1}Hg$) and this subgroup is invariant.

The following are subgroups of order 2:

$$H_1 = \{e, b\}, \quad H_2 = \{e, ba\}, \quad H_3 = \{e, ba^2\}.$$

By working out the elements it is easily seen that

$$e^{-1}H_1e = b^{-1}H_1b = H_1, \quad (a^2)^{-1}H_1a^2 = (ba^2)^{-1}H_1(ba^2) = H_2,$$
$$a^{-1}H_1a = (ba)^{-1}H_1(ba) = H_3.$$

Hence these three subgroups form a conjugacy class and none is invariant.

The following theorem sometimes enables us to discover all invariant subgroups of a given group, particularly in the case of groups of fairly small order.

Theorem 6.4.6. *A subgroup H is invariant in G if and only if it is the union of conjugacy classes of elements of G.*

Suppose first that H is invariant. Then if $x \in H$, so is $g^{-1}xg \ \forall \ g \in G$. Hence H contains all elements conjugate to x and so must be the union of conjugacy classes.

Conversely suppose that H is the union of conjugacy classes, so that if $x \in H$ all elements conjugate to x are in H, i.e. $g^{-1}xg \in H \ \forall \ g \in G$. Hence H is invariant by theorem 6.2.3.

Applying this to D_6, H is invariant if and only if it is the union of some or all of the classes

$$C_1 = \{e\}, \quad C_2 = \{a, a^2\}, \quad C_3 = \{b, ba, ba^2\}.$$

This gives us the invariant subgroups

$$C_1, \quad C_1 \cup C_2 \cup C_3 = D_6, \quad C_1 \cup C_2 = \{e, a, a^2\},$$

as before, whereas of course $\{e, b\}$, $\{e, ba\}$, $\{e, ba^2\}$ are not invariant.

(Note that the union of classes must be a *subgroup* for it to give an invariant subgroup: thus $C_1 \cup C_3$ does not give us one, not being itself a subgroup.)

6.5. Examples of invariant subgroups

Abelian groups

In an Abelian group every subgroup is invariant: since $gh = hg$ for all g and h, we certainly have $gH = Hg$. Although every subgroup is invariant the concept is still exceedingly important in the Abelian case and the identification of the quotient group is by no means a trivial process. For example we need only consider the example of the subgroup $\{\lambda n\}$ in the (Abelian) group of the integers to form the group of residues modulo n.

Trivial subgroups

It is seen at once that $\{e\}$ and G are always invariant subgroups of G. The cosets of $\{e\}$ each have only one element and the quotient group is G (strictly speaking, is isomorphic to G). There is just one coset of the subgroup G and the quotient group has just one element, being $\{e\}$.

Subgroups of vectors

Groups of vectors are Abelian and so every subgroup is invariant.

In the group of two-dimensional vectors let H be the subgroup consisting of all vectors of the form $(x, \lambda x)$ where λ is a fixed scalar. Suppose that the co-ordinates are elements of a field F (which may be the reals, rationals, complex numbers or any field). A typical coset is the set $\{(a, 0)+(x, \lambda x)\}$ for varying x, where a is an element of F, and there is a distinct coset for each $a \in F$, provided $\lambda \neq 0$. The sum of the cosets $(a, 0)+H$ and $(b, 0)+H$ is $(a+b, 0)+H$, and so the quotient group is isomorphic to F, the coset $(a, 0)+H$ corresponding to the element a in the isomorphism. (In geometrical terms the cosets are lines parallel to the line which represents the subgroup.)

Similarly in three dimensions if H is the subgroup of vectors of the form $(x, \lambda x, \mu x)$ the quotient group is the group of two-dimensional vectors, a typical coset being $(a, b, 0)+H$, unless $\mu = 0$. The subgroup consisting of the vectors (x, y, z) where $\lambda x+\mu y+\nu z = 0$ (a plane through O) has a quotient group isomorphic to the field of scalars F.

Subgroups of finite cyclic groups

If m is a factor of n so that $n = mr$ say, the elements

$$e, g^r, g^{2r}, \dots g^{(m-1)r} \quad \text{of } C_n,$$

generator g, form a subgroup C_m. The cosets are $g^i H$ (where we call the subgroup H), for $i = 0, 1, \dots, (r-1)$, and it is easily seen since $(g^i H)(g^j H) = (g^{i+j} H)$ and, of course, $g^r H = H$, the neutral element of the quotient group, that the quotient group is merely C_r.

Subgroups of index 2

Suppose H is a subgroup of a group G of index 2, so that there are just two cosets of H in G. Then the left cosets are H itself and one other, which must consist of all elements of G not in H. Similarly there are two right cosets, which must be the same. Thus in this case right and left cosets certainly coincide and H is invariant. The quotient group has two elements and so must be C_2. It is worth putting these results in the form of a theorem.

Theorem 6.5.1. *If H is a subgroup index 2 of a group G then H is always invariant in G and $G/H \cong C_2$.*

From this theorem we know at once that the subgroup we have previously used, that of $\{e, a, a^2\}$ in D_6, is invariant.

As another example, the alternating group of degree n (the subgroup of all even permutations) has index 2 in S_n and thus is always invariant.

Some examples of non-invariant subgroups

The reader may gain the impression that practically all subgroups are invariant. This idea is helped by the fact that there are no non-invariant subgroups of Abelian groups. In the non-Abelian case however very few subgroups are invariant in general, and we will consider a few examples here. We have already met some, the subgroups $\{e, b\}$, $\{e, ba\}$ and $\{e, ba^2\}$ in D_6.

In the general Dihedral group D_{2n}, defined by generators a and b with $a^n = b^2 = e$ and $ab = ba^{n-1}$, the subgroup of powers of a has index 2 and so is invariant by theorem 6.5.1. The subgroup $\{e, b\}$ is never invariant for $n > 2$, for

$$a^{-1}ba = ba^2 \neq e \text{ or } b.$$

Hence by theorem 6.2.3 $\{e, b\}$ is not invariant. (This is usually the best way of disproving invariancy: we need find just one example where $g^{-1}hg \notin H$ to show that H is not invariant.)

In the symmetric group S_n let t be the permutation that interchanges two of the objects and leaves the others unaltered (t is a *transposition*). For simplicity let t interchange the objects 1 and 2, so that $t = \begin{pmatrix} 1 & 2 & \dots & n \\ 2 & 1 & \dots & n \end{pmatrix}$. Then $t^2 = e$ and so $\{e, t\}$ is a subgroup. But it is not invariant for $n \geqslant 3$, as may be seen by finding $g^{-1}tg$ where $g = \begin{pmatrix} 1 & 2 & 3 & \dots & n \\ 2 & 3 & 1 & \dots & n \end{pmatrix}$. ($g^{-1}tg = \begin{pmatrix} 1 & 2 & 3 & \dots & n \\ 1 & 3 & 2 & \dots & n \end{pmatrix}$ and is not in $\{e, t\}$.)

Similarly let H be the subgroup of S_n which leaves $n-r$ objects unaltered, so that a typical element of H permutes the objects $1, 2, \dots, r$ say, where $r < n$. (H is clearly a subgroup.) Then if $g = \begin{pmatrix} 1 & 2 & \dots & n \\ n & 2 & \dots & 1 \end{pmatrix}$, $h = \begin{pmatrix} 1 & 2 & \dots & r & \dots & n \\ 2 & 1 & \dots & r & \dots & n \end{pmatrix}$ we see that $g^{-1}hg = \begin{pmatrix} 1 & 2 & \dots & n \\ 1 & n & \dots & 2 \end{pmatrix}$ and $\notin H$, so that H is not invariant.

Finally we give an infinite example. Let G be the group of transformations of a sphere and H be the subgroup of all rotations about a given diameter, XX' in figure 19. Let h be a rotation of 90° about XX', clockwise from X to X', and let g be a rotation of 90° about YY', clockwise from Y to Y'. Then we see from figure 19 that $g^{-1}hg \notin H$, since it does not leave the point P unchanged. (Note that if we had taken g as before but with a rotation of 180° then $g^{-1}hg$ *would* be in H, but for H to be invariant this must be the case *for all* g.)

If H and K are subgroups of G, then although $H \cup K$ may not even be a subgroup, we know that $H \cap K$ is always a subgroup of G (theorem 1.8.5) and in fact is always invariant if both H and K are invariant. This is easily proved.

Theorem 6.5.2. *If both H and K are invariant subgroups of G then $H \cap K$ is an invariant subgroup of G.*

Clearly $g(H \cap K) = gH \cap gK$ and $(H \cap K)g = Hg \cap Kg$. But $gH = Hg$ and $gK = Kg$, and the result follows.

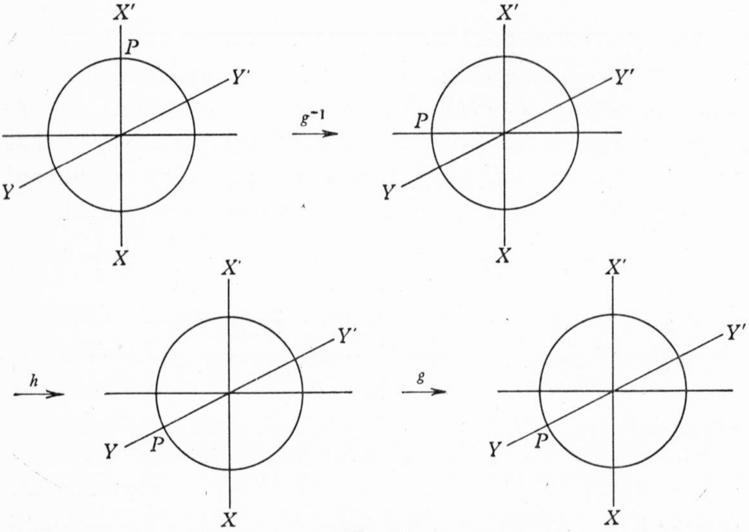

Fig. 19

Corollary. *If H_i is invariant for $i = 1, ..., n$ then $\bigcap_{i=1}^{n} H_i$ is invariant.*

It is important to note that invariancy is a relative term: thus strictly we must speak of H being an invariant subgroup *of* G, the group G being an essential part of the definition. It is possible for H to be invariant considered as a subgroup of one group but not invariant when considered as a subgroup of a larger group. As a trivial example of this H is always invariant considered as a subgroup of itself but may not be when considered as a proper subgroup of a group G. For a more interesting example consider the subgroup H consisting of the three permutations $\begin{pmatrix}1 & 2 & 3 & 4\\1 & 2 & 3 & 4\end{pmatrix}$, $\begin{pmatrix}1 & 2 & 3 & 4\\2 & 3 & 1 & 4\end{pmatrix}$ and $\begin{pmatrix}1 & 2 & 3 & 4\\3 & 1 & 2 & 4\end{pmatrix}$. This is a subgroup index 2, and hence invariant, of the group K of the six permutations in S_4 which leave the last object fixed. But we can soon check $\left(\text{take } g = \begin{pmatrix} 1 & 2 & 3 & 4 \\ 4 & 2 & 3 & 1 \end{pmatrix}, h = \begin{pmatrix} 1 & 2 & 3 & 4 \\ 2 & 3 & 1 & 4 \end{pmatrix}\right)$ that H is not invariant in S_4.

6.6. The centre of a group. Inner automorphisms

We saw in §6.4 that an element x of a group G is *self-conjugate* if its conjugacy class contains just the one element, i.e. if $g^{-1}xg = x \;\forall\; g \in G$, which implies that x commutes with all elements of G. The set of all self-conjugate elements is called the *centre* of G: it may equally well be defined as the set of all elements that commute with every element of G.

The centre may be thought of as being the Abelian part of G, but it is not in general the largest Abelian subgroup of G, or even the largest invariant Abelian subgroup. It is the whole group G if and only if G is Abelian. It may contain the single element e (as in the case of D_6) or may be a subset of G (for example the group of quaternions, where the centre is the subset of reals). In fact the centre is always a subgroup and is invariant.

Theorem 6.6.1. *The centre of any group G is an invariant subgroup of G.*

Suppose x and y are in the centre. Then for any $g \in G$,

$$g^{-1}xg = x, \quad g^{-1}yg = y$$

and so
$$g^{-1}xyg = (g^{-1}xg)(g^{-1}yg) = xy.$$

Hence xy is in the centre. If x is in the centre, $g^{-1}xg = x$ and so $x^{-1} = (g^{-1}xg)^{-1} = g^{-1}x^{-1}g$ and x^{-1} is in the centre. Finally e is clearly in the centre and so the centre is a subgroup. It is invariant since, if C is the centre, $g^{-1}Cg = C$ since

$$g^{-1}xg = x \;\forall\; x \in C.$$

Note that the centre is always Abelian.

Inner automorphisms

In chapter 2 we proved that the mapping $\theta: G \to G$ given by $g\theta = x^{-1}gx$, where x is a fixed element of G, is an automorphism of G, known as an *inner* automorphism (see theorem 2.5.1). We also saw that *all* automorphisms of G form a group.

Theorem 6.6.2. *The inner automorphisms form an invariant subgroup of the group of automorphisms of any group G.*

Denote the inner automorphism $g \to x^{-1}gx$ by θ_x. Then

$$g\theta_x\theta_y = y^{-1}x^{-1}gxy = (xy)^{-1}g(xy) = g\theta_{xy},$$

so that $\theta_x\theta_y = \theta_{xy}$ and is an inner automorphism.

Also since $g\theta_{x^{-1}} = xgx^{-1}$,

$$\begin{aligned}(x^{-1}gx)\,\theta_{x^{-1}} &= (x^{-1}\theta_{x^{-1}})\,(g\theta_{x^{-1}})\,(x\theta_{x^{-1}}) \\ &= (xx^{-1}x^{-1})\,(xgx^{-1})\,(xxx^{-1}) = g,\end{aligned}$$

so that $(\theta_x)^{-1} = \theta_{x^{-1}}$ and is an inner automorphism.

$g\theta_e = e^{-1}ge = g$ and thus θ_e is the identity automorphism.

Hence the set of inner automorphisms forms a subgroup of the group of automorphisms.

To show that it is invariant, let ϕ be any automorphism of G and consider $\phi^{-1}\theta_x\phi$.

$$\begin{aligned}g\phi^{-1}\theta_x\phi &= (g\phi^{-1})\,\theta_x\phi = [x^{-1}(g\phi^{-1})\,x]\,\phi = (x^{-1}\phi)\,(g\phi^{-1}\phi)\,(x\phi) \\ &= (x\phi)^{-1}\,g(x\phi) = g\theta_{x\phi}.\end{aligned}$$

Hence $\phi^{-1}\theta_x\phi = \theta_{x\phi}$ and is an inner automorphism, so that the subgroup of inner automorphisms is invariant.

Note that if G is Abelian $x^{-1}gx = g$ and so $\theta_x = I$, the identity automorphism, for all x.

An inner automorphism always sends an element into a conjugate element. In particular it leaves every element in the centre invariant, for $x^{-1}gx = g$ if g is in the centre.

6.7. The natural homomorphism

As we have already hinted, the subjects of invariant subgroups and homomorphisms of groups are closely connected: there is in fact a 1–1 correspondence between the invariant subgroups of a given group G and the essentially different homomorphisms which have G as object space, so that the study of either topic is akin to that of the other. We show first how an invariant subgroup gives rise to a related homomorphism; the converse problem will be dealt with in the next section.

Suppose that H is an invariant subgroup of G, so that the quotient group G/H is defined. We consider the mapping $\theta: G \to G/H$ defined by $g\theta = gH$, the coset containing g, which is of course an element of G/H.

Theorem 6.7.1. *The mapping $\theta: G \to G/H$ defined by $g\theta = gH$ is a homomorphism.*

The proof is almost immediate, since

$$\begin{aligned}(gg')\,\theta &= gg'H \\ &= (gH)(g'H), \quad \text{the product being as defined in } G/H, \\ &= (g\theta)(g'\theta).\end{aligned}$$

Corollary. *The kernel of θ is H and the image is G/H, so that θ is an epimorphism.*

If g is in the kernel, $g\theta = H$, the neutral element of G/H, and so $gH = H$, so that $g \in H$. Similarly if $g \in H$, $g\theta = H$ and g is in the kernel. That the image is G/H is obvious.

The homomorphism θ is called the *natural homomorphism* associated with H. (It is of course essential for H to be invariant, for otherwise G/H is not defined and we cannot speak of the product of cosets $(gH)(g'H)$.)

6.8. The quotient group associated with a given homomorphism

We now show that any homomorphism $\theta: G \to G'$ is closely related to the natural homomorphism associated with some invariant subgroup of G. This is the converse of the last section and consideration of theorem 6.7.1 and its corollary leads us to thinking of the kernel of θ as a subgroup of G; for if θ is to be a natural homomorphism then its kernel must be the invariant subgroup with which it is associated.

By theorem 2.3.1 we know that the kernel H of θ *is* a subgroup of G.

Theorem 6.8.1. *The kernel of $\theta: G \to G'$ is an invariant subgroup of G.*

If h is in the kernel H and g is any element of G,

$$\begin{aligned}(g^{-1}hg)\,\theta &= (g^{-1}\theta)\,(h\theta)\,(g\theta)\\ &= (g\theta)^{-1} f(g\theta), \quad \text{where } f \text{ is the neutral element of } G',\\ &= (g\theta)^{-1}\,(g\theta) = f.\end{aligned}$$

Hence $g^{-1}hg \in H$ and so H is invariant by theorem 6.2.3.

We now wish to show that θ is related to the natural homomorphism associated with its kernel H. If this natural homomorphism be denoted by ϕ, we know that $\phi: G \to G/H$. Since $\theta: G \to G'$ the most that we can prove is that the image $G\theta$ (a subgroup of G' by theorem 2.3.3) is isomorphic to the group G/H, since there is no reason why these should be *the same group*, nor indeed why G' should not be a larger group than G/H, as it will be if θ is not an epimorphism. To prove that θ and ϕ are essentially the same we must also show that $g\theta \sim g\phi$, the correspondence being that of the required isomorphism between $G\theta$ and G/H. But $g\phi = gH$. Hence we require to show that the mapping of G/H into $G\theta$ given by $gH \to g\theta$ is an isomorphism. We first prove it to be well-defined.

Theorem 6.8.2. *The mapping defined by $gH \to g\theta$ is a well-defined mapping of G/H onto $G\theta$.*

We must show that if g_1 and g_2 are in the same coset then $g_1\theta = g_2\theta$. If g_1 and g_2 are in the same coset then $g_2^{-1}g_1 \in H$. Hence $(g_2\theta)^{-1}(g_1\theta) = f$, the neutral element in G', since H is the kernel of θ. Thus $g_2\theta = g_1\theta$.

Also any element of $G\theta$ is the image of some $g \in G$ and so the mapping is *onto* $G\theta$, although not necessarily of course onto G'.

Theorem 6.8.3. *The mapping defined by $gH \to g\theta$ is an isomorphism between G/H and $G\theta$.*

We first show that it is a homomorphism. This is so, since

$$(g_1H)(g_2H) = g_1g_2H,$$

$$(g_1\theta)(g_2\theta) = (g_1g_2)\,\theta,$$

since θ is a homomorphism, and $g_1g_2H \to (g_1g_2)\,\theta$.

We have already seen that this mapping is onto $G\theta$. Hence it is an isomorphism provided it is 1–1, i.e. provided its kernel is the neutral element of G/H. Now suppose $gH \to f$. Then $g\theta = f$ and so $g \in H$, so that $gH = H$, the neutral element of G/H.

Notice that the sets of elements of G which map into the particular elements of $G\theta$ are merely the cosets of G relative to the kernel H.

We have proved that to any invariant subgroup H of G there corresponds a natural homomorphism from G onto G/H, and conversely that any homomorphism of G is equivalent to (is isomorphic to, to speak loosely) a natural homomorphism, namely to that associated with its kernel. Thus there is a 1–1 correspondence between the invariant subgroups and the essentially different homomorphisms of G. By essentially different homomorphisms we mean those having different kernels: two homomorphisms having the same kernel are both equivalent to the same natural homomorphism (their images are isomorphic and elements always map into elements of the image spaces which correspond under this isomorphism)—they are, in effect, the same homomorphism and exhibit the same structure. Note that the image *space* G' is largely irrelevant: the subgroup $G\theta$ of G' is isomorphic to G/H but G' may be larger, and of course there is no reason why $G\theta$ should be invariant in G'.

To help fix our ideas consider the following example. Let θ be the homomorphism of the group of integers onto the group of residues modulo n defined by $r\theta =$ the residue class containing r. (That this is a homomorphism was shown in example 2 on p. 32.) The kernel of θ is the subgroup of multiples of n, invariant of course since we are dealing with Abelian groups, and θ maps an integer into its coset, which is an element of the group of residues.

Isomorphisms

The kernel of an isomorphism is $\{e\}$ and the image is of course $\cong G$, i.e. is $\cong G/\{e\}$.

Zero homomorphisms

If $G\theta = \{f\}$, the kernel of θ is G and the image $\cong G/G$.

Homomorphisms from finite cyclic groups

In §6.5 we saw that for any factor m of n there is a subgroup of C_n consisting of the elements $e, g^r, g^{2r}, \ldots, g^{(m-1)r}$, where g is a generator of C_n and $r = n/m$. This is invariant and the quotient group is C_r.

These are the only subgroups of C_n and hence the homomorphisms of C_n are given by mapping $g^{\lambda r}$ to the neutral element and obtaining an image $\cong C_r$.

As a further example we return to the topics of §6.6, and explore the connection between inner automorphisms and the centre of a group. We have seen that the centre C of a group G is an invariant subgroup and we wish to investigate the quotient group G/C. We start with a lemma.

Lemma 1. *The inner automorphism θ_x given by $g \to x^{-1}gx$ is the identity automorphism if and only if $x \in$ the centre C.*

If $x \in C$, $x^{-1}gx = x^{-1}xg = g$ and so $\theta_x = I$, the identity. If $\theta_x = I$, $x^{-1}gx = g\ \forall\ g$ and so $x \in C$.

Now consider the mapping of G into the group of all automorphisms of G given by $x \to \theta_x$.

Lemma 2. *The mapping $x \to \theta_x$ of G into the group of automorphisms of G is a homomorphism.*

In the proof of theorem 6.6.2 we proved that $\theta_{xy} = \theta_x\theta_y$; the result follows.

By lemma 1 the kernel of this homomorphism is the centre C. The image is the subgroup of inner automorphisms. We obtain the following theorem.

Theorem 6.8.4. *The group of inner automorphisms of a group $G \cong G/C$, where C is the centre of G.*

6.9. Direct products

At the beginning of this chapter we used the direct product $A \times B$ to help us in understanding the ideas behind the concept of an invariant subgroup. We discuss it more formally here.

Theorem 6.9.1. *In the group* $G = A \times B$ *the subgroup of elements of the form* (e, b) *is invariant and its quotient group* $\cong A$.

To show that the subgroup is invariant, take a general element (x, y) of G and consider the product

$$\begin{aligned}(x, y)^{-1}(e, b)(x, y) &= (x^{-1}, y^{-1})(e, b)(x, y)\\ &= (x^{-1}ex, y^{-1}by)\\ &= (e, y^{-1}by)\end{aligned}$$

and is in the subgroup.

A typical coset is the set (a, b) for a fixed $a \in A$ and variable $b \in B$. Denote this by aB. Since $(a, b)(a', b') = (aa', bb')$ we see that the product of the cosets aB and $a'B$ is $aa'B$ and so the quotient group $\cong A$.

We may of course interchange the roles played by A and B in the above and obtain an invariant subgroup $\cong A$ and an associated quotient group $\cong B$.

The natural homomorphism for the invariant subgroup $\{(e, b)\}$ is given by mapping (a, b) into a, thus giving a homomorphism of $A \times B$ onto A. This is the *projection* mentioned in §2.5.

In the above example the group G is isomorphic to $B(G/B)$. This is not the case for a general group G and invariant subgroup H. Examples given earlier were that of the integers with $H = \{\lambda n\}$ (where the group of integers is *not* $H \times C_n$), that where $G = C_4$ and $H = C_2$, and where $G = D_6$, $H = \{e, a, a^2\}$ (where

$$D_6 \not\cong C_3 \times C_2).$$

6.10. Simple groups

A group G always possesses the invariant subgroups $\{e\}$ and G. If it has no others, i.e. no proper invariant subgroups, then G is called a *simple* group.

Theorem 6.10.1. *If* p *is prime,* C_p *is a simple group, and these are the only Abelian simple groups.*

By Lagrange's theorem the order of any subgroup of C_p must be a factor of p and hence is 1 or p, so that C_p has no proper subgroups, and hence no proper invariant subgroups. Conversely if G is Abelian and simple it has no proper subgroups (since all subgroups are invariant in the Abelian case). Let x be any non-neutral element. Then the subgroup of powers of x must be G and so G is generated by x and is cyclic. But the infinite cyclic group and finite cyclic groups of composite order have proper subgroups and hence $G \cong C_p$ for a prime p.

Thus all simple groups of composite order, or infinite order, are non-Abelian. Such groups are relatively rare, but we must mention one important example, which is of importance in the proof of the insolubility of the general quintic. The alternating group of degree n (the group of all *even* permutations on n objects) has order $\frac{1}{2}n!$ and this is not prime if $n \geqslant 4$. If $n = 4$ the alternating group has a proper invariant subgroup (the elements $\begin{pmatrix}1 & 2 & 3 & 4\\ 2 & 1 & 4 & 3\end{pmatrix}$, $\begin{pmatrix}1 & 2 & 3 & 4\\ 3 & 4 & 1 & 2\end{pmatrix}$ and $\begin{pmatrix}1 & 2 & 3 & 4\\ 4 & 3 & 2 & 1\end{pmatrix}$ can be seen to form a conjugacy class and these, together with the neutral element $\begin{pmatrix}1 & 2 & 3 & 4\\ 1 & 2 & 3 & 4\end{pmatrix}$, form a subgroup, which must be invariant by theorem 6.4.6). If $n \geqslant 5$ however, the alternating group is simple. The proof is not difficult but depends on the idea of 'cycles' and is too long for inclusion here (see any standard work, for example *Ledermann—'Introduction to the Theory of Finite Groups'*). We will however give an indication of the proof for the special case $n = 5$, leaving the reader to fill in the tedious details if he wishes. (This is a proof by enumeration of elements and is not the same as the general proof for any $n \geqslant 5$.)

Theorem 6.10.2. *The alternating group of degree* 5 *is simple.*

We divide the 60 elements into conjugacy classes, omitting the details of the proof.

(i) The neutral element $\begin{pmatrix}1 & 2 & 3 & 4 & 5\\ 1 & 2 & 3 & 4 & 5\end{pmatrix}$ is self-conjugate and forms a class of 1 *element*.

(ii) There are 15 *elements* in the class containing $\begin{pmatrix}1 & 2 & 3 & 4 & 5\\ 2 & 1 & 4 & 3 & 5\end{pmatrix}$, viz. all elements obtained by transposing two distinct pairs of the 5 objects.

(iii) There are 20 *elements* in the class containing

$$\begin{pmatrix}1 & 2 & 3 & 4 & 5\\ 2 & 3 & 1 & 4 & 5\end{pmatrix},$$

all obtained by permuting three objects cyclically (each changing its position by either 1 or 2 places) and leaving the other two objects invariant.

(iv) The class containing $\begin{pmatrix} 1 & 2 & 3 & 4 & 5 \\ 2 & 3 & 4 & 5 & 1 \end{pmatrix}$ has 12 *elements*, consisting of half those elements which are given by a single 'chain' of interchanges (object a going to b, b to c, c to d, d to e and e to a, where a, b, c, d, e are the 5 objects in some order).

(v) The remaining 12 *elements* given by a single chain of interchanges form a conjugacy class, a typical one being

$$\begin{pmatrix} 1 & 2 & 3 & 4 & 5 \\ 3 & 1 & 4 & 5 & 2 \end{pmatrix}.$$

Now by theorem 6.4.6 an invariant subgroup must be the union of complete conjugacy classes. But by Lagrange's theorem its order must be a factor of 60. It must also contain the neutral element. It is easily seen that we cannot total some of the integers, 1, 15, 20, 12 and 12 (these being the numbers of elements in the conjugacy classes), including the integer 1 for the neutral element, to obtain a proper factor of 60, and so the alternating group has no proper invariant subgroups and hence is simple.

Worked exercises

1. Prove that a subgroup H of a group G is invariant if and only if it is unchanged (as a whole) by *any* inner automorphism of G.

H is invariant if and only if $gH = Hg \; \forall \; g \in G$, i.e. if and only if

$$H = g^{-1}Hg \; \forall \; g \in G.$$

But the image of H under the inner automorphism given by $x \to g^{-1}xg$ is $g^{-1}Hg$; the result follows.

Note. We see from this result the reason for the term 'invariant': an invariant subgroup is one that is invariant under any inner automorphism.

2. Find all invariant subgroups of S_4 and identify the corresponding quotient groups.

We divide the 24 elements into conjugacy classes, using theorem 6.4.2 that conjugate elements have the same order. The details are tedious but may be verified quite simply—we omit the verification. We obtain the following classes.

(i) $\begin{pmatrix}1&2&3&4\\1&2&3&4\end{pmatrix}$. 1 element, order 1.

(ii) $\begin{pmatrix}1&2&3&4\\2&1&4&3\end{pmatrix}$, $\begin{pmatrix}1&2&3&4\\3&4&1&2\end{pmatrix}$, $\begin{pmatrix}1&2&3&4\\4&3&2&1\end{pmatrix}$. 3 elements, order 2.

(iii) $\begin{pmatrix}1&2&3&4\\2&1&3&4\end{pmatrix}$, $\begin{pmatrix}1&2&3&4\\3&2&1&4\end{pmatrix}$, $\begin{pmatrix}1&2&3&4\\4&2&3&1\end{pmatrix}$, $\begin{pmatrix}1&2&3&4\\1&3&2&4\end{pmatrix}$,

$\begin{pmatrix}1&2&3&4\\1&4&3&2\end{pmatrix}$, $\begin{pmatrix}1&2&3&4\\1&2&4&3\end{pmatrix}$. 6 elements, order 2.

(iv) $\begin{pmatrix}1&2&3&4\\1&3&4&2\end{pmatrix}$, $\begin{pmatrix}1&2&3&4\\1&4&2&3\end{pmatrix}$, $\begin{pmatrix}1&2&3&4\\3&2&4&1\end{pmatrix}$, $\begin{pmatrix}1&2&3&4\\4&2&1&3\end{pmatrix}$,

$\begin{pmatrix}1&2&3&4\\2&4&3&1\end{pmatrix}$, $\begin{pmatrix}1&2&3&4\\4&1&3&2\end{pmatrix}$, $\begin{pmatrix}1&2&3&4\\2&3&1&4\end{pmatrix}$, $\begin{pmatrix}1&2&3&4\\3&1&2&4\end{pmatrix}$.

8 elements, order 3.

(v) $\begin{pmatrix}1&2&3&4\\2&3&4&1\end{pmatrix}$, $\begin{pmatrix}1&2&3&4\\3&4&2&1\end{pmatrix}$, $\begin{pmatrix}1&2&3&4\\4&1&2&3\end{pmatrix}$, $\begin{pmatrix}1&2&3&4\\2&4&1&3\end{pmatrix}$,

$\begin{pmatrix}1&2&3&4\\3&1&4&2\end{pmatrix}$, $\begin{pmatrix}1&2&3&4\\4&3&1&2\end{pmatrix}$. 6 elements, order 4.

Now the invariant subgroups are those subgroups which are the union of conjugacy classes (theorem 6.4.6). Thus we search for unions of conjugacy classes whose total number of elements is a factor of 24 (by Lagrange's theorem), and which contain class (i).

The only possibilities are as follows:

(*a*) (i) alone. The trivial subgroup $\{e\}$, quotient group S_4.

(*b*) The trivial subgroup S_4, quotient group $\{e\}$.

(*c*) (i)+(ii). The set has order 4, and is easily seen to be in fact a subgroup, the Vierergruppe $C_2 \times C_2$.

The cosets are:

(α) $\begin{pmatrix}1&2&3&4\\1&2&3&4\end{pmatrix}$, $\begin{pmatrix}1&2&3&4\\2&1&4&3\end{pmatrix}$, $\begin{pmatrix}1&2&3&4\\3&4&1&2\end{pmatrix}$, $\begin{pmatrix}1&2&3&4\\4&3&2&1\end{pmatrix}$.

(β) $\begin{pmatrix}1&2&3&4\\2&1&3&4\end{pmatrix}$, $\begin{pmatrix}1&2&3&4\\1&2&4&3\end{pmatrix}$, $\begin{pmatrix}1&2&3&4\\4&3&1&2\end{pmatrix}$, $\begin{pmatrix}1&2&3&4\\3&4&2&1\end{pmatrix}$.

(γ) $\begin{pmatrix}1&2&3&4\\3&2&1&4\end{pmatrix}$, $\begin{pmatrix}1&2&3&4\\4&1&2&3\end{pmatrix}$, $\begin{pmatrix}1&2&3&4\\1&4&3&2\end{pmatrix}$, $\begin{pmatrix}1&2&3&4\\2&3&4&1\end{pmatrix}$.

(δ) $\begin{pmatrix}1&2&3&4\\4&2&3&1\end{pmatrix}$, $\begin{pmatrix}1&2&3&4\\3&1&4&2\end{pmatrix}$, $\begin{pmatrix}1&2&3&4\\2&4&1&3\end{pmatrix}$, $\begin{pmatrix}1&2&3&4\\1&3&2&4\end{pmatrix}$.

(ϵ) $\begin{pmatrix}1&2&3&4\\1&3&4&2\end{pmatrix}$, $\begin{pmatrix}1&2&3&4\\2&4&3&1\end{pmatrix}$, $\begin{pmatrix}1&2&3&4\\3&1&2&4\end{pmatrix}$, $\begin{pmatrix}1&2&3&4\\4&2&1&3\end{pmatrix}$.

(η) $\begin{pmatrix}1&2&3&4\\1&4&2&3\end{pmatrix}$, $\begin{pmatrix}1&2&3&4\\2&3&1&4\end{pmatrix}$, $\begin{pmatrix}1&2&3&4\\3&2&4&1\end{pmatrix}$, $\begin{pmatrix}1&2&3&4\\4&1&3&2\end{pmatrix}$.

We check that

$$\alpha = e,\ \epsilon^2 = \eta,\ \epsilon^3 = e,\ \beta^2 = e,\ \beta\epsilon = \delta,\ \beta\epsilon^2 = \gamma \text{ and } \epsilon\beta = \gamma = \beta\epsilon^2,$$

so that the factor group is S_3.

(*d*) (i)+(ii)+(iv). This contains all the even permutations and so is a subgroup, being the alternating group of order 12. The quotient group has order 2 and so must be C_2.

3. Give the kernel and a typical coset, and identify the quotient group of the kernel for the homomorphism $\theta: C_{24} \to C_{12}$ defined by $a\theta = b^4$ and $a^n\theta = b^{4n}$ where a is a generator of C_{24} and b a generator of C_{12}.

The kernel is the set of elements of C_{24} whose images are e, i.e. is

$$\{e, a^3, a^6, a^9, a^{12}, a^{15}, a^{18}, a^{21}\},$$

being isomorphic to C_8.

The other two cosets are $\{a^{3r+1}\}$ and $\{a^{3r+2}\}$, and the quotient group is isomorphic to C_3.

4. Find all possible homomorphic images of the Vierergruppe $C_2 \times C_2$.

Any homomorphic image is isomorphic to the factor group G/H where H is the kernel of the homomorphism (theorem 6.8.3). Thus we investigate all invariant subgroups (that is, all subgroups since $C_2 \times C_2$ is Abelian) and their quotient groups.

Apart from the trivial subgroups $\{e\}$ and $C_2 \times C_2$, giving the quotient groups, and hence images, $C_2 \times C_2$ and $\{e\}$, any subgroup must have order 2 and hence be C_2. Thus its quotient group has order 2 and is also C_2.

Thus the only possible images are $\{e\}$, C_2 and $C_2 \times C_2$. (Compare this method with that used in worked exercise 4 of chapter 2.)

5. Prove that the elements of G that commute with a given element g form a subgroup N_g. Prove that in a finite group G the number of elements in the conjugacy class containing g is equal to the index of N_g, and deduce that if a conjugacy class in a finite group G has s elements then s is a factor of the order of G.

Suppose x and y commute with g, so that $xg = gx$ and $yg = gy$. Then $xyg = xgy = gxy$ and so xy commutes with g. Also $x^{-1}xgx^{-1} = x^{-1}gxx^{-1}$, i.e. $gx^{-1} = x^{-1}g$ and x^{-1} commutes with g. But e commutes with g, and so those elements that commute with g form a subgroup. (This is called the *normaliser* or *centraliser* of g.)

Now suppose that $x^{-1}gx = y^{-1}gy$. Then

$$yx^{-1}g = gyx^{-1}, \quad \text{so that} \quad yx^{-1} \in N_g.$$

Conversely if $yx^{-1} \in N_g$, $x^{-1}gx = y^{-1}gy$.

Hence the conjugates $x^{-1}gx$ and $y^{-1}gy$ are the same element if and only if x and y are in the same right cosets of N_g in G, and thus each coset gives rise to one and only one distinct conjugate of g. Thus the number of elements in the conjugacy class containing g is equal to the number of such cosets, that is to the index of N_g.

If a class has s elements, s is the index of N_g for any g in the class, and thus s is a factor of the order of G.

Exercises 6A

Identify the quotient group of the subgroup H in the Abelian group G in **1–10**.

1. G = reals under addition, H = integers.

2. G = Gaussian integers under addition, H = integers.

3. $G = C_2 \times C_2 \times C_2$ (generators a, b, c), $H = \{e, abc\}$.

4. G = polynomials over reals, H = those polynomials of degree $\leqslant 3$.

5. G = polynomials over reals of degree $\leqslant 4$, H = those polynomials of degree $\leqslant 2$.

6. G = complex numbers under addition, H = real numbers under addition.

7. G = complex numbers except 0 under multiplication, H = real numbers except 0 under multiplication.

8. G = residues modulo 11 except 0 under multiplication, $H = \{1, 10\}$.

9. G = residues modulo 15 prime to 15 under multiplication $H = \{1, 14\}$.

10. G = residues modulo 15 prime to 15 under multiplication, $H = \{1, 4\}$.

11. Give an example to show that an Abelian subgroup H of a non-Abelian group G is not necessarily invariant.

12. If $(gH)(g'H) \subseteq gg'H \;\forall\, g, g' \in G$, where H is a subgroup of G, prove that $(gH)(g'H) = gg'H$.

Give the left and right cosets of H in G in **13–18**. Hence or otherwise determine whether H is invariant in G. If so, identify its quotient group.

13. $G = S_4$, H = the set of permutations in S_4 that leave the first object unaltered.

14. $G = S_7$, H = subgroup of even permutations in S_7.

15. $G = D_8$ (generators a, b with $a^4 = b^2 = e$ and $ab = ba^3$), $H = \{e, b\}$.

16. G = quaternion group (generators a, b with $a^4 = e, a^2 = b^2, ab = ba^3$), $H = \{e, a^2\}$.

17. $G = D_{12}$ (generators a, b with $a^6 = b^2 = e$ and $ab = ba^5$), $H = \{e, a^3\}$.

18. G = alternating group of degree 4,

$$H = \left\{\begin{pmatrix}1 & 2 & 3 & 4\\ 1 & 2 & 3 & 4\end{pmatrix}, \begin{pmatrix}1 & 2 & 3 & 4\\ 2 & 1 & 4 & 3\end{pmatrix}, \begin{pmatrix}1 & 2 & 3 & 4\\ 3 & 4 & 1 & 2\end{pmatrix}, \begin{pmatrix}1 & 2 & 3 & 4\\ 4 & 3 & 2 & 1\end{pmatrix}\right\}.$$

19. If H is an invariant subgroup of A prove that the set $\{(h,f)\}$, where $h \in H$, is an invariant subgroup of $A \times B$, and that its quotient group $\cong (A/H) \times B$.

20. If H is a subgroup of G prove that H is invariant if and only if $HX = XH$ for *any* complex X of G.

21. Give a counter-example to show that the converse of theorem 6.4.2 is not true, i.e. that two elements of the same order are not necessarily conjugate.

22. In which of **1–10** is $G \cong H \times (G/H)$?

23. Divide the elements of D_8 into conjugate classes.

24. Repeat **23** for the alternating group of degree 4.

25. Prove that the conjugate subgroups H and $g^{-1}Hg$ are isomorphic.

26. List the subgroups conjugate to $\{e, b\}$ in D_8.

27. Repeat **26** for the subgroup of permutations in S_4 that leave the first object unaltered.

28. If H is a subgroup of K, which is itself a subgroup of G, and if H is invariant in G prove that H is invariant in K.

29. If H and K are invariant subgroups of G and if HK is a subgroup prove that it is invariant.

30. Prove that $B \times C$ is an invariant subgroup of $A \times B \times C$ and the quotient group $\cong A$.

31. Prove that C is an invariant subgroup of $A \times B \times C$ and the quotient group $\cong A \times B$.

For the homomorphisms in **32–36** give the kernel and a typical coset of this and identify the quotient group of the kernel.

32. θ: group of integers under addition $\rightarrow D_6$, defined by $r\theta = a^r$, where a is a generator of D_6 and $a^3 = e$.

33. θ: group of two-dimensional vectors over the reals $\rightarrow$ the complex numbers under addition, defined by $(x, y)\,\theta = iy$.

34. θ: group of three-dimensional vectors over the reals $\rightarrow$ group of two-dimensional vectors over the reals, defined by $(x, y, z)\,\theta = (0, y-2z, 2y)$.

35. θ: integers under addition $\rightarrow$ complex numbers (except 0) under multiplication, defined by $r\theta = e^{(2\pi ir/n)}$ for some positive integer n.

36. $\theta: C_{15} \rightarrow C_9$, defined by $a\theta = b^3$ and $a^n\theta = b^{3n}$, where a is a generator of C_{15} and b a generator of C_9.

37. Show that a homomorphism from C_p where p is prime is either the trivial homomorphism or a monomorphism.

38. If H is an invariant subgroup of G and $\theta: H \rightarrow G$ is defined by $h\theta = h$ (considered as an element of G) with $\phi: G \rightarrow G/H$ being the natural homomorphism, prove that the homomorphism $\theta\phi$ is trivial.

Exercises 6B

1. Find the centre of D_8 and identify the group of inner automorphisms of D_8. Find the group of all automorphisms and identify the quotient group of the subgroup of inner automorphisms in this.

2. K is an *invariant complex* of G if $gK = Kg \; \forall \, g \in G$. Prove that

(i) if $h = k_1 k_2 \ldots k_n$, $g^{-1}hg = (g^{-1}k_1 g)(g^{-1}k_2 g) \ldots (g^{-1}k_n g)$;

(ii) if H is the *subgroup* generated by the elements of K, $g^{-1}Hg$ is the subgroup generated by the elements of $g^{-1}Kg$;

(iii) if K is an invariant complex, H is an invariant subgroup.

3. $\theta : G \to G'$ is a homomorphism onto G' and H' is an invariant subgroup of G'. Prove that

(i) $\theta^{-1}(H')$ is a subgroup of G;

(ii) if $\phi : G' \to G'/H'$ is the natural homomorphism, $\theta^{-1}(H')$ is the kernel of $\theta\phi$;

(iii) $\theta^{-1}(H')$ is an *invariant* subgroup of G.

4. Divide the elements of the alternating group of degree 4 into conjugacy classes and hence show that the subgroup

$$\left\{\begin{pmatrix}1 & 2 & 3 & 4\\1 & 2 & 3 & 4\end{pmatrix}, \begin{pmatrix}1 & 2 & 3 & 4\\2 & 1 & 4 & 3\end{pmatrix}, \begin{pmatrix}1 & 2 & 3 & 4\\3 & 4 & 1 & 2\end{pmatrix}, \begin{pmatrix}1 & 2 & 3 & 4\\4 & 3 & 2 & 1\end{pmatrix}\right\}$$

is the only proper invariant subgroup.

5. If H and K are invariant subgroups of G and $H \cap K = \{e\}$ prove that $hk = kh \; \forall \, h \in H, k \in K$. (*Hint*. Consider the element $h^{-1}k^{-1}hk$.)

6. Prove that in a finite group the conjugacy class containing x and that containing x^{-1} have the same number of elements and deduce that in a group of even order there is at least one pair of such classes (other than that given by $x = e$) that coincide.

7. Prove that the intersection of all subgroups conjugate to a given subgroup is invariant.

8. If H is an invariant subgroup of G apply the result of **3** (iii) to $\theta : G \to G/H$ to prove that if K is an invariant subgroup of G/H then $\theta^{-1}(K)$ is an invariant subgroup of G which contains H. Deduce that if no such subgroup exists other than G and H then G/H is a simple group. (In this case H is said to be a *maximal* invariant subgroup of G.)

9. Use the result of worked exercise 5 (that the number of elements in a conjugacy class for a finite group is a factor of the order of the group) to prove that in a group of order p^m, where p is prime, the number of self-conjugate elements is a positive multiple of p. Deduce that the centre contains non-neutral elements (if $p > 1$) and deduce that such a group, if non-Abelian, cannot be simple.

10. In a group of order p^2, where p is prime, prove that if the centre has order p then its quotient group is cyclic. By considering the cosets of the centre prove that the group is Abelian in this case (and so the centre must in fact have order p^2). Deduce by using exercise **9** above that a group of order p^2 *must* be Abelian.

7

IDEALS

7.1. Invariant subgroups of an Abelian group

In this chapter we discuss the analogue in a ring of an invariant subgroup of a group. We will see that the structures with which we will be concerned are a particular type of subring, and hence are certainly subgroups of the additive group of the ring. Since we also have multiplicative structure we must use the additive notation, and we first wish to see how the ideas of invariant subgroups and factor groups are dealt with in this notation. We assume that the groups are Abelian, for this will always be true when they are the additive group of a ring.

Suppose then that H is a subgroup of an Abelian group G and that we use the additive notation. We first notice that H is always invariant. The coset containing g is written in the normal way as $g+H$ (or, as a right coset, $H+g$), so that g and k are in the same coset if and only if $-k+g$, or $g-k$ because of commutativity, is in H. H is invariant and we have

$$g+H = H+g$$

or, in the conjugacy form, $-g+H+g = H$.

The cosets form a group, the addition process being written now in the form $(g+H)+(g'+H) = g+g'+H$. Note that the fact that H is a subgroup is expressed by the equations

$$H+H = H \quad \text{and} \quad -H = H.$$

The group of cosets, the quotient group of the invariant subgroup H in G, is sometimes called the *difference group* and written $G-H$, although we will usually retain the notation G/H.

Finally we recollect that the notation of conjugacy is trivial in the Abelian case and will therefore not enter into the work of this chapter.

7.2. Ideals

In searching for an analogy in ring theory to the concept of invariant subgroup for groups we are concerned with two structures: namely the additive and the multiplicative. We are looking for subrings which can be 'divided' into the whole ring to give a 'quotient ring'. We wish to be able to work 'modulo' the subring and still obtain a ring.

A subring is inevitably a subgroup, so let us first take any *subgroup* S of the additive structure of the ring R. Then S is invariant and the cosets $x+S$ form a group, the quotient group R/S. Thus far we may proceed with any subgroup, but we wish further to obtain a ring: thus we wish to be able to multiply cosets as well as add them. We investigate the conditions for this to be possible.

Suppose that the cosets $x+S$, $y+S$ can be multiplied. We clearly require the product to be $xy+S$ and, since x and y must be arbitrary elements of the cosets, we wish this coset $xy+S$ to be independent of the choice of x and y. Thus we need the product of the cosets, i.e. the set of elements of the form $(x+s_1)(y+s_2)$; $s_1, s_2 \in S$, to be contained in $xy+S$. Let us see what this means. We bear in mind that although addition is commutative, multiplication is not necessarily so, nor need R have a unity.

If $(x+s_1)(y+s_2) \in xy+S$ we have, multiplying out

$$xy+xs_2+s_1y+s_1s_2 \in xy+S.$$

We wish this to be true for all x and y in R. Putting $x = y = 0$ we see that we require $s_1s_2 \in S$. But this is merely the condition for S to be a subring, not only a subgroup. Thus, as expected, we are searching for a type of subring.

Now put $y = 0$. Then we require $xs_2+s_1s_2 \in S$ or $xs_2 = s-s_1s_2$ for some $s \in S$.

Now $s-s_1s_2 \in S$, S being a subring, and so we require $xs_2 \in S$. Similarly by putting $x = 0$ we also require $s_1y \in S$, and these conditions must be true for any x and $y \in R$.

Hence necessary conditions for a meaningful multiplication to be given for cosets is that S is a subring and

$$xs \in S,\ sx \in S\ \forall\ x \in R.$$

It is at once seen that these conditions are also sufficient. For if $xs \in S$, $sy \in S$ and S is a subring $xs_2 + s_1 y + s_1 s_2 \in S \,\forall s_1, s_2 \in S$ and we have

$$\begin{aligned}(x+s_1)(y+s_2) &= xy + xs_2 + s_1 y + s_1 s_2 \\ &= xy + (xs_2 + s_1 y + s_1 s_2) \in xy + S.\end{aligned}$$

We are led to the definition of a subring that satisfies the conditions and thus enables cosets to be both added and multiplied and ensures that the quotient group is a ring (this latter fact will be formally proved in the next section: we need to prove the Associative and Distributive Laws, although this proof is easy once the possibility of multiplication of cosets is assured). A subring with the required properties is called an *ideal.*

Definition. *A subring I of a ring R is an ideal (strictly speaking a two-sided ideal) if for all $x \in R$ and $i \in I$, xi and ix are both in I.*

Note that since R may not be commutative we need *both* xi and ix in I.

An important example of an ideal is the subring of multiples of n in the ring of integers. An example of a subring that is not an ideal is the subring of all polynomials of degree 0 in the ring of polynomials $R[x]$.

Here if we multiply any polynomial of degree $\geqslant 1$ with $(a, 0, 0, \ldots)$ where $a \neq 0$ we obtain an element not in the subring. Thus even when R is commutative not all subrings are ideals.

Theorem 7.2.1. *If R has a unity then*

$$\{xi : x \in R, i \in I\} \subseteq I \Rightarrow \{xi\} = I \text{ and similarly for } \{ix\}.$$

$\{xi\} \supseteq \{1i\} = I$ and the result follows.

Thus for a ring with unity the conditions are equivalent to the conditions that $\{xi\}$ and $\{ix\} = I$. For a ring without a unity equality does not follow. Thus if R is the ring of even integers and I is the subring of all multiples of 4, $\{ix\}$ is certainly contained in I but the inclusion is strict, since the integers ± 4, ± 12, ... are in I but not in $\{ix\}$. Note also in this connection that for S to be a general subring in a ring R without a unity we must always have $\{s_1 + s_2 : s_1 \in S, s_2 \in S\} = S$, but $\{s_1 s_2\} \subseteq S$, and the inclusion may be a strict inclusion.

As in the case of invariant subgroups, the concept of an ideal is fundamental to advanced work in ring theory. We have the same idea of the cosets forming a ring, and there is a very similar connection with homomorphisms. There are also some important types of ideals, some of which will be considered later in the chapter.

7.3. Quotient rings

We have seen that the condition for a multiplicative structure to exist in the set of cosets relative to a subring I of R is that I is an ideal, as defined in the last section. Once such a multiplication is established it is a simple matter to prove that the cosets form a ring. The complete result is proved below.

Theorem 7.3.1. *If I is an ideal of a ring R the cosets $x+I$ form a ring under the sum $(x+I)+(y+I) = (x+y)+I$ and product $(x+I)(y+I) = xy+I$.*

I is an invariant subgroup of the additive structure of R and hence the cosets certainly form an Abelian group under the sum defined.

Let x', y' be any elements of $x+I$, $y+I$ respectively. Then $x' = x+i_1$, $y' = y+i_2$ where $i_1, i_2 \in I$. Thus

$$\begin{aligned} x'y' &= (x+i_1)(y+i_2) \\ &= xy+xi_2+i_1y+i_1i_2. \end{aligned}$$

But xi_2 and $i_1y \in I$ since I is an ideal and, since I is a subring, $xi_2+i_1y+i_1i_2 \in I$. Hence $x'y' \in xy+I$ and so

$$x'y'+I \subseteq xy+I+I \subseteq xy+I.$$

But $x = x'-i_1 \in x'+I$ and similarly $y \in y'+I$ and so, as above, $xy+I \subseteq x'y'+I$. Hence these cosets are the same.

Thus the definition of product is independent of choice of members x and y of the given cosets, and thus is uniquely defined, and is of course a coset. (This is the vital part of the proof: the rest is easy.)

The Associative Law of multiplication is true, since both

$$\{(x+I)(y+I)\}(z+I) \quad \text{and} \quad (x+I)\{(y+I)(z+I)\}$$

are, by the definition and the Associative Law in R, equal to $xyz+I$.

Similarly for the Distributive Law:

$$\{(x+I)+(y+I)\}(z+I) = (x+y)z+I = (xz+yz)+I$$
$$= (x+I)(z+I)+(y+I)(z+I)$$

and

$$(x+I)\{(y+I)+(z+I)\} = x(y+z)+I = (xy+xz)+I$$
$$= (x+I)(y+I)+(x+I)(z+I).$$

Hence the cosets form a ring.

This ring of cosets is called the *quotient ring of I in R* and is written R/I. Note that we still use the quotient notation although the cosets, being cosets of the additive subgroup I, are written in the additive notation. The reason is that we are still in effect 'dividing' R by I. Some writers do use $R-I$ and call the ring the *difference ring*, but our notation and terminology is the more common one.

If the order of R is finite and equals n, with the order of I being m, then of course the order of R/I is n/m, as in the case of quotient groups.

We may naturally wonder what happens to R/I when R is a specialised type of ring.

***Theorem* 7.3.2.** *If R is commutative so is R/I.*

For $(x+I)(y+I) = xy+I = yx+I = (y+I)(x+I)$.

***Theorem* 7.3.3.** *If R has a unity* 1, *R/I has a unity* $1+I$.

For $$(1+I)(x+I) = 1x+I = x+I$$

and $$(x+I)(1+I) = x1+I = x+I.$$

The converses of these theorems are not true. A counter-example of the first is given by the trivial case where $I = R$ so that $R/I = \{0\}$ and is certainly commutative, whereas R is arbitrary. For a counter-example for the second, let I be the ideal of multiples of 6 in the ring R of multiples of 2. Then R has no unity, but R/I contains 3 elements, $0+I$, the zero, $2+I$ and $4+I$, which is easily seen to be a unity.

If we further restrict R we do not always get the expected results. If R is an integral domain then R/I need not be: if R is the domain of integers and I the ideal of multiples of 4, then

$(2+I)(2+I) = 4+I = I$, the zero, but $2+I$ is certainly not the zero coset.

If R is a field there are no proper ideals, as is shown in §7.7.

The cases where R/I is an integral domain or a field are dealt with later. At present all we will say is that these occurrences depend on the nature of I rather than on R.

7.4. Examples of ideals

Trivial ideals

In any ring R the whole ring R is obviously an ideal, the quotient ring R/R being $\cong \{0\}$.

The subring $\{0\}$ is always an ideal, since $x0 = 0x = 0$ for any x in R. The quotient ring $R/\{0\}$ is of course $\cong R$.

Since an ideal is a subgroup of the additive structure of R, it must contain 0. If R has a unity then a non-trivial ideal will never contain this, by the following theorem.

Theorem 7.4.1. *If an ideal I of a ring R with unity contains* 1, *then $I = R$.*

If x is any element of R, $x1 = x \in I$. Hence $I = R$.

By means of this theorem we can easily discover subrings that are not ideals: indeed most subrings that spring to mind are in this category. Ideals in fact are much rarer than invariant subgroups: all subgroups in Abelian groups and some in non-Abelian ones are invariant, but for a subring to be an ideal is the exception rather than the rule. Easy examples of those that are not ideals are the subring of the reals in the complex number ring, that of the integers in the ring of reals, and the subring of even integers in the ring of reals: the two former examples contain 1 but the latter does not.

If I and J are ideals then $I \cup J$ may not be a subring and hence certainly need not be an ideal. $I \cap J$, as we expect, is however always an ideal.

Theorem 7.4.2. *If I and J are ideals in R, $I \cap J$ is an ideal.*

$I \cap J$ is certainly a subring by theorem 3.7.2.

Let $x \in R$ and $i \in I \cap J$. Then since I and J are ideals, $xi \in I$ and $\in J$ and so $\in I \cap J$. Similarly $ix \in I \cap J$ and so $I \cap J$ is an ideal.

Ideals in the ring of integers

The subring of all multiples of n, for any positive integer n, is an ideal, since $m.\lambda n$, for any integers m and λ, is a multiple of n. Let us call this ideal $[n]$. It is easy to prove that there are no other proper ideals in the ring of integers.

Theorem 7.4.3. *If I is an ideal in the ring of integers, $I = [n]$ for some positive integer n, or $I = \{0\}$.*

If $I \neq \{0\}$ it contains a non-zero element. If I contains a negative integer $-k$ then it also contains k, since it is a subgroup. Hence I certainly contains a positive integer. Let n be the smallest such.

Since I is a subgroup it contains all multiples of n. Suppose it contains an integer m that is not a multiple of n. Then by the division algorithm $m = nq+r$ where $0 < r < n$. But $m \in I$ and $nq \in I$ and so $m-nq = r \in I$, contrary to the hypothesis that n is the smallest positive member of I.

Hence I contains no integer that is not a multiple of n and so $I = [n]$.

The cosets of $[n]$ are the residue classes modulo n hence $R/[n]$ is the ring of residues modulo n. Note that this is a field if n is prime, otherwise it is not even an integral domain.

Ideals in finite rings of residues

Corresponding to each factor m of n, the subring of multiples of m is an ideal in the ring of residues modulo n, and it is easily seen that the quotient ring is the ring of residues modulo m.

Polynomials with given zeros

Let R be commutative and with a unity. Then in chapter 3 we saw that the set of polynomials in $R[x]$ which have given zeros form a subring. This subring is in fact an ideal, since if c is one of the zeros and P any polynomial in the ideal so that $P(c) = 0$, then if X is any polynomial in R, $X(c)\,P(c) = 0$ and so XP, and similarly PX, has the given zeros and so is in the ideal.

If the given zeros form a finite set $c_1, c_2, \ldots, c_n$ then by the factor theorem any polynomial with these zeros is

$$(x-c_1)\,(x-c_2)\ldots(x-c_n)\;Q(x),$$

where Q is some polynomial. Hence the ideal consists of all multiples of $(x-c_1)(x-c_2)\ldots(x-c_n)$.

Similarly in the ring of polynomials in more than one indeterminate. For example in $R[x, y]$ the polynomials that have a zero at (2, 1) form an ideal, while in $R[x, y, z]$ those that have zeros at all points of intersection of, say, $z = 1$ and $x^2 + y^2 = 1$ also form an ideal.

Direct sums

We recollect that if A and B are groups then the set of elements of the form (e, b) is an invariant subgroup of $A \times B$, its quotient group being isomorphic to A. An analogous result holds for rings. We remember that we use the notation $R \oplus S$ for the direct *sum* of rings R and S.

Theorem 7.4.4. *If R and S are rings the set of elements I of the form $(0, s)$ is an ideal in $R \oplus S$, and $(R \oplus S)/I \cong R$.*

I is a subring.

Since $(x, y)(0, s) = (x0, ys) = (0, ys) \in I \, \forall \, (x, y)$, and similarly $(0, s)(x, y) \in I$, I is an ideal. The cosets are of the form $(r, 0)+I$, a typical element of this being (r, x) for some $x \in S$. These cosets are distinct, since if $(r, 0)+I = (r', 0)+I$ then $(r, x) = (r', x')$ for some $x' \in S$, and so $r = r'$.

Since $(r, x)+(r', x') = (r+r', x+x')$ we see that

$$((r, 0)+I)+((r', 0)+I) = (r+r', 0)+I$$

and similarly their product is $(rr', 0)+I$. Hence $(R \oplus S)/I \cong R$.

Note that $I \cong S$ and so we have, in this case, if

$$T = R \oplus S, \quad T \cong (T/I) \oplus I.$$

As for invariant subgroups, this is not true generally.

7.5. Principal ideals

In the ring of integers we have seen that the set of multiples of the integer n forms an ideal, and that in polynomial rings with coefficients in a ring that is commutative and with a unity, the set of multiples of the polynomial

$$(x-c_1)(x-c_2) \ldots (x-c_n)$$

also gives an ideal. This is true in general: in any commutative ring with unity the set of multiples of any element is an ideal.

Theorem 7.5.1. *If R is commutative and with a unity, and a is a fixed element of R, the set of elements of the form ra, for $r \in R$, is an ideal in R. It is contained in any ideal which contains a, and in fact is the intersection of all ideals that contain a.*

$ra - r'a = (r-r')a$ and is in the set, which is therefore a subgroup. $(ra)(r'a) = (rar')a$ and is in the set: hence it is a subring. For any $x \in R$, $x(ra) = (xr)a$ and $(ra)x = x(ra)$ since R is commutative. Hence both products are in the set, which is therefore an ideal.

If any ideal I contains a, then it contains ra for any $r \in R$, by definition of ideal.

By an extension of theorem 7.4.2, the intersection of all ideals that contain a *is* an ideal and must contain the set of elements ra by the previous part of the proof. But this is one ideal that contains $1.a = a$ and hence contains the intersection. Thus it is the intersection.

The ideal we have been discussing, the set $\{ra\}$ for a fixed $a \in R$, is called a *principal ideal* of R, and is denoted by $[a]$. (The notation (a) is often used, but the square brackets cause less confusion.) There will of course be a principal ideal corresponding to each $a \in R$, but these may not all be distinct. Thus in the ring of integers, $[n] = [-n]$, and in the ring of residues modulo 10, $[2] = [4]$.

The idea of a principal ideal as the smallest ideal containing a given element a is not limited to the case where R is commutative and with a unity, although this case is the most important in practice, since principal ideals are most significant in the case where R is an integral domain. Let us suppose first that R is commutative but has no unity. Then the whole of theorem 7.5.1 is valid except for the last part. The set $\{ra\}$ is an ideal, it is contained in any ideal that contains a, and hence in their intersection, but it is not necessarily the intersection in this case, since in general a itself cannot be put in the form ra (there is no unity). Thus to obtain the smallest ideal containing a we need to extend our set. We must clearly include all elements of the form

na, where n is an integer (*not* an element of the ring), and so all sums $na+ra$. The result then is as follows.

***Theorem* 7.5.2.** *If R is commutative but has no unity, and a is a fixed element of R, the set of elements of the form $na+ra$, where n is an integer, positive negative or zero, and r is an element of R, is an ideal in R. It is contained in any ideal which contains a, and is the intersection of all ideals that contain a.*

$$(na+ra)-(n'a+r'a) = (n-n')a+(r-r')a$$

and is in the set,

$$(na+ra)\,(n'a+r'a) = nn'a+(nar'+n'ra+rar')a$$

and is also in the set, since $n(ar')+n'(ra)+rar'$ is an element of R. The remainder of the proof is as in theorem 7.5.1.

When R has no unity but is commutative the set $\{na+ra\}$ is called a principal ideal and is denoted by $[a]$.

Examples. If R is the ring of even integers and $a = 4$, the set $\{ra\}$ consists of all multiples of 8. The principal ideal [4], however, is the set of multiples of 4 and is given by $\{n4+r4\}$, in fact in this case merely by $\{n4\}$.

If R is the ring of polynomials over the ring of even integers and $a = 4$, $[a]$ consists of all polynomials with constant term a multiple of 4, and every other coefficient a multiple of 8.

If now R is non-commutative but with unity any ideal containing a must contain ra and ar for any $r \in R$ and hence also rar' for any r and $r' \in R$. It must then contain any finite sum of such elements as rar' (which in general cannot be combined to give a single element of this form). If in addition R has no unity we must add a term na as before. The result, giving the principal ideal $[a]$ in these cases, is stated in theorem 7.5.3 and the proof, which is similar to those of theorems 7.5.1 and 7.5.2, will be omitted.

***Theorem* 7.5.3.** *If R is non-commutative with unity, the elements of the form $ra+ar'+\sum_{i=1}^{n} s_i a s_i'$, where r, r', s_i, s_i' are elements of R, give an ideal which is the intersection of all ideals which contain a. If R has no unity such an ideal is given by elements of the form $na+ra+ar'+\sum_{i=1}^{n} s_i a s_i'$.*

Ideals generated by a subset of elements

The ideal $[a]$ is the smallest ideal containing a and all its elements are obtained from a, in the commutative case where R also has a unity merely by multiplying a by an element of R, and by more elaborate methods in the other cases. We say that $[a]$ is *the ideal generated by a*. The idea may be extended to any subset of elements of R, which for convenience we will take as a finite subset, though this is not in fact necessary

Suppose then that we have an ideal I containing $a_1, a_2, \ldots, a_n$. If R is commutative and with a unity I must contain all elements of the form $\sum_{i=1}^{n} r_i a_i$, and it is a straightforward matter to show that these form an ideal. As before we can easily show that it is the intersection of all ideals containing $a_1, a_2, \ldots, a_n$. It is called *the ideal generated by* $a_1, a_2, \ldots, a_n$ and denoted by $[a_1, a_2, \ldots, a_n]$.

Note that all the a_i's need not be independent, in the sense that the ideal generated by a subset of them may be the same as that generated by the complete set. On the other hand, if $a_{k+1} \notin [a_1, a_2, \ldots, a_k]$ for any $1 \leqslant k \leqslant n-1$, we say that the a_i's form an *independent set of generators*. An ideal may of course contain more than one distinct independent set of generators.

We obtain similar ideals in the cases where R is non-commutative or without unity.

Example. In the ring of polynomials $R[x, y, z]$ over the reals the ideal $[z-1, x^2+y^2-1]$ consists of all polynomials of the form $(z-1)P+(x^2+y^2-1)Q$ and is in fact the ideal consisting of all polynomials which have zeros at all points of intersection of $z = 1$ and $x^2+y^2 = 1$.

Principal ideals in integral domains

It is in the theory of integral domains that principal ideals have most of their importance. This is because they are closely connected with factorisation properties.

We first consider the domain of integers. The principal ideal $[n]$ consists of multiples of n. Suppose m is a factor of n. Then $[m]$ is the set of multiples of m, and contains the ideal $[n]$ as a subset (for example let $n = 6$ and $m = 3$: then the set of multiples of 3 contains all multiples of 6). Conversely let $[m] \supseteq [n]$. Then $n \in [m]$ and so is a multiple of m. Hence $m/n \Leftrightarrow [m] \supseteq [n]$.

This result is a general one, valid for all integral domains R, and in fact for all commutative rings with unity.

Theorem 7.5.4. *If R is an integral domain, $b/a \Leftrightarrow [b] \supseteq [a]$.*

If b/a, $a = xb$ for some $x \in R$. Hence for any $r \in R$, $ra = (rx)b$ and $rx \in R$. Hence $[a] \subseteq [b]$.

Conversely let $[b] \supseteq [a]$. Then certainly $a \in [b]$ and so $a = xb$ for some $x \in R$. Thus b/a.

It is now a simple matter to discover the conditions under which $[a]$ and $[b]$ are the same.

Theorem 7.5.5. $[a] = [b]$ *if and only if a and b are associates.*

$[a] = [b]$ if and only if $[a] \supseteq [b]$ and $[b] \supseteq [a]$, i.e. if and only if a/b and b/a, or by theorem 5.3.3, if and only if a and b are associates.

The idea of a principal ideal is important in simplifying the discussion of unique factorisation in an integral domain. We will first investigate its connection with H.C.F. If a and b have an H.C.F. h we know that h/a and h/b; so that $[h] \supseteq [a]$ and $[h] \supseteq [b]$ and hence $[h] \supseteq [a, b]$. Since h is the *highest* common factor, we might expect $[h]$ to be the smallest ideal containing a and b, i.e. to be *equal* to $[a, b]$. Unfortunately however this is not always the case. For consider the domain of polynomials over the integers, and let $a = 2$, $b = x$. Then $h = 1$ and $[h]$, being the set of multiples of 1, is the whole ring. But $[2, x]$ is certainly not this, since the element 1 is not a member, not being expressible in the form $2P + xQ$ for any P and Q in the ring.

In an important case however our surmise is correct, and that is when every ideal of R is a principal one.

Definition. *A principal ideal domain is an integral domain in which every ideal is a principal ideal.*

Theorem 7.5.6. *If R is a principal ideal domain, any two non-zero elements a and $b \in R$ have an* H.C.F. *h, where $[h] = [a, b]$.*

Since the domain is a principal ideal domain, the ideal $[a, b] = [h]$ for some h. Since $[a, b] \supseteq [a]$, h/a and similarly h/b.

Now suppose c/a and c/b. Then $[c] \supseteq [a]$ and $[b]$ and hence $[c] \supseteq [a, b]$: i.e. $[c] \supseteq [h]$ and so c/h. Thus the conditions for an H.C.F. are satisfied by h.

Note 1. We must be able to state that $[a, b]$ is a principal ideal, and thus require that R is a principal ideal domain.

Note 2. The step from $[c] \supseteq [a]$ and $[c] \supseteq [b]$ to $[c] \supseteq [a, b]$ is straightforward, for if an ideal contains all elements ra and all $r'b$ then it must contain $ra+r'b$.

Corollary. *The set of elements $\{ra+r'b\}$ is the set of multiples of h, although more than one member of the first set may give rise to the same multiple of h.*

We recall theorem 5.4.5, that if a domain has an H.C.F. and also satisfies criterion 1 of p. 130 then it is a unique factorisation domain. It turns out that a principal ideal domain always satisfies criterion 1.

Theorem 7.5.7. *In a principal ideal domain there exists no infinite sequence of elements $\{a_i\}$ such that a_{i+1} is a proper factor of a_i for all i.*

The fact that a_{i+1} is a *proper* factor of a_i is equivalent to the statement that $[a_{i+1}] \supset [a_i]$, the inclusion being strict.

Suppose then that we have an infinite sequence of ideals $[a_1] \subset [a_2] \subset [a_3] \subset \dots$. Let $U = \bigcup_{i=1}^{\infty} [a_i]$. We prove first that U is an ideal. For let x and y be members of U, with $x \in [a_m]$ and $y \in [a_n]$. Let $m \leqslant n$. Then since $[a_m] \subset [a_n]$, $x \in [a_n]$ and so $x-y \in [a_n]$ and $xy \in [a_n]$, and so both $x-y$ and xy are $\in U$. We also have, since $x \in [a_m]$, rx $(= xr) \in [a_m]$ and so $\in U$. Thus U is an ideal.

Hence since the domain is a principal ideal domain, $U = [b]$ for some b.

Now b, being an element of $U = \bigcup_{i=1}^{\infty} [a_i]$, $\in [a_N]$ for some particular N, and so $[b] \subseteq [a_N]$. But $[b] = U \supseteq [a_N]$ and so

$$U = [a_N].$$

Hence $[a_{N+1}] \supset U$ and must $\subseteq U$, since $U = \bigcup_{i=1}^{\infty} [a_i]$.

Thus we have a contradiction and so no such infinite sequence exists.

We at once have the important corollary that *a principal ideal domain is a unique factorisation domain.*

We see now that theorem 7.4.3 proved that the domain of integers is a principal ideal domain. In a similar way we can prove that any Euclidean domain is necessarily a principal ideal domain.

Theorem 7.5.8. *A Euclidean domain is a principal ideal domain.*

Let the Euclidean domain R have an ideal I. Either $I = \{0\}$, in which case it is $[0]$, or I contains elements x for which $\delta(x) > 0$. Let a be an element of I for which $0 < \delta(a) \leqslant \delta(x)$ for all non-zero $x \in R$. Then certainly $I \supseteq [a]$. We will prove that $I \subseteq [a]$, i.e. that if $b \in I$ then $b = ya$ for some $y \in R$. By the definition of a Euclidean domain, $b = ya+z$ for some y and $z \in R$ and $\delta(z) < \delta(a)$. But $z = b-ya \in I$ and so, by the hypothesis for a, $\delta(z) = 0$ and so $z = 0$. The result follows.

Thus the polynomials over a field form a principal ideal domain. The domain of polynomials over the integers does not: the ideal generated by 2 and x is not principal, as may easily be verified.

The above work enables us to show that a Euclidean domain is a principal ideal domain, and hence a unique factorisation domain, without using the Euclidean algorithm and thus avoiding the rather tedious work which that algorithm involves. But of course the algorithm is useful in giving a straightforward method of finding the H.C.F. in particular cases.

7.6. The natural homomorphism

There is a connection between the ideals of a ring R and the ring-homomorphisms from R similar to that between the invariant subgroups of a group G and the group-homomorphisms from G. Thus to any ideal there corresponds a homomorphism, and all homomorphisms are related to ideals in this way. The work is very like that of §6.7 and §6.8 and we will deal briefly with the details.

Theorem 7.6.1. *If I is an ideal of a ring R then the mapping $\theta: R \to R/I$ defined by $x\theta = x+I$ is a ring-homomorphism, with* *kernel I and image R/I. It is called the* natural homomorphism *associated with I.*

$$(x+y)\,\theta = (x+y)+I$$
$$= (x+I)+(y+I),$$

by theorem 7.3.1, the sum being as defined in R/I,

$$= x\theta + y\theta.$$

$$(xy)\,\theta = xy+I$$
$$= (x+I)(y+I),$$

again by theorem 7.3.1, the product being as defined in R/I,

$$= (x\theta)(y\theta).$$

Hence θ is a homomorphism. x is in the kernel of θ if and only if $x+I = I$, the zero of R/I; i.e. if and only if $x \in I$. Hence I is the kernel. That the image is R/I is obvious.

To prove the converse theorem we show first that, given any homomorphism $\theta: R \to R'$, the kernel of θ is an ideal of R.

Theorem 7.6.2. *The kernel of $\theta: R \to R'$ is an ideal of R.*

By considering the group structure the kernel is a subgroup.

If i is in the kernel and x is any element of R,

$$(xi)\,\theta = (x\theta)(i\theta) = (x\theta)\,0,$$

since i is in the kernel,

$$= 0,$$

and so xi is in the kernel.

Similarly ix is in the kernel and the kernel is an ideal. (That the kernel is a *subring*, i.e. that ii' is in it whenever i and i' are, follows from the fact that ix is in the kernel.)

We now prove that θ is closely related to the natural homomorphism associated with its kernel I in exactly the same way as was done for groups. Thus we prove the homomorphisms $x \to x+I$ and $x \to x\theta$ are equivalent, by showing that the mapping $x+I \to x\theta$ is a well-defined isomorphism between R/I and $R\theta$.

Theorem 7.6.3. *The mapping defined by $x+I \to x\theta$ is well-defined and is an isomorphism between R/I and $R\theta$.*

We first show that it is well-defined, i.e. that if x_1 and x_2 are in the same coset then $x_1\theta = x_2\theta$. If x_1 and x_2 are in the same coset, $x_1 - x_2 \in I$ and so $(x_1 - x_2)\theta = 0$, the zero of $R\theta$, since I is the kernel of θ. Hence $x_1\theta - x_2\theta = 0$ and so $x_1\theta = x_2\theta$.

We now prove that $x+I \to x\theta$ is a homomorphism. The sum $(x+I)+(y+I) = (x+y)+I$ and goes to $(x+y)\theta = x\theta + y\theta$. The product $(x+I)(y+I) = xy+I$ and goes to $(xy)\theta$, the product of $x\theta$ and $y\theta$. Thus the mapping *is* a homomorphism.

It is onto $R\theta$, since any element of $R\theta$ is the image $x\theta$ for some $x \in R$. Finally we have to prove the mapping 1–1, i.e. that its kernel is the zero of R/I.

If $x\theta$ is the zero element of $R\theta$, x is in the kernel of θ, i.e. is in I, and then $x+I$ *is* I, the zero of R/I. This completes the proof.

Thus as in the case of groups there is a 1–1 correspondence between the ideals of a ring R and the essentially different ring-homomorphisms of R: homomorphisms are essentially different if they have different kernels, homomorphisms with the same kernel being equivalent to the same natural homomorphism and being in effect the same, exhibiting the same homomorphism structure and having the same image, although their image *spaces* may of course be different.

As an example, the ideal $[n]$ in the ring of integers corresponds to the homomorphism which maps an integer into its residue modulo n.

Trivial ideals

The trivial ideal $\{0\}$ is associated with isomorphisms of R, and the trivial ideal R is associated with zero homomorphisms.

Direct sums

The natural homomorphism associated with the ideal consisting of the elements of $R \oplus S$ of the form $(0, s)$ is the projection of $R \oplus S \to R$ defined by $(r, s) \to r$.

7.7. Ideals in a field

Let us apply the theory of the previous section to the case where R is a field F: then to each ideal I of F there corresponds a natural homomorphism of F, with kernel I. But from theorem

4.5.1 all homomorphisms of F are either monomorphisms (with kernel $\{0\}$) or are zero-homomorphisms (with kernel F). It follows that the only ideals in F are the trivial ones. It is easy to give a direct proof of this.

Theorem 7.7.1. A field F has no proper ideals.

Suppose an ideal I of F contains a non-zero element x, so that, since F is a field, x^{-1} exists. Then if y is any element of F, $(yx^{-1})x$ is in I, and so $y \in I$. Hence $I = F$. Thus either I contains no non-zero element and is $\{0\}$, or is the trivial ideal F.

As a consequence of the above theorem we see that the theory of ideals and quotient rings becomes trivial for fields: it is essentially a concept of *ring* theory and we have no possibility of 'dividing' a field by a proper subfield to obtain a 'quotient field' (although when R is a ring we may still have R/I as a field, as will be seen later). It is interesting in this connection to recollect theorem 3.5.1 that the direct sum of two non-trivial fields is never a field.

7.8. Simple rings

A ring is said to be *simple* if it has no non-trivial ideals (compare the definition of a simple group in §6.10). It follows at once that there are no homomorphisms from a simple ring except zero-homomorphisms and monomorphisms.

By theorem 7.7.1 any field is a simple ring.

We now investigate the converse of this: are there any simple rings other than fields? We restrict ourselves to commutative rings.

We may find an example quite easily. Consider the simple group C_p, and the zero-ring R with this as additive structure and $xy = 0$ for all x and y. Then any ideal of R is a subgroup of C_p, and hence is trivial since C_p has no proper subgroups. However R is not a field.

If we do not allow zero-rings there are no commutative simple rings other than fields. This result is not difficult to prove, and is sometimes a useful method of proving that a given ring (usually obtained theoretically so that its structure may not be known in detail) is a field: we will use it in §7.10.

Theorem 7.8.1. *If R is a commutative simple ring, with non-trivial multiplication, then R is a field.*

We need to prove that R contains a unity and inverses. It is convenient to prove first that R has no zero divisors, and in order to do this to consider the set of elements x such that $ax = 0$, where a is any fixed element of R.

Lemma 1. *If $a \in R$, and $I_a = \{x : ax = 0\}$, then I_a is an ideal.*

If $ax_1 = ax_2 = 0$, $a(x_1 - x_2) = ax_1 - ax_2 = 0$ and so $x_1 - x_2 \in I_a$ and I_a is a subgroup.

If $ax = 0$, $a(yx) = yax = 0$ and so $yx \in I_a$ for any $y \in R$. $xy = yx \in I_a$ since R is commutative. Hence I_a is an ideal.

Lemma 2. *R has no zero divisors.*

Since R is simple and I_a is an ideal, $I_a = \{0\}$ or R, for each element $a \in R$. Suppose $I_a = R$ for some non-zero a. Then $ax = 0 \;\forall\; x$ and so I_x contains a. But $a \neq 0$ and so $I_x = R$ for all x and the ring has trivial multiplication, contrary to hypothesis. Therefore $I_a = 0$ for *all* non-zero a and so $ax = 0 \Rightarrow x = 0$ or $a = 0$, so that R has no zero divisors.

To prove that R contains a unity

If a is any non-zero element of R consider the set

$$J_a = \{ax : x \in R\}.$$

J_a is an ideal, as was proved in theorem 7.5.1. (Theorem 7.5.1 assumes that R has a unity, but the proof is valid when this is not assumed, except that in this case we have no guarantee that J_a contains a. Thus in claiming merely that J_a is an ideal we are not begging the question of the existence of a unity.) Hence since R is simple $J_a = \{0\}$ or R. But since $a \neq 0$ and there are no zero divisors, $a^2 \neq 0$ and is an element of J_a. Thus $J_a \neq \{0\}$ and so is R: thus there is an element 1_a such that $a.1_a = a$. Now let y be any element of R and let $y.1_a = z$. Then

$$\begin{aligned} az &= a1_a y \quad (R \text{ is commutative}) \\ &= ay. \end{aligned}$$

Hence $a(z-y) = 0$ and so, by lemma 2, $z - y = 0$ and $z = y$, so that $y.1_a = y$ and 1_a is a unity.

To prove that R contains inverses

If $a \neq 0$, we have proved that $J_a = R$ and so contains the unity 1. Hence there exists an element a^* such that $aa^* = 1$, and a^* is the inverse of a.

7.9. Prime ideals

We now come to an investigation of the circumstances under which the quotient ring R/I is an integral domain. We have already shown that this need not be the case whenever R is an integral domain, and conversely it is possible for R/I to be an integral domain even when R is not. (For example if R is the ring of residues modulo 6 and I the ideal of multiples of 3, R/I is the ring of residues modulo 3 and is an integral domain, although R is not in this case.) The property depends in fact on the nature of I and its relationship to R rather than on R itself.

If R/I is an integral domain then it must be commutative and have a unity, and this is assured if we assume R to be commutative and with unity. Since this includes most important cases, we will assume throughout this section that *all our rings are commutative and have a unity.*

Assuming this condition, R/I is an integral domain if and only if it has no zero divisors, i.e. if and only if $(x+I)(y+I) = I$ (the zero in R/I) $\Rightarrow x+I = I$ or $y+I = I$. But since

$$(x+I)(y+I) = xy+I$$

this condition merely states that $xy \in I \Rightarrow$ either $x \in I$ or $y \in I$. We are thus led to the following definition and theorem.

Definition. *An ideal I of a commutative ring R with unity is said to be a* prime ideal *if $xy \in I \Rightarrow$ either $x \in I$ or $y \in I$.*

Theorem 7.9.1. *R/I is an integral domain if and only if I is prime.*

Examples of prime ideals

In the ring of integers consider the ideal $[n]$. For this to be prime we must have that $xy = \lambda n \Rightarrow$ either x or y is a multiple of n, i.e. that n is a prime. R/I is then of course an integral domain, being the finite field of residues modulo n.

In the ring of residues modulo n consider the ideal $[m]$, where m is a factor of n. Then by a similar argument to that in the previous example $[m]$ is prime if and only if m is a prime number, and R/I is again the field of residues modulo m.

Now consider the ring of polynomials $F[x]$ over a field F. The principal ideal $[P]$ is prime if and only if the fact that XY is a multiple of P implies either X or Y is a multiple of P. But $F[x]$ is a unique factorisation domain and so the condition is that P is an irreducible polynomial, by theorem 5.4.1.

Prime ideals in integral domains with unique factorisation

In the examples of the integers and polynomials above, we saw that a principal ideal is prime if and only if its generator is an irreducible element. This is true generally.

***Theorem* 7.9.2.** *The principal ideal $[p]$ in a unique factorisation domain D is a prime ideal if and only if p is an irreducible element of D.*

If p is irreducible then by theorem 5.4.1 $p/xy \Rightarrow$ either p/x or p/y: i.e. $xy \in [p] \Rightarrow$ either $x \in [p]$ or $y \in [p]$, and so $[p]$ is prime.

Conversely suppose p is not irreducible, so that $p = xy$ where neither x nor y are units. Then xy is certainly in $[p]$, but neither x or y are, since if say $x \in [p]$, $x = zp$ and so

$$p = xy = yzp, \quad 1 = yz$$

and y is a unit, giving a contradiction. Hence $[p]$ is not prime.

Corollary. *If D is a principal ideal domain then the prime ideals are the ideals $[p]$, for all irreducible elements p of D.*

This follows at once, since a principal ideal domain is a unique factorisation domain (theorem 7.5.7).

7.10. Maximal ideals

We continue with the ideas which led to the work of the previous section and investigate the conditions under which the quotient ring R/I is a field. This again depends on the nature of I rather than on that of R, since if R is a field then I must be trivial.

As in §7.9 it is convenient to restrict our rings R to be

commutative and with unity, and we assume this throughout the present section.

We will use the characteristic property of a field, that it possesses no proper ideals, in connection with the natural homomorphism $R \to R/I$, and we need first to establish the fact that ideals are preserved 'both ways' under homomorphisms. This is proved formally in the two theorems below.

Theorem 7.10.1. *If $\theta: R \to S$ is an epimorphism (i.e. is* onto *S), and I is an ideal of R, then $I\theta$ is an ideal of S.*

If y_1 and $y_2 \in I\theta$, so that $y_1 = x_1\theta$ and $y_2 = x_2\theta$ for some $x_1, x_2 \in I$, $y_1 - y_2 = (x_1 - x_2)\theta$ and $y_1 y_2 = (x_1 x_2)\theta$ and are both in $I\theta$, since $x_1 - x_2$ and $x_1 x_2$ are in I, I being an ideal and so a subring. Hence $I\theta$ is a subring of S.

Now let $y \in I\theta$ and s be any element of S. Since θ is an epimorphism $\exists$ an element $r \in R$ for which $r\theta = s$. $\exists$ also $x \in I$ for which $x\theta = y$. Then $(rx)\theta = sy$. But $rx \in I$ since I is an ideal, and so $sy \in I\theta$. Similarly $ys \in I\theta$ and $I\theta$ is an ideal.

Note. It is essential to the above proof that θ is an epimorphism: if this is not the case then $I\theta$ need not be an ideal. An example is given by letting R be the ring of integers, S be the ring of real numbers and I the ideal of even integers, with θ the injection defined by $n\theta = n$. Then $I\theta$ is the set of even integers, and this is not an ideal in the ring of reals, since these form a field and hence have no proper ideals.

Theorem 7.10.2. *If $\theta: R \to S$ is a homomorphism and J is an ideal of S, then the set of elements of R that map into elements of J is an ideal of R.*

Let $\phi: S \to S/J$ be the natural homomorphism. Then $\theta\phi: R \to S/J$ is a homomorphism, as may be easily proved, and hence its kernel is an ideal of R, by theorem 7.6.2. But the kernel of $\theta\phi$ is the set of elements r such that $r\theta\phi = J$, and is the set in question.

Now consider the natural homomorphism $R \to R/I$. Then under this mapping ideals of R/I arise from ideals of R *which contain I.* Hence the condition that R/I has no proper ideals is equivalent to the condition that R has no proper ideals containing I. We are led to the definition and theorem below.

Definition. *An ideal I of a commutative ring R with unity is said to be a* maximal ideal *if there exists no ideal J other than I or R such that $I \subseteq J \subseteq R$.*

(Compare this with the definition of maximal invariant subgroup given in exercise 6B, no. 8.)

Theorem 7.10.3. *R/I is a field if and only if I is maximal.* (*R is assumed commutative and with unity.*)

Suppose first that R/I is a field: we will prove I maximal. Suppose J is such that $I \subseteq J \subseteq R$. Then if θ is the natural homomorphism $R \to R/I$, $J\theta$ is an ideal of R/I by theorem 7.10.1, since θ is certainly an epimorphism. Hence, since R/I is a field, $J\theta$ is either R/I or the trivial ideal I. In the latter case $J = I$. We will prove that if $J\theta = R/I$ then $J = R$.

Let r be any element of R. Then since $J\theta = R/I$, $\exists\, j \in J$ such that $r\theta = j\theta$ and so $r \in j+I$. But $I \subseteq J$ and so $j+I \subseteq J$ since J is an ideal. Hence $r \in J$ and $J = R$. Thus $J = I$ or R and so I is maximal.

We now assume I maximal and prove that R/I is a field. We first show that R/I is simple. Let K be an ideal of R/I. Then by theorem 7.10.2 the set of elements of R that map into elements of K under the natural homomorphism θ is an ideal of R, and contains I since K contains the zero I of R/I. Hence since I is maximal, this set is either I or R. In the first case $K = I$, the zero ideal of R/I, and in the second $K = R/I$. Thus R/I is a simple ring. Also, since R is commutative, R/I is commutative. It remains to be shown that multiplication in R/I is non-trivial.

If it is trivial then $(x+I)(y+I) = I$ for any cosets and so $xy \in I$ for all x, y in R. Putting $y = 1$, the unity, which we have assumed to exist in R, we see that $x \in I$ for all x, i.e. I is the whole ring R. Thus, except in the trivial case when $I = R$ and R/I has just one element, multiplication in R/I is non-trivial and so, by theorem 7.8.1, R/I is a field.

Note that although intuitively the above theorem is fairly

obvious, the details of the proof are rather harder to establish than in the corresponding theorem for prime ideals.

Note also that we may define maximal ideals in *any* ring R, and R/I will still be a simple ring, but that to prove R/I a field we need further criteria, which we have taken to be the commutativity and existence of unity in R.

Examples of maximal ideals

In the ring of integers $[n]$ is not maximal if n is composite, for if m is a proper factor of n, $[m] \supset [n]$. If however n is prime then $[n]$ is maximal. For suppose an ideal J contains $[n]$ and a further integer r. Then r is prime to n, so there exist integers t and u such that $tr+un = 1$ and so J contains 1 and hence is the whole ring by theorem 7.4.1.

Similarly in the ring of residues modulo n the ideal $[m]$, where m is a factor of n, is maximal if and only if m is a prime integer.

In both the above cases we could prove that the ideals in question are maximal by considering whether or not the quotient rings are fields: in these cases the finite fields of residues modulo a prime.

By applying the argument given above (for the ring of integers) to the ring of polynomials over a field we can see that the principal ideal $[P]$ is maximal if and only if P is irreducible.

The relationship between prime and maximal ideals

That a maximal ideal is necessarily prime is almost trivial.

Theorem 7.10.4. *If I is maximal then it is prime.*

For R/I is a field, and hence is an integral domain.

In general the converse is not true: all prime ideals are not necessarily maximal. For example consider the ring of polynomials over the integers. Then the ideal generated by 2 is prime (by theorem 7.9.2 since the ring is a U.F.D. and 2 is an irreducible element). But this ideal is *not* maximal, for the ideal generated by 2 and x contains it properly and is properly contained in the complete ring.

In the case of the rings of integers and of polynomials over a field however the converse is true: for a prime ideal is generated by an irreducible element and so is maximal by the work above. In fact the converse is always true for principal ideal domains, as will now be proved directly.

Theorem 7.10.5. *If R is a principal ideal domain then every prime ideal is maximal.*

Suppose I is a prime ideal. Then $I = [p]$ where p is irreducible, by the corollary to theorem 7.9.2. Now let $J \supseteq I$. Since R is a principal ideal domain, $J = [q]$ for some $q \in R$. Hence by theorem 7.5.4, $q|p$. But p is irreducible and hence q is either a unit or an associate of p. If q is a unit, $J = [q] = R$ (for $vq = 1$ for some v and then any $x = xv.q$ and $\in [q]$), whereas if q is an associate of p, $J = I$ by theorem 7.5.5. Thus for any ideal $J \supseteq I$ we have $J = R$ or I and so I is maximal.

Corollary. *If R is a principal ideal domain then $[p]$ is maximal if and only if p is irreducible.*

This follows at once from the above work and the corollary to theorem 7.9.2.

Worked exercises

1. Show that the set $S = \{m+ni\}$, where m and n are multiples of 4, is an ideal in the ring $R = \{m+ni\}$, where m and n are both even. Identify R/S.

If $s_1 = m_1+n_1 i$ and $s_2 = m_2+n_2 i$ where m_1, n_1, m_2, n_2 are all multiples of 4, $s_1-s_2 = (m_1-m_2)+(n_1-n_2)\, i$ and m_1-m_2 and n_1-n_2 are multiples of 4. Hence S is a subgroup.

If $r = \mu+\nu i$, where μ and ν are even, $rs_1 = (\mu m_1-\nu n_1)+(\nu m_1+\mu n_1)\, i$ and both $\mu m_1-\nu n_1$ and $\nu m_1+\mu n_1$ are multiples of 4, so that $rs_1 \in S$. Since R is commutative, $s_1 r = rs_1 \in S$ and, putting $s_2 = r$ we also have $s_1 s_2 \in S$ so that S is a subring and thus by the above an ideal.

The cosets are $0+S$, $2+S$, $2i+S$ and $2+2i+S$. We easily see that the additive structure is the Vierergruppe and that multiplication is trivial, so that R/S is the zero-ring obtained from the Vierergruppe.

2. If R is commutative and with a unity show that the natural homomorphism associated with the ideal $[(x-b)]$ in $R[x]$ is that which maps $P(x)$ into $P(b)$.

Since R is commutative and with a unity the remainder theorem applies and so $P \equiv (x-b)\, Q+P(b)$. Thus P is in the coset $P(b)+[(x-b)]$ and the result follows. (The quotient ring being of course R.)

3. If the integral domain D is *not* a U.F.D. prove that if $[p]$ is a prime ideal then p is an irreducible element of D but, by considering $D = I[\sqrt{-5}]$ prove that the converse is not necessarily true. (Compare theorem 7.9.2.)

As in the proof of theorem 7.9.2 suppose that p is not irreducible: we will show that $[p]$ is not a prime ideal, and thus prove the first part.

If p is not irreducible it has a factor x which is neither a unit nor an associate of p, so that $p = xy$ where neither x nor y are units. Then $xy \in [p]$ but neither x nor y are, since $x \in [p] \Rightarrow x = zp$ and $p = yzp$, $1 = yz$ and y is a unit. Hence $[p]$ is not prime.

Conversely considering $I[\sqrt{-5}]$ we recollect §5.4 (p. 129) where we showed that $9 = 3.3 = (2+\sqrt{-5})(2-\sqrt{-5})$ and that all these factors are irreducible. Now consider the irreducible 3. Then $9 = 3.3 \in [3]$, but neither $2+\sqrt{-5}$ nor $2-\sqrt{-5} \in [3]$, since as both are irreducible they cannot be multiples of 3, not being associates of 3. Hence $[3]$ is not a prime ideal.

4. Prove theorem 7.10.3 by consideration of elements and cosets.

Suppose first that I is maximal. Since R is assumed commutative and with unity, R/I is certainly commutative and has a unity $1+I$. We require to prove that it has inverses.

Take an element x of any non-zero coset $x+I$. Then the set of elements $i+rx$, for variable $i \in I$ and $r \in R$, can easily be shown to be an ideal of R (in fact it is the ideal generated by the elements of I plus the additional element x). Since $x+I$ is non-zero, $x \notin I$ and so the ideal $\{i+rx\}$ properly contains I. Hence, as I is maximal, this ideal must be R and hence contain the unity 1, i.e. $1 = i+tx$ for some $t \in R$. Hence $1+I = (t+I)(x+I)$ and so $x+I$ has the inverse $t+I$. Thus any non-zero coset has an inverse and R/I is a field.

Conversely let R/I be a field and suppose J is an ideal that properly contains I, with $j \in J$ but $\notin I$. Then $j+I$ is non-zero and hence

$$(j+I)(x+I) = r+I$$

has a solution for all $r \in R$. Hence $jx = r+i$ for some $i \in I$ and so

$$r = jx-i \in J$$

since J is an ideal and both j and i are in J (since $I \subset J$). Thus as r is arbitrary $J = R$ and so I is maximal.

Exercises 7A

1. Prove that a subset I of a ring R is an ideal if and only if

$$I-I = I \quad \textit{and} \quad xi \in I,\ ix \in I, \quad \forall i \in I,\ x \in R.$$

Which of the subrings S of the rings R in **2–8** are ideals? If S is an ideal identify the quotient ring R/S.

2. R is the Gaussian integers, S the set $\{m+0i\}$ for variable m.

3. R is the Gaussian integers, S the set $\{m+ni\}$ where m and n are both even.

4. R is the ring of polynomials over the integers, S the set

$$\{a_0+a_1x+...\} \text{ where } a_0 \text{ is even and } a_1, a_2, ... \text{ are arbitrary.}$$

5. R is as in **4**, S the set $\{a_0+a_1x+...\}$ where a_0 and a_1 are both even and $a_2, a_3, ...$ are arbitrary.

6. R is as in **4**, S the set $\{a_0+a_1x+...\}$ where a_1 is even and $a_0, a_2, ...$ are arbitrary.

7. R is the Gaussian integers, S the set $\{m+ni\}$ where m is a multiple of 4 and n a multiple of 6.

8. R is the ring of even integers, S the set of multiples of 4.

9. Show that the set of elements of the form $(r, 0, 0)$ is an ideal in $R \oplus S \oplus T$ and that its quotient ring $\cong S \oplus T$.

10. Show that the set of elements of the form $(r, s, 0)$ is an ideal in $R \oplus S \oplus T$ and that its quotient ring $\cong T$.

11. Let R be the ring of integers and I the ideal of even integers. If $x = y = 0$ show that $\{(x+i_1)(y+i_2): i_1, i_2 \in I\} \subset xy+I$, and that the inclusion is a *strict* inclusion; i.e. that $\{(x+i_1)(y+i_2)\} \neq xy+I$.

12. If R is the ring of even integers and I the principal ideal [6], identify R/I.

13. Repeat **12** where $I = [8]$.

14. In any ring prove that [0] is the trivial ideal $\{0\}$.

15. In any ring R with unity prove that $[1] = R$.

16. If R is an integral domain prove that $[x] = R \Leftrightarrow x$ is a unit.

17. Give independent sets of generators for those subrings in **2–8** which are ideals.

18. In the ring of polynomials over the integers prove that

$$[3, 2x, x^2] = [3, 2x] = [3, x].$$

19. In the ring of polynomials over the rationals, identify P if

$$[x^3+1, x^2-1] = [P].$$

20. In the ring R of polynomials over the reals prove that $[x^3+1, x^2+1]$ is the trivial ideal R.

21. Give the form of the general element of the ideal $[a_1, ..., a_n]$ in the cases where

(i) R has a unity but is not commutative;
(ii) R is commutative but has no unity;
(iii) R is not commutative and has no unity.

22. In a principal ideal domain prove that $[a, b] = R \Leftrightarrow a$ and b are co-prime.

23. Prove that if $\theta: R \to R'$ is a ring-homomorphism then the set of elements of R that map into a given element y of R' form a coset of the kernel of θ.

24. Prove that any homomorphic image of the ring of integers is isomorphic either to the ring of residues modulo n for some n, or to the ring of integers.

25. Prove that in *any* integral domain (not necessarily a U.F.D.), $[p]$ is a prime ideal if and only if $p/xy \Rightarrow p/x$ or p/y.

26. Prove that the trivial ideal R is always prime.

27. Prove that the trivial ideal $\{0\}$ is prime in R if and only if R is an integral domain.

28. Which of the subrings in **2–8**, which are ideals, are prime ideals? Which are maximal ideals?

29. If R is a commutative ring with non-trivial multiplication whose additive structure is a simple group prove that R is a field.

30. Prove that *any* subgroup of a zero ring (i.e. a ring where $xy = 0 \; \forall \; x$ and y) is an ideal. Deduce that the only simple zero rings are those with additive structure C_p, for prime p.

31. If D_1 and D_2 are integral domains prove that the set $\{(d_1, 0)\}$ is a *prime* ideal in the ring $D_1 \oplus D_2$.

32. If F_1 and F_2 are fields prove that the set $\{(f_1, 0)\}$ is a *maximal* ideal in the ring $F_1 \oplus F_2$.

33. If I is prime and R/I finite prove that I is maximal.

34. By considering the nature of the ideal [2] or otherwise prove that if $Z[x]$ is the ring of polynomials over the integers then $Z[x]/[2]$ is an integral domain but *not* a field.

35. Prove that R/I is a zero-ring $\Leftrightarrow \{xy: x, y \in R\} \subseteq I$. Deduce that if R has a unity then R/I is a zero-ring if and only if $I = R$.

36. If $\theta: I \to I$ is an endomorphism of the ring I of integers, prove that θ is either the identity automorphism or the zero endomorphism (with $I\theta = \{0\}$).

37. If I, I' are ideals in R, R' respectively prove that the set

$$\{(i, i'): i \in I, \; i' \in I'\}$$

is an ideal in $R \oplus R'$ and that its quotient ring $\cong R/I \oplus R'/I'$.

38. Give an example of a U.F.D. that is not a principal ideal domain.

Exercises 7B

1. The *centre* of a ring R is the set of elements that commute with *all* elements of R. Show that

(i) the centre is always a subring;

(ii) if R has no unity and is not commutative then the centre is not necessarily an ideal;

(iii) if R has a unity but is not commutative then the centre can never be an ideal;

(iv) if R is commutative the centre is the trivial ideal R.

2. If I and J are ideals in R prove that $I+J = \{i+j : i \in I, j \in J\}$ is an ideal and is contained in every ideal containing I and J. Show further that if $I = [a_1, ..., a_m]$, $J = [b_1, ..., b_n]$ then $I+J = [a_1, ..., a_m, b_1, ..., b_n]$.

3. If R is the ring of integers and $I = [m]$, $J = [n]$ show that $I \cap J = [l]$, $I+J = [h]$ where l, h are the L.C.M. and H.C.F. of m and n.

4. If I and J are ideals in R, where R is commutative and has a unity, define $I * J$ to be the set of all finite sums

$$i_1 j_1 + ... + i_n j_n, \quad \text{where} \quad i_1, ..., i_n \in I, \quad j_1, ..., j_n \in J.$$

Prove that $I * J$ is an ideal. Prove further that if $I = [p]$ and $J = [q]$ then $I * J = [pq]$.

5. If R is commutative and has a unity prove that

(i) $I * J \subseteq I \cap J$;

(ii) $I(*J*K) = (I*J)*K$;

(iii) $I*(J+K) = I*J+I*K$.

6. Identify the general ideal in the ring of Gaussian integers.

7. Give examples to show that the *subring* generated by an element of a ring (see p. 78) and the *ideal* generated by the same element may or may not be the same.

8. If R is commutative and with a unity show that the set of all polynomials in $R[x, y]$ that vanish at each of a given set C of points (where C is either a finite or infinite set) form an ideal in $R[x, y]$. Give a set of independent generators for this ideal in the cases where R is the ring of reals and

(i) C is the set of points on the circle $x^2+y^2 = 1$;

(ii) C is the pair of points $(1, 1)$ and $(-1, -1)$.

9. Give an example where $R \not\cong (R/I) \oplus I$.

10. Let R be a ring with a unity e and which is generated by e. Prove that

(i) the mapping θ: ring I of integers $\rightarrow R$ defined by $n\theta = ne$ is a ring-homomorphism;

(ii) the image $I\theta$ is a subring of R containing e, and θ is *onto* R;

(iii) $R \cong I$ or the ring of residues R_m for some m.

8

EXTENSIONS OF STRUCTURES

8.1. Introduction and examples

A very common procedure in abstract algebra is that of obtaining new structures from ones which are already known. We have used this idea many times in the previous chapters and in volume 1—see the examples below.

There are two main purposes in investigating such extensions. The first is so that we can study a structure in terms of its components, which are usually simpler. Thus we are interested in the substructures of a given group, ring or field and in the relationship that the whole structure bears to them. We have used this idea in expressing sets as the direct products of two or more of their subsets.

The second purpose is to obtain new structures, by applying certain more or less standard methods to sets that we already know. We may thus obtain structures that possess some required additional property: for example the integers do not admit of division and we extend these to the field of rationals so as to obtain a set that does so, while still containing a subset isomorphic to the integers. In many cases we use the extension methods in order to obtain rigorously definitions of sets that we already know about intuitively: the complex numbers give a good example, where the rigorous development by number pairs is somewhat tedious but gives us a secure foundation for working with complex numbers, defining them in terms of real numbers, which are already known. In other cases we do obtain quite new structures: the Galois fields $GF[2^3]$ and $GF[3^2]$ of §4.8 were obtained by such methods, which will be explained more fully in §8.5.

An example can often be looked at from both points of view given above. Thus the ring of polynomials $R[x]$ over a ring R with a unity may be thought of as being generated by the subring of coefficients (i.e. all polynomials of degree 0) and the element x, so that a general element is given by a linear combination of the coefficients and positive powers of x (thus: $a_0+a_1x+a_2x^2+\ldots+a_nx^n$). Alternatively we may start with the ring R and *form* all such linear combinations, proving that they give a ring under the usual definitions. Either point of view is valid although for some work one will be more useful, and for some the other will be best.

We now give the main examples that we have used so far, both earlier in the present volume and in volume 1. The work that follows will be independent of this previous work, though a knowledge of it will help in fixing the ideas by showing concrete examples. The remainder of the present chapter will be fairly abstract as it is concerned with generalising the particular examples so as to apply them more widely.

Extensions of the number system

The basic set of the natural numbers does not admit of subtraction in general and, in order to allow of this, is extended to the complete set of integers. This is done by considering the ordered pairs (a, b) of the natural numbers (i.e. the positive integers) and setting up an equivalence relation by $(a, b)\ R(c, d)$ if and only if $a+d = b+c$. We prove this to be an equivalence relation and then define the integers to be the equivalence classes. Addition of classes is defined in terms of addition of arbitrary elements of them, the negative of the class containing (a, b) is defined as the class containing (b, a) and the usual laws are obeyed. The details are given in §4.2 of volume 1, but note that the basic idea is that of forming an equivalence relation on ordered pairs of natural numbers.

A similar method is used to define the rationals in terms of the integers. Here we take ordered pairs (a, b) of integers with $b \neq 0$, and set up a relation $(a, b)\ R(c, d)$ if and only if $ad = bc$. We prove this to be an equivalence relation and define the classes to be our rational numbers, proceeding to define addition and the other rules and to verify that the fundamental laws apply. The details were given in §5.1 of volume 1. This method is generalised in §8.2.

The extension of the field of rationals to that of the reals, in order to solve equations of the form $x^2 = 2$, belong to analysis rather than algebra since it essentially involves the idea of the limit of an infinite series.

The real numbers are extended to the field of complex numbers by another application of the idea of ordered pairs: thus a complex number is an ordered pair (a, b) where a and b are real. We do not need to form an equivalence relation here, since the complex number is given by the *unique* pair (a, b). We do of course need to define addition and the other rules and to verify the fundamental laws, and this was done in outline in §5.3 of volume 1. An alternative method of defining complex numbers is given in §8.4.

Polynomials

In §3.6 we defined a polynomial over a ring R as an ordered set of elements of R and defined addition and multiplication of polynomials. We showed how such a polynomial could be written in the customary way as $\sum_0^n a_r x^r$ and pointed out that if R has a unity then this notation may be obtained systematically in terms of the polynomial x. Polynomial extensions will be dealt with further in §8.3.

Rational functions

The field of rational functions is obtained from the integral domain of polynomials (both being over a coefficient set that is itself an integral domain) in the same way as the field of rational numbers is obtained from the integral domain of integers. It gives another example of the general process of the next section.

Direct products and direct sums

We have discussed direct products (for groups) and direct sums (for rings) in detail in the relevant chapters and will not extend this treatment here: the reader is merely reminded of their importance as a method of obtaining new structures in terms of simpler ones.

Quotient groups and rings

Corresponding to each invariant subgroup H of a group G we form the new group G/H and corresponding to each ideal I of a ring R we obtain the new ring R/I. §§8.4 and 8.5 give further uses of this idea.

Extensions of fields

In §4.3 we showed that the field of rationals could be extended by 'adjoining' the irrational number $\sqrt{n}$, where n is rational and not a square of a rational number. The resulting set of numbers of the form $a+b\sqrt{n}$ forms a field. Similarly we can 'adjoin' the complex number ω, where ω is a complex cube root of unity. By a similar method applied to the field of residues modulo 2 we obtained the Galois field $GF[2^2]$ and later, in §4.8, the fields $GF[2^3]$ and $GF[3^2]$. Such methods of 'adjunction' will be investigated in §8.4.

From the above examples we see that methods of extension are both varied and fruitful. The purpose of the present chapter is to generalise some of these methods and to explain the reasoning which leads us to them. Most of the work will be in connection with fields, extending and illuminating some of the topics of chapter 4 and culminating in a brief discussion of the general countable field.

8.2. The field of fractions of an integral domain

In this section we generalise the method we used to form the field of rationals from the integral domain of the integers, and show how any integral domain I may be extended to give a field F, called the *field of fractions of I*, which contains a subset isomorphic to I. Applied to the domain of polynomials over an integral domain it gives us a field of rational functions over the integral domain. The work is very similar to that in volume 1 where the result was proved for the special case of the integers.

We consider the set of ordered pairs (a, b) of elements of the integral domain I, with $b \neq 0$.

We then set up a relation $(a, b)R(c, d) \Leftrightarrow ad = bc$ and prove this to be an equivalence relation. (In our minds we identify (a, b) with the 'quotient' a/b or ab^{-1}, which must exist in our fields of fractions, and this gives us the reason for our definition of R and later of addition and multiplication. But of course we must not use this idea in the formal work since at present we do not have a field and must not beg the question of its existence: all our work is in terms of the *domain* I and its properties.)

Reflexive. Since $ab = ba$ we have $(a, b)R(a, b)$.

Symmetric. If $(a, b)R(c, d)$ then $ad = bc$ and so $cb = da$, giving $(c, d)R(a, b)$.

Transitive. If $(a, b)R(c, d)$ and $(c, d)R(e, f)$ then $ad = bc$ and $cf = de$. Hence $adf = bcf = bde$. But $d \neq 0$ and I *is an integral domain*, so that $af = be$ and $(a, b)R(e, f)$.

Hence R is an equivalence relation. We consider the equivalence classes under R and define addition and multiplication so that these form a field F.

Addition. Take typical elements (a, b) and (c, d) of two classes and consider the element $(ad+bc, bd)$. Since $b \neq 0$ and $d \neq 0$ we certainly have $bd \neq 0$ and so the element is admissible. We will show that whatever elements we choose from the two given classes, the element $(ad+bc, bd)$ will always be in the same class, and thus we may define the sum of classes without ambiguity to be the class containing this element.

Thus suppose $(a, b)R(a', b')$ and $(c, d)R(c', d')$ so that $a'b = b'a$ and $c'd = d'c$. We wish to show that

$$(ad+bc, bd)R(a'd'+b'c', b'd'),$$

i.e. that $(ad+bc)\,b'd' = (a'd'+b'c')\,bd$, which is easily seen to be the case.

Multiplication. Consider the element (ac, bd), where again $bd \neq 0$ and so the element is admissible. Again if $(a, b)R(a', b')$ and $(c, d)R(c', d')$ we have $a'b = b'a$ and $c'd = d'c$ and so $acb'd' = a'c'bd$, or $(ac, bd)R(a'c', b'd')$. Thus we define the product of classes to be the class containing (ac, bd) where (a, b) and (c, d) are any members of the given classes, and this definition again gives a well-defined product.

We must now verify the field axioms. We take particular elements of the classes in all cases.

F is a commutative group under addition

$$(a, b)+(c, d) = (ad+bc, bd) = (cb+da, db)$$

by commutativity of both addition and multiplication in I,

$$= (c, d)+(a, b).$$

$$\begin{aligned}((a, b)+(c, d))+(e,f) &= (ad+bc, bd)+(e,f)\\ &= (adf+bcf+bde, bdf)\\ &= (a, b)+(cf+de, df)\\ &= (a, b)+((c, d)+(e,f)).\end{aligned}$$

The zero is the class containing $(0, a)$ and the negative of that containing (a, b) is that containing $(-a, b)$, the latter being easily seen to be well-defined.

F (except for 0) is a commutative group under multiplication

The Commutative and Associative Laws of multiplication are immediate. The unity is the class containing (a, a) and the inverse of the class containing (a, b) with both a and b non-zero is that containing (b, a): both these concepts are easily seen to be well-defined.

F satisfies the Distributive Law

This is immediate, the proof being left to the reader.

Hence F is a field. We now show that it contains a subset isomorphic to I.

Consider the mapping $\theta: I \to F$ defined by $a\theta = (a, 1)$, where the RHS is taken to mean the class containing $(a, 1)$. Then

$$\begin{aligned}(a+b)\,\theta &= (a+b, 1)\\ &= (a, 1)+(b, 1)\\ &= a\theta+b\theta.\end{aligned}$$

$$\begin{aligned}(ab)\,\theta &= (ab, 1)\\ &= (a, 1).(b, 1)\\ &= (a\theta)\,(b\theta).\end{aligned}$$

Hence θ is a homomorphism. It is 1–1 since suppose $a\theta = b\theta$. Then $(a, 1) = (b, 1)$. Thus they are in the same class and $a = b$.

Hence I is isomorphic to $I\theta$, a subset of F.

8.3. Transcendental extensions

As we saw in §3.6, one way of considering the polynomial ring $R[x]$ in the case where R has a unity (which covers most important examples) is as being the set of expressions of the form $a_0 + a_1x + a_2x^2 + \ldots + a_nx^n$, where x is itself an element of $R[x]$. Addition and multiplication of such polynomials obey the same rules as ordinary ring multiplication provided x is taken to commute with the elements of the coefficient set R, and the fact that $\sum_0^n a_rx^r = 0 \Rightarrow a_r = 0 \;\forall\, r$ may be interpreted to mean that x satisfies no polynomial equation whose coefficients are in R.

Thus to any ring R with unity we may form an extension ring $R[x]$ by 'adjoining' an element x which commutes with all elements of R and which satisfies no polynomial equation with coefficients in R. Such an extension is called a *transcendental extension* of R.

We may consider this process in a slightly different way. Let R be a subring of a ring R^* and let x be an element of R^* which is not in R. Suppose further that R^* has a unity which is in R and that x commutes with all elements of R. Then the subring of R^* generated by the elements of R plus the element x consists of all elements of the form $\sum_0^n a_rx^r$ where all a_r are in R. If x satisfies no polynomial equation with coefficients in R (i.e. $\sum_0^n a_rx^r = 0 \Rightarrow a_r = 0 \;\forall\, r$) then x is said to be *transcendental over* R and in this case all such elements $\sum_0^n a_rx^r$ can easily be seen to be different. The subring generated by R and x is isomorphic to the polynomial ring $R[x]$, and we see the reason for the name 'transcendental extension'.

Two such extensions, given by transcendental elements x and y, are clearly isomorphic, the isomorphism being given by

$\sum_0^n a_r x^r \sim \sum_0^n a_r y^r$. Note also that the transcendental extension contains a subring isomorphic to R, consisting of the polynomials of degree 0.

As an example of the above let R^* be the ring (in fact field) of the real numbers and let R be the subring of the integers. Then the real number π is transcendental over R, and we obtain a subring by considering the set of all polynomials in π with integral coefficients. The subring of polynomials in the transcendental element π^2 is isomorphic to this, as the subring of polynomials is in, say, e.

Simple transcendental extensions of a field

If F is a field the polynomials with coefficients in F form an integral domain $F[x]$, which is however *not* a field. But we may now form the field of fractions of the domain $F[x]$ and obtain a field that has a sub-domain isomorphic to $F[x]$, which in turn has a subfield isomorphic to F, so that the field of fractions has a subfield isomorphic to F. This field of fractions is called a *simple transcendental extension of* F, and of course all such are isomorphic.

Again, if F is a subfield of a field F^* and x is an element of F^* which is not in F and which is transcendental over F then the field of fractions of $F[x]$ is the *subfield* generated by the elements of F plus x (that is, the smallest subfield containing them).

8.4. Algebraic extensions of a field

In the previous section we considered the subring generated by a subring R of a ring R^* and one additional element x where x is transcendental over R. We now consider the effect of the same procedure when x is not transcendental. For the full theory to work it is necessary for R and R^* to be fields and so we will assume this at the outset. It is also convenient to replace x by u reserving x for use as an indeterminate.

Thus suppose we have a subfield F of a field F^* and an element u of F^* which is not in F and suppose that u is not transcendental over F. This means that u satisfies a polynomial equation with coefficients in F, and in such a case we say that

u is *algebraic over* F (thus giving us the name *algebraic extension* for the type of extension that we are considering). As an example, to fix the ideas, let F^* be the field of real numbers, F the subfield of rationals and u the real number $\sqrt{2}$, which satisfies the polynomial equation $x^2-2 = 0$.

Now the above equation is not the only one with rational coefficients which is satisfied by $\sqrt{2}$, other examples being $\frac{1}{2}x^2-1 = 0$, $x^4-4 = 0$, etc. But it is the only irreducible polynomial with leading coefficient 1. This uniqueness property is true generally, and it is important to establish this theorem before we proceed further.

Theorem 8.4.1. *If u is algebraic over F then it is a zero of a unique irreducible polynomial $f(x)$ with coefficients in F and leading coefficient* 1, *and if u satisfies any polynomial equation $g(x) = 0$ then $g(x)$ is merely a multiple of $f(x)$.*

Since u is algebraic over F it certainly satisfies some polynomial equation $p(x) = 0$. But it was shown in §5.6 that the polynomial domain $F[x]$, for F a field, is a U.F.D. Hence $p(x)$ may be expressed as the product of irreducible polynomials, say $p(x) \equiv p_1(x).p_2(x) \ldots p_n(x)$. Also, $F[x]$ is an integral domain, and so the fact that $p(u) = 0$ implies that $p_i(u) = 0$ for some i. Let $p_i(x)$ have leading coefficient c, so that $p_i(x) \equiv cf(x)$, where $f(x)$ has leading coefficient unity and, since $p_i(u) = 0$ and $c \neq 0$, $f(u) = 0$. Thus u is a zero of the irreducible polynomial $f(x)$.

To show this unique, let u be also a zero of a second irreducible polynomial $f'(x)$, with leading coefficient unity, so that f and f' are certainly not associates and hence must be co-irreducible as both are irreducible. Thus f and f' have H.C.F. 1 and so polynomials s and t exist with $1 \equiv sf+tf'$. (This follows since $F[x]$ is a Euclidean domain, F being a field, and Euclid's algorithm may be used to give the H.C.F. in the form $sf+tf'$—see exercises 5B, no. 6.) Since $f(u) = f'(u) = 0$ we have a contradiction at once, since 1 is certainly not equal to zero.

Now suppose $g(u) = 0$. Then applying the division algorithm, valid for polynomials over a field, we obtain $g \equiv qf+r$ with degree of $r <$ degree of f. Thus $r(u) = 0$. But r and f must be co-irreducible since r has degree less than that of

f and f is irreducible (any common non-unit factor must be an associate of f and this cannot be a factor of r by consideration of degrees) and a repetition of the argument of the preceding paragraph leads to a contradiction, unless $r \equiv 0$, so that g is a multiple of f, as was to be proved.

We now consider the subring of F^* generated by F and u. In the case where F is the field of rationals and $u = \sqrt{2}$ this subring is the set of real numbers of the form $a + b\sqrt{2}$, with a and b rational and manipulated according to the usual rules: it is in fact the field discussed on p. 98, not merely a ring.

In the general case the subring consists of all elements (of F^*) of the form $\sum_0^n a_r u^r$, where all a_r are in F. So far this is the same as in the transcendental case, but we have the important difference that not all such elements are different, since $f(u) = 0$. In fact, two such polynomials in u are the same if they differ by a multiple of $f(u)$: we are in effect working modulo $f(u)$.

Thus the subring is no longer isomorphic to the polynomial ring $F[x]$: indeed we suspect (as we are working modulo $f(u)$) that it is isomorphic to the quotient ring $F[x]/[f(x)]$ where of course $[f(x)]$ is the principal ideal consisting of multiples of $f(x)$. We will now prove this fundamental fact.

Theorem 8.4.2. *If u is algebraic over F and $f(x)$ is as in theorem* 8.4.1, *then the subring generated by F and u is isomorphic to the quotient ring $F[x]/[f(x)]$.*

Consider the mapping $\theta: F[x] \to F^*$ defined by

$$\left(\sum_0^n a_r x^r\right)\theta = \sum_0^n a_i u^r.$$

The image of θ is the subring required. θ is a ring-homomorphism since both addition and multiplication are defined identically in the polynomial ring $F[x]$ and in the field F^*.

A polynomial $p(x)$ is in the kernel of θ if $p(u) = 0$ and so, by the second part of theorem 8.4.1, p is a multiple of f and so is in $[f(x)]$. Conversely any member of $[f(x)]$ is a multiple of f and so has u as a zero. Hence the kernel of θ is the principal ideal $[f(x)]$.

Now by theorem 7.6.3 the image of $F[x]$ under θ, that is the subring generated by F and u, is the quotient ring $F[x]/[f(x)]$, and the theorem is proved.

We have obtained the quotient ring as the *subring* generated by F and u, but it is in fact a *subfield* of F^*, since $f(x)$ is irreducible, $F[x]$ is a principal ideal domain and hence $[(fx)]$ is maximal by the corollary to theorem 7.10.5, so that $F[x]/[f(x)]$ is a field by theorem 7.10.3. It is called a *simple algebraic extension of F*, formed by 'adjoining' a root of $f(x) = 0$ on to F. The degree of $f(x)$ is called the *degree of the extension.*

The extension is obtained in practical cases by considering the polynomials of $F[u]$ and putting $f(u) = 0$: expressing a general polynomial g in the form $g = qf+r$ by the division algorithm and replacing $g(u)$ by $r(u)$. Notice from this that all polynomials of degree $< \deg f$ will be unchanged, but any polynomial of degree $\geqslant \deg f$ will be put 'equal' to one of degree $< \deg f$.

Thus in the case where $u = \sqrt{2}$ we do not change polynomials $a+b\sqrt{2}$, but for those of degree $\geqslant 2$ we put $x^2 = 2$ and obtain polynomials of degree 0 or 1.

In the general case the elements of our extension field are those of the form $g(u)$ with $\deg g < \deg f$, and addition is straightforward. But multiplication of such polynomials will introduce various convolutions, by virtue of the fact that $f(u) = 0$, and we will obtain exciting fields, often different for different irreducible polynomials f.

Theorem 8.4.3. *If u and v are both zeros of the same irreducible polynomial f(x) then the simple algebraic extensions of F obtained by adjoining u and v are isomorphic.*

For each is isomorphic to $F[x]/[f(x)]$.

As an example of this, the field of the set $\{a+b\sqrt{2}\}$ is isomorphic to that of the set $\{a-b\sqrt{2}\}$, the isomorphism being the mapping which interchanges these two typical elements.

In all the above work of this section the field F^* hardly enters and is not used specifically: its only function is to provide us with an element u and a field of which both F and its extension field are subfields. Thus we could merely start with a given field F and adjoin to it the fictitious root of an irreducible polynomial $f(x)$ with coefficients in F, fictitious in the sense that it is extraneous to the known field F. We may thus, provided irreducible polynomials of degree > 1 exist, obtain new fields from our known field F. Thus from the field of rationals we obtain the fields of §4.3: $\{a+b\sqrt{n}\}$ by adjoining a root of $x^2-n = 0$ and $\{a+b\sqrt[3]{3}+c\sqrt[3]{9}\}$ by adjoining a root of $x^3-3 = 0$ (strictly speaking the root $\sqrt[3]{3}$, but if we choose one of the other two roots we obtain an isomorphic field by theorem 8.4.3), and $\{a+b\omega\}$ by adjoining a root of $x^2+x+1 = 0$ (in this last case F^* if introduced would need to contain complex numbers: note also that we do not use the reducible equation $x^3-1 = 0$, the quotient ring obtained from the principal ideal $[x^3-1]$ not being even a field).

Extensions of finite fields will be dealt with in the following section.

There is one very important application of the above work: Cauchy's method of defining complex numbers. We form the quotient field $F[x]/[x^2+1]$ where F is the field of real numbers, over which x^2+1 is of course irreducible. We obtain elements of the form $a+bx$ and add and multiply them according to the usual rules by putting $x^2 = -1$ whenever it occurs. Thus x fulfils the role of i (or $-i$) and we obtain the field of complex numbers in a rigorous manner, although in a manner that is advanced and which depends on much of our work on ideals.

8.5. Finite fields as extensions

If F is finite then a transcendental extension of F is always infinite (there are an infinite number of polynomials, since the degree is arbitrary). But simple algebraic extensions will be finite.

Theorem 8.5.1. *If F is finite of order m then a simple algebraic extension of F of degree n will have order m^n.*

For the elements are merely the polynomials with coefficients in F of degree $< n$, and two such polynomials are the same if and only if all corresponding coefficients are the same. Since there are m possibilities for each coefficient (including 0) there are m^n such polynomials.

Thus if we start from the simplest finite fields, those of the residues modulo a prime, and adjoin roots of irreducible polynomials to these we will obtain further finite fields. This was the method used in chapter 4, although we could not then give the reasoning behind it.

On p. 100 we took the field of residues modulo 2 and considered the irreducible equation $x^2+x+1 = 0$. We obtained a field with 4 elements, the Galois field $GF[2^2]$ in line with theorem 8.5.1. We did not prove that the given equation *is* irreducible, but we can soon convince ourselves of this by forming all possible products of linear equations in pairs (modulo 2). If we do this we discover that the polynomial x^2+x+1 is the only irreducible polynomial of degree 2 over the field of residues modulo 2.

Later in chapter 4 we discovered the fields $GF[2^3]$ and $GF[3^2]$ by the same method, and we can now *prove* that we do in fact obtain fields, provided we can show the equations irreducible. In fact for $GF[2^3]$ we could choose either of the polynomials x^3+x+1 or x^3+x^2+1. We did in fact use the former and put $x^3 = x+1$, remembering that we are working modulo 2. If we have chosen the other we would still have obtained a field with 2^3 elements and, by theorem 4.8.3, this must be the same field.

The method can be used, in theory, to construct any Galois field. To form $GF[p^n]$ we find an irreducible polynomial $f(x)$ of degree n over the field F of residues modulo p and then $GF[p^n] \cong F[x]/[f(x)]$. It can be shown that such an irreducible polynomial always exists: often there will be more than one and the choice of which to take is arbitrary.

In practice of course the method is impossibly tedious unless p and n are small, but we have the interesting result that any finite field is a simple algebraic extension of a field of residues.

8.6. The general countable field

It should be obvious that the field axioms are highly restrictive: the variety of possible fields is far less than that of rings or groups. Already in §4.8 we have classified completely the finite (commutative) fields, and we are now in a position to extend this classification. We can still not do this completely, but will restrict ourselves to the case of countable (commutative) fields. The study of non-countable fields raises very deep questions, which do not enter into the countable case.

A countable set is one which can be put into 1–1 correspondence with the integers 1, 2, 3, ... It was proved in volume 1 that the field of rationals is countable, whereas the field of the reals is not (nor of course is that of the complex numbers).

We have found two ways of extending fields: a simple transcendental extension of F is formed by constructing the field of fractions of $F[x]$, and a simple algebraic extension by forming the quotient field $F[x]/[f(x)]$ where f is irreducible over F. We also know (theorems 4.6.3 and 4.6.4) that any field contains a prime field as subfield, either the field of rationals of that of the residues modulo p, for a prime p. We will prove that any finite or countably infinite field can be formed from its prime subfield by a succession (finite or countably infinite) of simple extensions, transcendental or algebraic. This is not difficult to prove, being done by adjoining the elements in turn.

Theorem 8.6.1. *If F is finite or countably infinite it can be constructed from its prime subfield by a succession (finite or countably infinite) of simple extensions, transcendental or algebraic.*

Let the elements of F be denumerated thus: $\phi_1, \phi_2, \phi_3, \ldots$. (It is here of course that we assume F countable.)

We start with the prime subfield F_0 of F and consider the subfield F_1 generated by F_0 and ϕ_1. If $\phi_1 \in F_0$ this subfield is F_0. If ϕ_1 is transcendental over F_0 the subfield F_1 is a simple transcendental extension of F_0. If ϕ_1 is algebraic over F_0 then F_1 is a simple algebraic extension of F_0.

Now consider the subfield F_2 generated by F_1 and ϕ_2: the results are as before. Continue forming subfields $F_3, F_4, \ldots$.

At each stage we have F_{i+1} either equal to F_i or a simple extension of it. Since all elements of F are in the sequence $\{\phi_i\}$, we obtain F by constructing the countable set $F_0, F_1, F_2, \ldots$, i.e. by performing a succession (at most countably infinite) of simple extensions on the prime subfield of F.

(If F is finite then there will be a finite number of extensions, but if F is infinite the number may be finite or infinite.)

Worked exercises

1. Modify the work of §8.2 to show how to form a field of fractions of a commutative ring R with no zero divisors *but with no unity*, having at least two elements. What is the field of fractions of the ring R of the even integers?

Since R contains at least two elements it contains non-zero elements and so ordered pairs (a, b) with $b \neq 0$ do exist. We may proceed exactly as for integral domains (R has all the properties of an integral domain except the possession of a unity) as far as proving that F is a field. Note that F has a unity still, viz. the class containing (a, a) for any a.

To show that F contains a subset isomorphic to R we consider the mapping $\theta: R \to F$ defined by $a\theta = (ab, b)$ for some $b \neq 0$, where the RHS means the class containing (ab, b). This class is independent of any particular choice of b, since (ab, b) and (ac, c) are in the same class since $(ab)c = b(ac)$. Thus θ is well-defined. Then

$$(a+a')\,\theta = (ab+a'b, b)$$
$$= (ab, b)+(a'b, b),$$

since this is $(ab^2+a'b^2, b^2)$ and is in the same class as $(ab+a'b, b)$,

$$= a\theta+a'\theta,$$

and
$$(aa')\,\theta = (aa'b, b)$$
$$= (aa'b^2, b^2), \quad \text{this being in the same class,}$$
$$= (ab, b).(a'b, b)$$
$$= (a\theta)\,(a'\theta).$$

Hence θ is a homomorphism. Now suppose $a\theta = a'\theta$. Then

$$(ab, b) = (a'b, b) \quad \text{so that} \quad ab^2 = a'b^2 \quad \text{and so} \quad a = a'.$$

Hence θ is 1–1 and as before $R \cong R\theta$, a subset of F.

The field of fractions of the ring of even integers is merely the field of the rationals, since any rational p/q may be put in the form $2p/2q$ and $2p$, $2q$ are certainly even.

2. Give a direct proof by considering cosets and using H.C.F. theory that $F[x]/[f(x)]$ is a field if $f(x)$ is irreducible.

The quotient ring is certainly a ring and is commutative since $F[x]$ is commutative as F is a field and thus is commutative. It has a unity since

$F[x]$ has a unity. Thus it remains to prove only that the quotient ring possesses inverses.

Let $g(x)$ be any member of $F[x]$ which is not in the zero coset, i.e. is not a multiple of $f(x)$. Then since f is irreducible g and f are co-prime, having H.C.F. 1.

Now $F[x]$ is a Euclidean domain and hence if h is the H.C.F. of f and g there exist polynomials s and t such that $h = sf + tg$. (This follows at once from Euclid's algorithm which gives h in this form; see also exercises 5B, no. 6.) Here $h = 1$ and so s and t exist with $1 = sf + tg$.

In terms of cosets with respect to the ideal $[f]$ this means that

$$1 + [f] = (t + [f])(g + [f])$$

and so the coset $g + [f]$ has an inverse, viz. the coset $t + [f]$.

3. Use the method of worked exercise 2 above to find the inverse of $x+2$ in $GF[3^2]$ given by $J_3[x]/[x^2+1]$, where J_3 is the field of residues modulo 3.

Using the Euclidean algorithm we obtain

$$x^2+1 = (x+1)(x+2)+2,$$
$$x+2 = (2x+1)\,2.$$

Hence $2 = (x^2+1)-(x+1)(x+2)$ and so, multiplying by 2, we obtain $1 = 2(x^2+1)-(2x+2)(x+2)$.

(This last step is necessary since the algorithm gives us merely *an* H.C.F. and we recollect that we may always multiply an H.F.C. by any unit to obtain another. In the present example any constant polynomial is a unit and thus an H.C.F. and we must multiply by the inverse of 2, which is also 2, to obtain the particular H.C.F. 1 in terms of f and g.)

Hence the inverse of $x+2$ (strictly speaking of the coset containing $x+2$) is $-(2x+2)$, i.e. is $x+1$ (i.e. the coset containing $x+1$).

Exercises 8A

1. In the work of §8.2 prove that F satisfies the Distributive Law.

2. Show that the field of fractions of a *field* F is isomorphic to F.

3. Let I be a subring of a field F^* which contains the unity 1 of F^*. Prove that I is an integral domain.

4. In 3 prove that if F is the intersection of all subfields of F^* which contain I then $F \cong$ the field of fractions of I. (F is called the *subfield generated by I* and is the smallest subfield containing I. See exercises 4B, no. 5.)

5. Show that if the integral domains I_1 and I_2 are isomorphic then this isomorphism can be extended in a *unique* way to give an isomorphism between their fields of fractions.

6. Prove that, if for any n, $\sum_0^n a_r x^r = 0 \Rightarrow a_r = 0 \;\forall\, r$ then

$$\sum_0^n a_r x^r = \sum_0^n b_r x^r \Rightarrow a_r = b_r \;\forall\, r.$$

(This is the assertion made in §8.3 that if x is transcendental over a ring R then all elements $\sum_0^n a_r x^r$ are different.)

7. Prove that the reals *cannot* be obtained by a finite sequence of algebraic extensions of the rationals.

8. Prove that a transcendental extension of R is always infinite, even if R is finite.

9. If $f(x)$ is linear, so that its zero u is in F, prove that

$$F[x]/[f(x)] \cong F.$$

10. Prove that any simple algebraic extension of the field of reals is either isomorphic to the reals or to the field of complex numbers. (*Hint.* Use the fact that there are no irreducible polynomials of degree > 1, over the complex numbers.)

11. Prove that the only *irreducible* polynomials of degree 3 over the field of residues modulo 2 are x^3+x+1 and x^3+x^2+1.

12. Working over the residues modulo 3 show that x^2+x+2 is irreducible, and use it to form $GF[3^2]$. Show that the field so obtained is isomorphic to that given on p. 110, where we used the irreducible polynomial x^2+1.

13. Show that the fields $\{a+b\sqrt[4]{2}\}$ and $\{a+bi\sqrt[4]{2}\}$ are isomorphic and give two different isomorphisms. (a and b are rational.)

14. Give a subfield of the field of complex numbers which is not a subfield of the reals and which is isomorphic to the field $\{a+b\sqrt[3]{2}\}$, where a and b are rational.

15. Working over the residues modulo p, if $f(x)$ is an irreducible polynomial of degree n, prove that there is at least one element $u \in GF[p^n]$ for which $f(u) = 0$. (*Hint.* Consider the field $J_p[x]/[f(x)]$.)

16. By considering $GF[p^n]$ as an algebraic extension of F, prove that any subfield F of $GF[p^n]$ must have order p^{n_1} where n_1 is a factor of n. (Cf. exercises 4B, no. 12.)

Exercises 8B

1. Prove that the field of rational functions over an integral domain I is isomorphic to the field of rational functions over the field of fractions of I. (Cf. exercises 4B, no. 6.)

2. A field F is said to be *algebraically closed* if every polynomial of degree $\geqslant 1$ in $F[x]$ has a zero in F. (The field of complex numbers is an example, by the Fundamental Theorem of Algebra.) Prove that in such a field every polynomial of degree $\geqslant 1$ factorises into linear factors and that there are no irreducible polynomials of degree > 1. Deduce that the only simple algebraic extensions of F are those isomorphic to F, and that there are no proper simple algebraic extensions of the field of complex numbers.

3. If u and v are zeros of the same irreducible polynomial $f(x)$ prove that the correspondence $g(u) \leftrightarrow g(v)$ between elements of the simple algebraic extensions obtained by adjoining u and v to F is 1–1. Prove further that this correspondence is an isomorphism and deduce theorem 8.4.3.

4. If R is a ring without a unity consider the set S of ordered pairs (m, r) with m an integer and $r \in R$. Define

$$(m, r)+(n, s) = (m+n, r+s),$$
$$(m, r).(n, s) = (mn, nr+ms+rs).$$

With this definition prove that S is a ring with a unity $(1, 0)$ and that S contains a subring $\tilde{R} \cong R$. Prove also that $\tilde{R}$ is an ideal in S.

(This gives a method of extending a ring R without unity to a ring S with unity.)

5. By considering R as the ring of even integers and R^* as the ring of rationals, or otherwise, show that if R is a subring without unity of a ring R^* with unity then S as defined in **4** is not necessarily isomorphic to the subring of R^* generated by R and the unity 1. Give an example where S *is* isomorphic to this subring.

6. Prove that if I is a sub-domain of a field F, then F contains a subfield isomorphic to the field of fractions of I.

VECTOR SPACES

9.1. Introduction

The study of vectors, particularly of their algebraic properties, is often best performed by considering the set of all vectors of a given type (for example that of all three-dimensional vectors with real co-ordinates) and the algebraic structure which these comprise. Such a structure is called a *vector space* and, as we expect, may be defined by taking as axioms a very few fundamental laws which we require it to satisfy. It turns out that, in the 'finite dimensional' case (we anticipate a little in the present section the ideas which will be developed formally later in the chapter), our basic axioms characterise completely sets of what we would naturally think of as 'vectors': in other words they lead to no structures that are not immediately recognizable as sets of vectors in the elementary sense. (This is quite different from the case in group theory, where there is an extraordinary richness of possible structure, and from fields, where we lay down a large collection of axioms and still obtain a fairly wide diversity of structure.) The advantage of approaching the subject axiomatically is, as always, that we know precisely what we may assume and with what we are dealing: the work follows rigorously from the axioms and we are made aware of the distinction between fundamental properties and derived ones.

We come now to an important point. Any advanced, or even fairly advanced, work on vectors leads inevitably to the study of matrices, and to the whole vast subject of what is known as 'linear algebra'. As expected, matrices develop naturally when we study vector spaces, and this is probably the most illuminating (though not the most elementary) method of developing their theory. But such work is essentially different from most of *abstract* algebra. It is analytical in that it deals with individual elements and individual homomorphisms, rather than with structures as a whole. Indeed, any use of matrices in connection with vector spaces depends on the choice of a particular 'base' (see §9.5), and different bases lead to different matrices, albeit connected.

This is not to decry the importance of linear algebra. This importance is very great and the subject has many theoretical and practical uses (for example in the solution of simultaneous equations and in its connection with geometrical transformations). It is a huge subject and there is now no difficulty in finding books on it which are suitable for students at any required level of mathematical knowledge. But the outlook, the 'feel' of the subject, are distinct from those of abstract algebra, though there is inevitably some overlap, as in the present chapter.

For these reasons we will deal here only very briefly with matrices. We will show how they arise naturally and indicate some basic properties, but will base this firmly on the abstract structural ideas of vector spaces with which we are chiefly concerned. Any detailed analytical work is foreign to our treatment and will not be attempted. For this the reader is referred to any of the standard works on linear algebra and, particularly if he knows some elementary matrix theory already, he may well find that this is illuminated by the indications that we give on the fundamental link between matrices and the basic structure of vector spaces

We are concerned then with a vector space as an example of an algebraic structure. As such it shows many similarities to the other structures with which we have already dealt: thus we will consider substructures and homomorphisms of vector spaces. The terminology will often be rather different, since the subject has grown up largely on its own, by virtue of its connection with linear algebra (homomorphisms, for example, are known as 'linear transformations').

Although most familiar examples are finite dimensional, and although elementary linear algebra is derived from the finite dimensional case, we will be concerned with both finite dimensional and infinite dimensional spaces, and will always indicate where the theory applies only to the former case.

9.2. Algebraic properties of vectors

We assume that the reader is familiar with vectors from the practical and intuitive points of view. A brief account is given in chapter 8 of volume 1, and in the present section we summarise the algebraic properties of vectors as given there, before defining an abstract vector space in the next section. As usual, vectors are written in bold type.

If we think of an n-dimensional vector as the ordered set $(\xi_1, \xi_2, \ldots, \xi_n)$, where the co-ordinates $\xi_1, \xi_2, \ldots, \xi_n$ are called *scalars* and may be real, complex or indeed members of any field, then the two basic processes of combination are addition of vectors and multiplication of a vector by a scalar, in both cases obtaining a vector as the result.

Addition

If $\mathbf{x} = (\xi_1, \xi_2, \ldots, \xi_n)$ and $\mathbf{y} = (\eta_1, \eta_2, \ldots, \eta_n)$ then

$$\mathbf{x}+\mathbf{y} = (\xi_1+\eta_1, \xi_2+\eta_2, \ldots, \xi_n+\eta_n)$$

and addition is commutative and associative. The zero vector $(0, 0, \ldots, 0) = \mathbf{0}$ and satisfies the equations $\mathbf{0}+\mathbf{x} = \mathbf{x}+\mathbf{0} = \mathbf{x}$ for any $\mathbf{x}$, whereas $-\mathbf{x} = (-\xi_1, -\xi_2, \ldots, -\xi_n)$ is a negative of $\mathbf{x}$ and satisfies $\mathbf{x}+-\mathbf{x} = -\mathbf{x}+\mathbf{x} = \mathbf{0}$.

Multiplication by a scalar

$\lambda\mathbf{x} = (\lambda\xi_1, \lambda\xi_2, \ldots, \lambda\xi_n)$ and satisfies

(i) $\lambda(\mathbf{x}+\mathbf{y}) = \lambda\mathbf{x}+\lambda\mathbf{y}$.

(ii) $(\lambda+\mu)\mathbf{x} = \lambda\mathbf{x}+\mu\mathbf{x}$.

(iii) $\lambda(\mu\mathbf{x}) = (\lambda\mu)\mathbf{x}$.

(iv) $1.\mathbf{x} = \mathbf{x}$.

The above laws are all obvious when vectors are defined in terms of co-ordinates: in the abstract treatment of the next section we take these laws as axioms, and avoid any discussion of co-ordinate representation, although this will arise naturally later.

Multiplication of vectors

There is no one definition, although various ideas are useful in certain cases and for certain work. Thus complex numbers may be multiplied, and scalar and vector products are often useful in practice. The scalar, or inner, product is important in defining a distance on a vector space, for details of which see §10.1.

9.3. Abstract vector spaces

In giving an axiomatic definition of a vector space, we wish the structure so defined to possess the usual properties of sets of vectors. It turns out that the properties given above in §9.2 are sufficient to ensure this (which is why they were chosen!) and thus are taken as our axioms.

We notice that the axioms involve the idea of multiplication by a scalar, which is not itself a member of the vector space. This is a completely new idea. In groups, rings, fields and integral domains all our processes were of combination between elements of the set, whereas here we seek to combine an element of the set with an element of another, quite distinct, set to give a product, which is of course itself a vector, a member of the space. The other set, that of the scalars, may in fact be any field and for full generality we take an arbitrary field, thus including the cases where it is the set of reals, that of the complex numbers or indeed a finite field, all these being important cases in practice. It is customary to borrow the word 'scalar' from the physical background to vectors to refer to a member of this field, though we must bear in mind that this is not necessarily a set of 'numbers' in any ordinary sense.

In our definition and indeed in all the work of this chapter we omit any reference to length or modulus of a vector. This is foreign to the basic abstract definition, and depends on the idea of an 'inner product' which will be dealt with in §10.1.

The axioms fall into two sets, those for addition and those for multiplication by a scalar. If F is the field of scalars we talk of a vector space *over* F: the specification of the field F is an integral part of the definition. Finally we note that as usual we represent vectors in bold type, while scalars are left in normal type.

The abstract definition is then as follows.

Definition of a vector space

A set V of elements forms a vector space over a given field F if to any two elements $\mathbf{x}$ and $\mathbf{y}$ of V there is associated a unique third element of V, called their sum and denoted by $\mathbf{x}+\mathbf{y}$, and if to any element $\mathbf{x}$ of V and any element λ of F there is associated an element of V called their product and denoted by $\lambda\mathbf{x}$, where sum and product satisfy the following laws:

V is an Abelian group under addition of vectors: specifically

A1 $\mathbf{x}+\mathbf{y} = \mathbf{y}+\mathbf{x}$ *for all* $\mathbf{x}$ *and* $\mathbf{y}$ *in* V.

A2 $(\mathbf{x}+\mathbf{y})+\mathbf{z} = \mathbf{x}+(\mathbf{y}+\mathbf{z})$ *for all* $\mathbf{x}$, $\mathbf{y}$ *and* $\mathbf{z}$ *in* V.

A3 *There is an element of V called the zero vector and denoted by* $\mathbf{0}$ *which has the property that* $\mathbf{x}+\mathbf{0} = \mathbf{0}+\mathbf{x} = \mathbf{x}$ *for all* $\mathbf{x}$ *in* V.

A4 *Corresponding to each* $\mathbf{x}$ *in* V *there is an element* $-\mathbf{x}$ *in* V *called the negative of* $\mathbf{x}$ *which has the property that*

$$\mathbf{x}+-\mathbf{x} = -\mathbf{x}+\mathbf{x} = \mathbf{0}.$$

The product $\lambda\mathbf{x}$ satisfies the following axioms:

V1 $\lambda(\mathbf{x}+\mathbf{y}) = \lambda\mathbf{x}+\lambda\mathbf{y}$ *for all* λ *in* F *and* $\mathbf{x}$, $\mathbf{y}$ *in* V.

V2 $(\lambda+\mu)\mathbf{x} = \lambda\mathbf{x}+\mu\mathbf{x}$ *for all* λ, μ *in* F *and* $\mathbf{x}$ *in* V.

V3 $\lambda(\mu\mathbf{x}) = (\lambda\mu)\mathbf{x}$ *for all* λ, μ *in* F *and* $\mathbf{x}$ *in* V.

V4 $1.\mathbf{x} = \mathbf{x}$ *where* 1 *is the unity of* F *and* $\mathbf{x}$ *is any element of* V.

The first four axioms do not introduce F: they merely state that V is an Abelian group under addition. Because of this all the properties of an Abelian group are true for a vector space and we do not repeat them here: for example subtraction is possible, the Cancellation Law holds for addition, $-(-\mathbf{x}) = \mathbf{x}$, and so on.

The last four axioms are new, and link V with the scalar field F. We have of course not stated the field axioms for scalars, but these hold: we have merely said that F is a field. V1 defines the effect of scalar multiplication with a sum of vectors—both addition signs refer to addition in V. V2 and V3 state the effect of scalar multiplication by sums and products in F: the addition sign on the LHS of V2 and the product on the RHS of V3 refer to F, whereas the addition on the RHS of V2 is addition in V, and the LHS of V3 is of course repeated scalar multiplication in V, since $\mu\mathbf{x}$ is an element of V.

V4 is of a different type. It, so to speak, fixes a 'starting point' for scalar multiplication and ensures that we do not get awkward convolutions of our space. For example, if $1.\mathbf{x} = \mathbf{y}$ we have $\lambda\mathbf{x} = (\lambda 1)\,\mathbf{x} = \lambda(1.\mathbf{x})$ by V3 $= \lambda\mathbf{y}$ and so, if $\mathbf{y} \neq \mathbf{x}$ we cannot 'cancel' λ from the equation $\lambda\mathbf{x} = \lambda\mathbf{y}$. The assumption of V4 avoids this kind of result, which is certainly not in accord with our everyday notions of how vectors ought to behave.

Note the distinction between 0, the zero of F, and $\mathbf{0}$, the zero vector. Note also that, since there is no proper multiplicative structure in V, the unit vector $\mathbf{1}$ has no obvious meaning.

We proceed to prove some elementary results. Those concerning only the additive structure are omitted, as being identical to those for a general Abelian group. The proofs of the others, given below, are shown only briefly, since the reader should by now be quite familiar with the type of argument used.

Theorem 9.3.1. $0\mathbf{x} = \mathbf{0}$.

$\lambda\mathbf{x} = (\lambda+0)\,\mathbf{x} = \lambda\mathbf{x}+0\mathbf{x}$ by V2.

Hence by the Cancellation Law of addition $\mathbf{0} = 0\mathbf{x}$.

Theorem 9.3.2. $(-1)\,\mathbf{x} = -\mathbf{x}$.

$\mathbf{0} = 0\mathbf{x} = (1+-1)\,\mathbf{x} = 1\mathbf{x}+(-1)\,\mathbf{x} = \mathbf{x}+(-1)\,\mathbf{x}$ by V4.

The result follows by the uniqueness property of the negative, true for any group.

Theorem 9.3.3. $\lambda\mathbf{0} = \mathbf{0}$.

$\lambda\mathbf{x} = \lambda(\mathbf{0}+\mathbf{x}) = \lambda\mathbf{0}+\lambda\mathbf{x}$ by V1.

The result follows by the Cancellation Law.

Theorem 9.3.4. $\lambda\mathbf{x} = \mathbf{0} \Rightarrow$ *either* $\lambda = 0$ *or* $\mathbf{x} = \mathbf{0}$.

If $\lambda \neq 0$, so that λ^{-1} exists in F, we have $\lambda^{-1}(\lambda\mathbf{x}) = \lambda^{-1}\mathbf{0}$ by hypothesis $= \mathbf{0}$ by theorem 9.3.3. But by V3,

$$\lambda^{-1}(\lambda\mathbf{x}) = (\lambda^{-1}\lambda)\,\mathbf{x} = 1\mathbf{x} = \mathbf{x} \text{ by V4.}$$

Hence $\mathbf{x} = \mathbf{0}$.

Note. It is here that V4 is essentially used, avoiding strange results like that given earlier.

Corollary 1. $\lambda\mathbf{x} = \lambda\mathbf{y} \Rightarrow \lambda = 0$ *or* $\mathbf{x} = \mathbf{y}$.

For $\mathbf{0} = \lambda\mathbf{x}-\lambda\mathbf{y} = \lambda\mathbf{x}+(-1)(\lambda\mathbf{y})$ by theorem 9.3.2,

$= \lambda\mathbf{x}+(-\lambda)\,\mathbf{y}$ by V3,

$= \lambda\mathbf{x}+\lambda(-1.\mathbf{y}) = \lambda\mathbf{x}+\lambda(-\mathbf{y})$ by theorem 9.3.2,

$= \lambda(\mathbf{x}-\mathbf{y})$ and the result follows.

Corollary 2. $\lambda\mathbf{x} = \mu\mathbf{x} \Rightarrow$ *either* $\lambda = \mu$ *or* $\mathbf{x} = \mathbf{0}$.

For $\mathbf{0} = \lambda\mathbf{x}-\mu\mathbf{x} = \lambda\mathbf{x}+(-\mu)\,\mathbf{x}$ by V3,

$= (\lambda-\mu)\,\mathbf{x}$ by V2 and the result follows.

9.4. Linear dependence

It turns out that many vector spaces (the so-called 'finite dimensional' ones) are isomorphic to the familiar sets of vectors, represented by co-ordinates. We now proceed to the series of results which establish this important fact. These results are not difficult and are fundamental to the theory of vector spaces. They are of wider applicability than the classification theorems for fields and are proved more easily.

In the next few sections we will take as examples the sets of n-dimensional co-ordinate vectors, over the fields of real numbers for convenience, and usually with $n = 3$. As we will prove, these examples are typical (with possibly different scalar fields and with general n) but we must not assume this at present. They are familiar, and a consideration of them provides us with the motivation for the work that follows.

In the set of three-dimensional co-ordinate vectors the 3 vectors $(1, 0, 0)$, $(0, 1, 0)$ and $(0, 0, 1)$ play a special role, in that any vector can be represented in terms of them, whereas none of these three can be given in terms of the others. They form a *base* for the set, and are not unique in this: any three non-coplanar vectors would do, and we can transform our co-ordinates accordingly.

These are two basic properties of sets of 3 non-coplanar vectors in the three-dimensional case. The first is that any vector can be given in terms of them: i.e. that any $\mathbf{x}$ can be expressed in the form $\lambda_1\mathbf{e}_1+\lambda_2\mathbf{e}_2+\lambda_3\mathbf{e}_3$, where $\mathbf{e}_1$, $\mathbf{e}_2$, and $\mathbf{e}_3$ are the non-coplanar vectors, and $\lambda_1, \lambda_2, \lambda_3$ are of course scalars. We say that $\mathbf{e}_1, \mathbf{e}_2, \mathbf{e}_3$ *span*, or *generate*, the whole space. The second property is that none of the three can be expressed in terms of the others, thus that we cannot express $\mathbf{e}_1$ in the form $\lambda_2\mathbf{e}_2+\lambda_3\mathbf{e}_3$, and similarly for the other two. We say that the vectors are *linearly independent*, and this is the idea with which we deal first for the general abstract case. The most convenient definition, a symmetrical one, is given below.

Definition. *In the abstract vector space V the finite set of vectors* $\mathbf{x}_1, \mathbf{x}_2, \ldots, \mathbf{x}_n$ *is said to be* linearly dependent *if scalars* $\lambda_1, \lambda_2, \ldots, \lambda_n$ *exist, not all zero, such that* $\lambda_1\mathbf{x}_1+\lambda_2\mathbf{x}_2+\ldots+\lambda_n\mathbf{x}_n = \mathbf{0}$. *If no such scalars exist, i.e. if* $\lambda_1\mathbf{x}_1+\lambda_2\mathbf{x}_2+\ldots+\lambda_n\mathbf{x}_n = \mathbf{0} \Rightarrow \lambda_i = 0$ $\forall i = 1, \ldots, n$ *then the set is* linearly independent.

Note. $\sum_1^n \lambda_i\mathbf{x}_i = \mathbf{0}$ is true if each λ_i is zero, by theorem 9.3.1: in a linearly independent set this is the only possibility for the λ_i's.

Examples for three-dimensional co-ordinate vectors

1. $\mathbf{x}_1 = (1, 1, 0)$, $\mathbf{x}_2 = (1, 0, 0)$ are linearly independent (i.e. form a linearly independent set), since

$$\lambda_1\mathbf{x}_1+\lambda_2\mathbf{x}_2 = \mathbf{0} \Rightarrow \lambda_1+\lambda_2 = 0 \quad \text{and} \quad \lambda_1 = 0,$$

so that both λ_1 and λ_2 are zero.

2. $\mathbf{x}_1 = (1, 0, 0)$, $\mathbf{x}_2 = (0, 1, 0)$, $\mathbf{x}_3 = (1, 1, 0)$ are linearly dependent since $\mathbf{x}_1+\mathbf{x}_2-\mathbf{x}_3 = \mathbf{0}$.

3. $\mathbf{x}_1 = (1, 0, 0)$, $\mathbf{x}_2 = (0, 1, 0)$, $\mathbf{x}_3 = (1, 1, 0)$, $\mathbf{x}_4 = (0, 0, 1)$ are linearly dependent, since $\mathbf{x}_1+\mathbf{x}_2-\mathbf{x}_3+0.\mathbf{x}_4 = \mathbf{0}$. In this example $\lambda_4 = 0$, but this is not precluded by the definition: the essential condition is that not all λ_i's are zero.

In examples 2 and 3 we see that $\mathbf{x}_1 = -\mathbf{x}_2+\mathbf{x}_3$, so that $\mathbf{x}_1$ is expressible in terms of $\mathbf{x}_2$ and $\mathbf{x}_3$. In general if

$$\mathbf{x}_1 = \lambda_2\mathbf{x}_2+\lambda_3\mathbf{x}_3+\ldots+\lambda_n\mathbf{x}_n$$

we say that $\mathbf{x}_1$ is a *linear combination* of $\mathbf{x}_2, \mathbf{x}_3, \ldots, \mathbf{x}_n$. This expression may not be unique: if $\mathbf{x}_1 = (1, 0, 0)$, $\mathbf{x}_2 = (2, 0, 0)$ and $\mathbf{x}_3 = (5, 0, 0)$ then $\mathbf{x}_3 = 3\mathbf{x}_1+\mathbf{x}_2$ and also $\mathbf{x}_1+2\mathbf{x}_2$.

If $\mathbf{x}_1$ is a linear combination of $\mathbf{x}_2, \ldots, \mathbf{x}_n$ then

$$\mathbf{x}_1 - \lambda_2\mathbf{x}_2 - \ldots - \lambda_n\mathbf{x}_n = \mathbf{0} \quad \text{so that} \quad \mathbf{x}_1, \mathbf{x}_2, \ldots, \mathbf{x}_n$$

are linearly dependent. But the converse is not necessarily true: thus in example 3 above $\mathbf{x}_1$, $\mathbf{x}_2$, $\mathbf{x}_3$ and $\mathbf{x}_4$ are linearly dependent but $\mathbf{x}_4$ cannot be expressed as a linear combination of the other three. However, in this case any of $\mathbf{x}_1$, $\mathbf{x}_2$ or $\mathbf{x}_3$ is expressible in terms of the remaining vectors (which include $\mathbf{x}_4$ although its coefficient will in fact be zero). For a general linearly dependent set some members are linear combinations of the others, but not necessarily all members have this property. This fact is expressed best in the basic theorem that follows, which is easily proved.

Theorem 9.4.1. *If $\mathbf{x}_1, \mathbf{x}_2, \ldots, \mathbf{x}_n$ are all non-zero then they form a linearly dependent set if and only if $\mathbf{x}_{k+1}$ is a linear combination of $\mathbf{x}_1, \mathbf{x}_2, \ldots, \mathbf{x}_k$ for some k with $1 \leqslant k \leqslant n-1$.*

If $\mathbf{x}_{k+1} = \lambda_1\mathbf{x}_1 + \lambda_2\mathbf{x}_2 + \ldots + \lambda_k\mathbf{x}_k$ then

$$\lambda_1\mathbf{x}_1 + \lambda_2\mathbf{x}_2 + \ldots + \lambda_k\mathbf{x}_k - \mathbf{x}_{k+1} + 0.\mathbf{x}_{k+2} + \ldots + 0.\mathbf{x}_n = \mathbf{0}$$

and the set is linearly dependent, since at least one coefficient (that of $\mathbf{x}_{k+1}$) is non-zero.

Conversely suppose the set is linearly dependent, so that $\lambda_1\mathbf{x}_1 + \lambda_2\mathbf{x}_2 + \ldots + \lambda_n\mathbf{x}_n = \mathbf{0}$, with not all λ_i's zero.

Let λ_{k+1} be the greatest non-zero λ. Then k is certainly $\leqslant n-1$, and is also $\geqslant 1$, for if $k = 0$ we have $\lambda_1\mathbf{x}_1 = 0$ with $\lambda_1 \neq 0$ and $\mathbf{x}_1 \neq \mathbf{0}$ by hypothesis, which contradicts theorem 9.3.4. Then $\lambda_{k+1}\mathbf{x}_{k+1} = -\lambda_1\mathbf{x}_1 - \lambda_2\mathbf{x}_2 - \ldots - \lambda_k\mathbf{x}_k$ and so, since

$$\lambda_{k+1} \neq 0, \quad \mathbf{x}_{k+1} = \left(-\frac{\lambda_1}{\lambda_{k+1}}\right)\mathbf{x}_1 + \ldots + \left(-\frac{\lambda_k}{\lambda_{k+1}}\right)\mathbf{x}_k$$

and is a linear combination of $\mathbf{x}_1, \mathbf{x}_2, \ldots, \mathbf{x}_k$.

Corollary. *Since any set is either linearly dependent or linearly independent, we see at once that $\mathbf{x}_1, \mathbf{x}_2, \ldots, \mathbf{x}_n$ are linearly independent if and only if no $\mathbf{x}_{k+1}$ is a linear combination of*

$$\mathbf{x}_1, \mathbf{x}_2, \ldots, \mathbf{x}_k \quad \textit{for} \quad 1 \leqslant k \leqslant n-1.$$

9.5. Bases

We recollect that if $\mathbf{e}_1$, $\mathbf{e}_2$, $\mathbf{e}_3$ are non-coplanar three-dimensional co-ordinate vectors and thus may be taken as basic co-ordinate vectors then they are linearly independent and also span the whole space, in that any vector may be expressed as a linear combination of them. We generalise this idea.

Definition. *The set of vectors* $\mathbf{x}_1, \mathbf{x}_2, \ldots, \mathbf{x}_n$ *are said to* span, *or* generate, *the space V if any vector* $\mathbf{x}$ *of V is expressible as a linear combination of them, i.e. if* $\mathbf{x} = \lambda_1\mathbf{x}_1 + \lambda_2\mathbf{x}_2 + \ldots + \lambda_n\mathbf{x}_n$ *for some scalars* $\lambda_1, \lambda_2, \ldots, \lambda_n$, *not necessarily unique.*

We are now in a position to define a base of V, by analogy with the co-ordinate case.

Definition. *The set* $\mathbf{e}_1, \mathbf{e}_2, \ldots, \mathbf{e}_n$ *is a* base *of V if (i) they are linearly independent and (ii) they span V.*

If V possesses a (finite) base it is said to be *finite dimensional*, if not then it is *infinite dimensional.*

Note. We may here be tempted to anticipate and say that if $\mathbf{e}_1, \mathbf{e}_2, \ldots, \mathbf{e}_n$ is a base of V then V is of dimension n. There is no guarantee however that V may not have other bases containing other than n members. That any two bases do in fact have the same number of members is true and is the subject of the next section, but at present we are merely making a distinction between spaces possessing a finite base and spaces that do not.

Examples. In the three-dimensional co-ordinate case, (1, 0, 0), (0, 1, 0), (0, 0, 1) form a base, as do (1, 0, 0), (1, 1, 0), (0, 0, 1), while (1, 0, 0), (0, 1, 0), (1, 1, 0), (0, 0, 1) do not since although they span the space they are not linearly independent.

Theorem 9.5.1. *If* $\mathbf{e}_1, \mathbf{e}_2, \ldots, \mathbf{e}_n$ *span V then they form a base if and only if any* $\mathbf{x}$ *in V is expressible* uniquely *as a linear combination of* $\mathbf{e}_1, \mathbf{e}_2, \ldots, \mathbf{e}_n$.

Suppose first that the $\mathbf{e}_i$'s form a base, i.e. that they are linearly independent. Then if $\mathbf{x} = \sum_1^n \lambda_i \mathbf{e}_i = \sum_1^n \mu_i \mathbf{e}_i$ we have, subtracting, that $\sum_1^n (\lambda_i - \mu_i)\,\mathbf{e}_i = \mathbf{0}$. Hence, since the $\mathbf{e}_i$'s are linearly independent $\lambda_i - \mu_i = 0 \;\forall\, i$, i.e. $\lambda_i = \mu_i \;\forall\, i$, so that the expression of $\mathbf{x}$ is unique.

Conversely suppose that the expression of any $\mathbf{x}$ as a linear combination of the $\mathbf{e}_i$'s is unique. In particular the expression of $\mathbf{0}$, a member of V, is unique and so if $\sum_1^n \lambda_i \mathbf{e}_i = \mathbf{0}$ we must have $\lambda_i = 0 \;\forall\; i$. Hence $\mathbf{e}_1, \mathbf{e}_2, \ldots, \mathbf{e}_n$ are linearly independent and so form a base of V.

The above theorem plays an important part in setting up a co-ordinate system in V, and will be used in §9.7.

The next two theorems prove for the general space the expected results that a spanning set must contain a base and a linearly independent set must (in the finite dimensional case) be contained in a base.

Theorem 9.5.2. *If* $\mathbf{x}_1, \mathbf{x}_2, \ldots, \mathbf{x}_m$ *span* V *then there is a subset of these which is a base of* V.

If $\mathbf{x}_1, \mathbf{x}_2, \ldots, \mathbf{x}_m$ are linearly independent they form a base and the theorem is proved. If they are linearly dependent then by theorem 9.4.1 some $\mathbf{x}_{k+1}$ is a linear combination of

$$\mathbf{x}_1, \mathbf{x}_2, \ldots, \mathbf{x}_k, \quad \text{say} \quad \mathbf{x}_{k+1} = \sum_1^k \mu_i \mathbf{x}_i.$$

Consider the set $\mathbf{x}_1, \mathbf{x}_2, \ldots, \mathbf{x}_k, \mathbf{x}_{k+2}, \ldots, \mathbf{x}_m$. Any vector $\mathbf{x} = \sum_1^m \lambda_i \mathbf{x}_i$ since the $\mathbf{x}_i$'s span V. Hence $\mathbf{x} = \sum_{\substack{i=1 \\ i \neq k+1}}^m \lambda_i \mathbf{x}_i + \lambda_{k+1} \sum_1^k \mu_i \mathbf{x}_i$ and is a linear combination of $\mathbf{x}_1, \ldots, \mathbf{x}_k, \mathbf{x}_{k+2}, \ldots, \mathbf{x}_m$, so that these also span V.

If this set is now linearly independent the theorem is proved, as they form a base. If not repeat the above argument and obtain a spanning set of $m-2$ vectors, which may or may not be linearly independent. Repeat until after at most $m-1$ steps we do obtain a base, which is a subset of the original set. (This must occur, since after $m-1$ steps we are left with a single vector which spans V, and this must be non-zero unless V is the trivial space $\{\mathbf{0}\}$ and hence is linearly independent, since $\lambda_1 \mathbf{x}_1 = \mathbf{0}$ with $\mathbf{x}_1 \neq \mathbf{0} \Rightarrow \lambda_1 = 0$.)

Corollary. *If* V *possesses a finite spanning set then* V *is finite dimensional.*

Theorem 9.5.3. *If* $\mathbf{x}_1, \mathbf{x}_2, \ldots, \mathbf{x}_p$ *are linearly independent and* V *is finite dimensional then there exists a base containing*

$$\mathbf{x}_1, \mathbf{x}_2, \ldots, \mathbf{x}_p.$$

Let $\mathbf{e}_1, \mathbf{e}_2, \ldots, \mathbf{e}_n$ be a base of V (existing since V is finite dimensional).

Consider the set $\mathbf{x}_1, \ldots, \mathbf{x}_p, \mathbf{e}_1, \ldots, \mathbf{e}_n$.

This certainly spans V. We now use the argument of theorem 9.5.2 on this set, but note that at each stage we must always obtain one of the $\mathbf{e}_i$'s as a linear combination of the preceding vectors, since no $\mathbf{x}$ can be a linear combination of the preceding $\mathbf{x}_i$'s by the corollary to theorem 9.4.1 as the $\mathbf{x}_i$'s form a linearly independent set.

Hence we are left always with the $\mathbf{x}_i$'s and the process must stop before (or exactly at the point when) the $\mathbf{e}_i$'s are exhausted. We thus obtain a base which contains $\mathbf{x}_1, \mathbf{x}_2, \ldots, \mathbf{x}_p$, together possibly with some $\mathbf{e}_i$'s.

9.6. Dimension of a vector space

We can now prove the expected fundamental theorem, that any two bases have the same number of elements. Most of the proof is contained in theorem 9.6.1, which is really a lemma for the main theorem 9.6.2.

Theorem 9.6.1. *If* $\mathbf{x}_1, \mathbf{x}_2, \ldots, \mathbf{x}_p$ *is a linearly independent set, and* $\mathbf{y}_1, \mathbf{y}_2, \ldots, \mathbf{y}_m$ *spans* V, *then* $p \leqslant m$.

Consider the set $\mathbf{x}_p, \mathbf{y}_1, \mathbf{y}_2, \ldots, \mathbf{y}_m$. Since the $\mathbf{y}_i$'s span V, $\mathbf{x}_p$ is a linear combination of them and so the set is linearly dependent. Hence some $\mathbf{y}_{k+1}$ is a linear combination of

$$\mathbf{x}_p, \mathbf{y}_1, \ldots, \mathbf{y}_k,$$

by theorem 9.4.1. As in the proof of theorem 9.5.2 this means that $\mathbf{x}_p, \mathbf{y}_1, \ldots, \mathbf{y}_k, \mathbf{y}_{k+2}, \ldots, \mathbf{y}_m$ span V.

Continue similarly, adding $\mathbf{x}_{p-1}$ to this set. We obtain as before a linearly dependent set and some $\mathbf{y}_{l+1}$ expressible as a linear combination of the others, so that we have a spanning set comprising $\mathbf{x}_{p-1}, \mathbf{x}_p$, and all $\mathbf{y}_i$'s except $\mathbf{y}_{k+1}$ and $\mathbf{y}_{l+1}$. Continue this process. After each step we obtain a spanning set containing one more $\mathbf{x}$ and one less $\mathbf{y}$. We always obtain a $\mathbf{y}_i$ as a linear combination of the $\mathbf{x}_i$'s and preceding $\mathbf{y}_i$'s since the $\mathbf{x}_i$'s are linearly independent by hypothesis, and so no $\mathbf{x}_{i+1}$ is a linear combination of the preceding $\mathbf{x}_i$'s that have been introduced.

The process must be possible until all $\mathbf{x}_i$'s are used, and we must at each stage still have some $\mathbf{y}_i$'s left, otherwise the set, consisting of $\mathbf{x}_i$'s alone, could not be linearly dependent. Hence we finally obtain a spanning set $\mathbf{x}_1, \mathbf{x}_2, \ldots, \mathbf{x}_p$, together with $(m-p)$ $\mathbf{y}_i$'s and thus $m-p$ must be $\geqslant 0$. Hence $p \leqslant m$.

Theorem 9.6.2. *If* $\mathbf{e}_1, \mathbf{e}_2, \ldots, \mathbf{e}_n$ *and* $\mathbf{f}_1, \mathbf{f}_2, \ldots, \mathbf{f}_m$ *are both bases of* V *then* $m = n$.

Both sets are linearly independent and both span V. Using theorem 9.6.1, since the $\mathbf{e}_i$'s are linearly independent and the $\mathbf{f}_i$'s span V, we have $n \leqslant m$. Since the $\mathbf{e}_i$'s span V and the $\mathbf{f}_i$'s are linearly independent, $m \leqslant n$. Hence $m = n$.

Thus is any vector space, V, one of two things must happen. Either V possesses a base, and in this case all bases of V have the same number of elements. We say that V is *finite dimensional* and the number of elements in any base is called the *dimension* of V. It is denoted by dim (V) or merely dim V. Or V possesses no finite base, in which case we say that V is *infinite dimensional.*

We will see in the next section that all finite dimensional vector spaces are isomorphic to co-ordinate spaces, and some of our theory will be concerned exclusively with this case. Many infinite dimensional spaces exist however, and some examples will be given in §9.8.

The following theorems are immediate.

Theorem 9.6.3. *If* V *has dimension* n *there cannot exist a set of more than* n *linearly independent vectors in* V.

By theorem 9.5.3 there exists a base containing the set, and this base has only n elements.

Theorem 9.6.4. *If* V *has dimension* n *there cannot exist a set of less than* n *vectors which spans* V.

By theorem 9.5.2 there exists a base contained in the set, and this base contains n elements.

9.7. Co-ordinate representation

This section will be exclusively concerned with the finite dimensional case.

Suppose V has dimension n and that $\mathbf{e}_1, \mathbf{e}_2, \ldots, \mathbf{e}_n$ is a base of V. Then any $\mathbf{x}$ in V is expressible uniquely in the form $\xi_1\mathbf{e}_1 + \xi_2\mathbf{e}_2 + \ldots + \xi_n\mathbf{e}_n$ by theorem 9.5.1, and of course every vector of this form is in V, since all $\mathbf{e}_i$'s are in V and V is closed under addition and scalar multiplication. Hence there is a 1–1 correspondence between the elements of V and the co-ordinate vectors of the form $(\xi_1, \xi_2, \ldots, \xi_n)$ where the ξ_i's are scalars, i.e. are elements of the scalar field F. (See §8.7 of volume 1 for the definition of co-ordinate vectors.) It is seen at once that this correspondence preserves addition and scalar multiplication and hence is an isomorphism. Hence we have the following important theorem.

Theorem 9.7.1. *Any finite dimensional vector space V is isomorphic to the space of n-dimensional co-ordinate vectors, the co-ordinates being members of the scalar field F and n being the dimension of V.*

Corollary. *If V and W are vector spaces of the same finite dimension and over the same field then they are isomorphic.*

Note the very important point that the isomorphism in theorem 9.7.1 depends on the choice of a particular base: there will be many possible bases and the corresponding co-ordinate representations of the same space will be different: the same element may have different co-ordinates under different bases. Thus *any co-ordinate representation depends on the choice of a particular base.*

This fact leads to another important point. All finite dimensional vector spaces are indeed isomorphic to co-ordinate spaces, but we will not restrict ourselves to work based on co-ordinates, since such work depends, for a particular space, on the base chosen and is thus not so general as an abstract approach. Thus we will continue to treat vector spaces from the abstract definition where possible, using the co-ordinate representation only where necessary. As mentioned in §9.1, the study of matrices depends on the choice of a particular base, and indeed any work which uses particular bases, or even interchange of bases, leads inevitably to the use of matrices. But we are more concerned with the structure of the space independent of a choice of base—by this means we will be able to prove general theorems very easily, which may then be applied to particular co-ordinate systems.

Geometrical examples

In three dimensions any set of three non-coplanar vectors forms a base and may be chosen as the basic co-ordinate vectors (1, 0, 0), (0, 1, 0), (0, 0, 1). (We do not define 'coplanar' as this is best done using the ideas of spanning sets and linear dependence—we are here considering the geometrical properties intuitively.)

Consider the vector $\mathbf{x} = (3, 4, 5)$ with respect to some given base. In terms of the base

$$\mathbf{e}_1 = (1, 0, 0), \quad \mathbf{e}_2 = (0, 1, 0), \quad \mathbf{e}_3 = (1, 1, 1),$$

we have $\mathbf{x} = -2\mathbf{e}_1 - \mathbf{e}_2 + 5\mathbf{e}_3$ and so, with respect to this base its co-ordinates are $(-2, -1, 5)$.

9.8. Examples of vector spaces

Finite dimensional

Since all are isomorphic to co-ordinate spaces we restrict ourselves to giving a very few examples.

The trivial space. The vector $\mathbf{0}$ by itself forms a vector space over any field. It may be said to have dimension 0.

Polynomials. All polynomials over a field of degree $\leqslant n-1$ form a vector space of dimension n, a typical element being $a_0 + a_1 x + \ldots + a_{n-1} x^{n-1}$, where some a_i's, even the last one, may be zero. This corresponds to the co-ordinate vector

$$(a_0, a_1, \ldots, a_{n-1})$$

in the obvious isomorphism.

Fields over themselves. Any field F may be considered as a vector space over itself, with scalar multiplication being merely the ordinary field multiplication. It has of course dimension 1.

The set $\{a+b\sqrt{2}\}$ over the rationals. The field $\{a+b\sqrt{2}\}$, where a and b are rational, forms a vector space over the rationals. It has dimension 2, possible bases being $\{1, \sqrt{2}\}$, $\{1+\sqrt{2}, \sqrt{2}\}$, etc.

Infinite dimensional

Polynomials. All polynomials, that is all expressions of the form $\sum_{i=0}^{n} a_i x^i$, where n may be any positive integer, form a vector space over the field of coefficients. This possesses no finite base, since if it had finite dimension m we would have $m+1$ linearly independent vectors $1, x, x^2, x^3, \ldots, x^m$, contradicting theorem 9.6.3.

The reals over the rationals. The real numbers form an Abelian group. We may also multiply a real by a rational number to obtain another real, and this multiplication satisfies axioms V1–4. Hence the reals form a vector space over the field of rationals. It has infinite dimension, since no finite set can span it. (Since all linear combinations of such a finite set of rationals form a countable set, and the reals are not countable.)

Vector spaces of functions. Consider for example all real valued functions $f(t)$ defined for $a \leqslant t \leqslant b$. Then the sum of any two such functions, given by $(f+g)(t) = f(t)+g(t)$, is in the set, as is the product of any with a real scalar (given by

$$(\lambda f)(t) = \lambda(f(t)),$$

and the axioms are obviously satisfied. Hence the set forms a vector space over the reals.

9.9. Subspaces

Whenever we meet a new type of structure we investigate its substructures. In the case of vector spaces we follow the same procedure, although the work is not so fruitful as in the case of many other structures, groups for example.

In the usual way we define a *vector subspace* M of a vector space V to be a subset of V which is closed under addition and scalar multiplication: if $\mathbf{x}$ and $\mathbf{y}$ are in M so is $\mathbf{x}+\mathbf{y}$, and so is $\lambda\mathbf{x}$ for any scalar λ. It follows at once that M is a vector space in its own right, over the same field of scalars as V, the inclusion of negatives and the zero vector being guaranteed by theorems 9.3.2 and 9.3.1.

Examples of vector subspaces

The space $\{\mathbf{0}\}$ and the whole space V are trivial subspaces.

If V is two-dimensional, the vectors on any line through O (speaking geometrically) form a subspace: analytically if the general element is (ξ_1, ξ_2) these consist of all elements with $\lambda_1\xi_1+\lambda_2\xi_2 = 0$ for some λ_1 and λ_2, not both zero.

If V is three-dimensional, all lines and planes through O are subspaces. (For a plane the co-ordinates must satisfy one linear equation, and for a line they must satisfy two such equations.)

For an infinite dimensional example, consider the vector space of real valued functions defined for $a \leqslant t \leqslant b$, discussed above. Then all such functions which have zeros at certain specified points form a vector subspace, for the sum of any two such and the product of any with a scalar both have the same property.

Subspaces of finite dimensional spaces

We expect that if V is a vector space of finite dimension n, and M a subspace of V, then M will have finite dimension $\leqslant n$. This is true, but it takes a little care to prove. Certainly *if* we know that M is finite dimensional, with a finite base, then this base is a set of linearly independent vectors, and hence by theorem 9.6.3 has not more than n members: the difficulty lies in showing that M does have a finite base.

Theorem 9.9.1. *If V has finite dimension n and M is a non-trivial subspace of V, then M has finite dimension m where* $0 < m < n$.

Take a non-zero vector $\mathbf{x}_1$ in M. If $\mathbf{x}_1$ spans M it is a base of M.

If $\mathbf{x}_1$ does not span M there exists $\mathbf{x}_2$ in M which is not expressible as a linear combination of $\mathbf{x}_1$. Then $\mathbf{x}_1$ and $\mathbf{x}_2$ form a linearly independent set. If they span M they form a base of M. If not we can choose $\mathbf{x}_3$ in M which is not expressible as a linear combination of $\mathbf{x}_1$ and $\mathbf{x}_2$, and as before $\mathbf{x}_1, \mathbf{x}_2, \mathbf{x}_3$ are linearly independent. Continue in this way. Since at each stage we obtain a linearly independent set, by theorem 9.6.3 the process must terminate and we must obtain a spanning set, and hence a base for M containing at most n vectors. Hence M has finite dimension $\leqslant n$. If $\dim M = n$ we have a base of M containing n linearly independent vectors and so, by theorem 9.5.3, they must be a base of V and so must span V, so that $M = V$, and is trivial. Hence if M is non-trivial (not V or $\{\mathbf{0}\}$) it has finite dimension $< n$ and > 0.

Theorem 9.9.2. *If M is a subspace of V, where V is finite dimensional, then any base of M is contained in a base of V.*

A base of M forms a linearly independent set, and the result follows at once from theorem 9.5.3.

We now give some general examples of subspaces.

The subspace spanned by a set of elements

If $\mathbf{x}_1, \mathbf{x}_2, \ldots, \mathbf{x}_n$ are elements of a vector space V, of finite or infinite dimension, where the $\mathbf{x}_i$'s are not necessarily linearly independent, then the set of elements of the form

$$\lambda_1\mathbf{x}_1 + \lambda_2\mathbf{x}_2 + \ldots + \lambda_n\mathbf{x}_n$$

is easily seen to be a vector subspace of V. Note that if the $\mathbf{x}_i$'s are linearly dependent two such representations may give rise to the same element. We call this subspace the subspace spanned by the set.

The intersection of subspaces

If M and N are subspaces of V, then $M \cap N$ is a subspace: the proof is trivial and is left to the reader. $M \cup N$ is not in general a subspace. Similarly the intersection of any set of subspaces is a subspace.

The linear sum of subspaces

If M and N are subspaces of V, the set of elements of the form $\mathbf{m}+\mathbf{n}$, where $\mathbf{m}$ is in M and $\mathbf{n}$ in N, is easily seen to be a subspace. It is called the *linear sum* of M and N and denoted by $M+N$. Similarly we may form the linear sum of any finite set of subspaces.

Direct sums

If V and W are vector spaces over the same scalar field, we define the direct sum $V \oplus W$ in the obvious way as the set of pairs $(\mathbf{v}, \mathbf{w})$, $\mathbf{v} \in V$ and $\mathbf{w} \in W$, with the sum and scalar product defined as

$$(\mathbf{v}_1, \mathbf{w}_1) + (\mathbf{v}_2, \mathbf{w}_2) = (\mathbf{v}_1 + \mathbf{v}_2, \mathbf{w}_1 + \mathbf{w}_2),$$
$$\lambda(\mathbf{v}, \mathbf{w}) = (\lambda\mathbf{v}, \lambda\mathbf{w}).$$

That this definition gives a vector space is easily proved.

Theorem 9.9.3. *If M and N are subspaces of V, $V \cong M \oplus N$ if and only if $M \cap N = \{\mathbf{0}\}$ and $M+N = V$.*

If $V \cong M \oplus N$ then M is the set of elements $(\mathbf{m}, \mathbf{0})$ and N the set $(\mathbf{0}, \mathbf{n})$, so that $M \cap N = \{(\mathbf{0}, \mathbf{0})\} = \{\mathbf{0}\}$. The general element $(\mathbf{m}, \mathbf{n}) = (\mathbf{m}, \mathbf{0}) + (\mathbf{0}, \mathbf{n})$ and so $M+N = V$.

Conversely suppose $M \cap N = \{\mathbf{0}\}$ and $M+N = V$. Set up a correspondence between V and $M \oplus N$ by $\mathbf{m}+\mathbf{n} \leftrightarrow (\mathbf{m}, \mathbf{n})$.

Given $(\mathbf{m}, \mathbf{n})$ the element $\mathbf{m}+\mathbf{n}$ is certainly unique. Conversely given an element of V this can certainly be expressed in the form $\mathbf{m}+\mathbf{n}$ since $M+N = V$, and this expression is unique since if $\mathbf{m}_1+\mathbf{n}_1 = \mathbf{m}_2+\mathbf{n}_2$, $\mathbf{m}_1-\mathbf{m}_2 = \mathbf{n}_2-\mathbf{n}_1$ and is in $M \cap N$, so that $\mathbf{m}_1-\mathbf{m}_2 = \mathbf{n}_2-\mathbf{n}_1 = \mathbf{0}$ and $\mathbf{m}_1 = \mathbf{m}_2$, $\mathbf{n}_1 = \mathbf{n}_2$. Thus the correspondence is 1–1.

It is trivially a homomorphism, and hence is an isomorphism between V and $M \oplus N$.

9.10. Linear manifolds

Since a vector space V is an Abelian group, and a subspace M is a subgroup, and is invariant since V is Abelian, we can form the cosets of M in V. A typical such coset will be $\mathbf{x}+M$, consisting of the vectors of the form $\mathbf{x}+\mathbf{m}$, for a fixed $\mathbf{x}$ and variable $\mathbf{m} \in M$.

Cosets in vector spaces are often called *linear manifolds.* They have an important geometrical significance. In two dimensions the manifold $\mathbf{x}+M$, where M is a line through O, gives the line through the point representing $\mathbf{x}$ parallel to M, and the cosets of M are all such lines. In three dimensions if M is a plane through O, the manifolds of M are the parallel planes, and if M is a line the manifolds are parallel lines. Similar ideas hold for spaces of higher dimension.

Quotient spaces

For any subspace M of V we can form the quotient group V/M. This is in fact a vector space, if we define scalar product of cosets by $\lambda(\mathbf{x}+M) = \lambda\mathbf{x}+M$. Such a definition is independent of choice of $\mathbf{x}$ in the coset, since $\lambda M = \{\lambda\mathbf{m} : \mathbf{m} \in M\} = M$, as M is a subspace. It is easily seen to satisfy the axioms for a vector space, and is called the *quotient space* V/M.

9.11. Linear transformations

As with previous types of structures, we now pass on to consider homomorphisms of vector spaces. Although subspaces were not on the whole very important, the opposite is the case now: homomorphisms have possibly even greater importance in the case of vector spaces than for groups and rings. The reason is that they lead naturally to the idea and definition of a matrix, and have a close connection with geometrical transformations and the solution of simultaneous linear equations. In the following sections we develop the basic ideas, while the study of geometrical transformations will be left until the next chapter.

As usual, a specialised terminology is used for homomorphisms in the case where we are dealing with vector spaces. Homomorphisms are called *linear transformations*, or sometimes *linear mappings*: they preserve linear relationships, that is relationships of the form $\mathbf{y} = \Sigma\lambda_i\mathbf{x}_i$. The definition is straightforward.

Definition. *A mapping θ of a vector space V into a vector space W, over the same field of scalars, is a* linear transformation *if $\theta(\mathbf{x}+\mathbf{y}) = \theta\mathbf{x}+\theta\mathbf{y}$, and $\theta(\lambda\mathbf{x}) = \lambda(\theta\mathbf{x})$, for all $\mathbf{x}$, $\mathbf{y}$ in V and λ in the scalar field F.*

Note. We are departing from our usual practice of denoting the image of $\mathbf{x}$ under the homomorphism θ by $\mathbf{x}\theta$. The author would like to keep the earlier notation as being more convenient when we consider the product of homomorphisms, but the notation $\theta\mathbf{x}$ is so well-established and still in such general use that to avoid it would cause a lot of trouble when the student refers to other books: it is also in line with the notation for matrices where the image of $\mathbf{x}$ under the matrix $\mathbf{A}$ is denoted by $\mathbf{Ax}$.

A linear transformation is of course a group-homomorphism of the Abelian group of V into that of W, and all the theory of homomorphisms for groups applies.

Examples of linear transformations

The zero transformation. The mapping that sends all elements of V into the zero of W is a linear transformation, since

$$\mathbf{0}+\mathbf{0} = \mathbf{0} \quad \text{and} \quad \lambda\mathbf{0} = \mathbf{0}.$$

The identity transformation. If $W = V$ the mapping that sends every element of V into itself is a linear transformation and is denoted by I, or sometimes I_V.

Transformations of finite dimensional spaces. As an example let V be three-dimensional and W two-dimensional over any field F. Define

$$\theta(\xi_1, \xi_2, \xi_3) \quad \text{by} \quad (\alpha_1 \xi_1 + \alpha_2 \xi_2 + \alpha_3 \xi_3, \beta_1 \xi_1 + \beta_2 \xi_2 + \beta_3 \xi_3),$$

where $\alpha_1, \alpha_2, \alpha_3, \beta_1, \beta_2, \beta_3$ are in F. This is clearly a linear transformation.

Differentiation of polynomials. If $V = W$ and is the infinite dimensional vector space of polynomials, the mapping given by $\theta \mathbf{P} = \mathbf{P}'$, the derivative of $\mathbf{P}$, is a linear transformation. It is an endomorphism, and is onto V but not 1–1, the kernel being the subspace of polynomials of degree 0, plus the zero polynomial.

Other examples appear in exercises 9A.

Combination of linear transformations

If $\theta: V \to W$ and $\phi: W \to X$ are linear transformations, the *product* of θ and ϕ is a linear transformation, as can easily be proved by the usual methods. It is a linear transformation of $V \to X$ and is denoted by $\phi\theta$, so that $\phi\theta\mathbf{x} = \phi(\theta\mathbf{x})$.

If θ and ϕ are both linear transformations of $V \to W$, the *sum* of θ and ϕ is defined as that mapping which sends $\mathbf{x}$ into $\theta\mathbf{x} + \phi\mathbf{x}$, and can easily be shown to be a linear transformation (remember that V and W are *Abelian* groups).

If θ is a linear transformation of $V \to W$ and λ is any scalar, the *scalar product* $\lambda\theta$ is defined as the mapping given by $\mathbf{x} \to \lambda(\theta\mathbf{x})$: this again is a linear transformation.

Theorem 9.11.1. *Under the definitions of sum and scalar product given above, the linear transformations of V into W form a vector space over the same field F.*

Sum is clearly commutative and associative. There is a zero transformation, and the mapping $\mathbf{x} \to -\theta\mathbf{x}$ is clearly a linear transformation and gives a negative.

Hence the set of linear transformations is an Abelian group.

To prove the other axioms:

$$V1 \quad [\lambda(\theta+\phi)]\,\mathbf{x} = \lambda[(\theta+\phi)\,\mathbf{x}] = \lambda(\theta\mathbf{x})+\lambda(\phi\mathbf{x}) = (\lambda\theta+\lambda\phi)\,\mathbf{x}.$$

$$V2 \quad [(\lambda+\mu)\,\theta]\,\mathbf{x} = (\lambda+\mu)\,[\theta\mathbf{x}] = \lambda(\theta\mathbf{x})+\mu(\theta\mathbf{x}) = (\lambda\theta+\mu\theta)\,\mathbf{x}.$$

$$V3 \quad [\lambda(\mu\theta)]\,\mathbf{x} = \lambda[\mu(\theta\mathbf{x})] = (\lambda\mu)\,\theta\mathbf{x}.$$

$$V4 \quad [1.\theta]\,\mathbf{x} = 1.(\theta\mathbf{x}) = \theta\mathbf{x}.$$

Theorem 9.11.2. *The endomorphisms, or set of linear transformations of V into itself, form a ring, under the definitions of sum and product given above.*

Since sum and product are always linear transformations, the result follows from theorem 3.9.1. The ring has a unity but is not in general commutative.

9.12. Representation of a linear transformation by a matrix

The whole of this section deals with finite dimensional vector spaces. We recollect that a vector space V of finite dimension n is isomorphic to the space of n-dimensional co-ordinate vectors, with co-ordinates in the field F of scalars.

Suppose that we have a linear transformation $\theta: V \to W$, where V has dimension n and W has dimension m, both over the same scalar field F, of course. Let $\mathbf{e}_1, \mathbf{e}_2, \ldots, \mathbf{e}_n$ form a base of V. Then the images of the $\mathbf{e}_j$'s under θ are elements of W. Let $\mathbf{f}_1, \mathbf{f}_2, \ldots, \mathbf{f}_m$ be a base of W, so that the image of $\mathbf{e}_j$ is a linear combination of the $\mathbf{f}_i$'s, say

$$\theta\mathbf{e}_j = a_{1j}\mathbf{f}_1 + a_{2j}\mathbf{f}_2 + \ldots + a_{mj}\mathbf{f}_m = \sum_{i=1}^{m} a_{ij}\mathbf{f}_i.$$

Here the mn scalars a_{ij} are elements of F of course, and are determined by θ.

Since θ is a linear transformation we may see its effect on a general element of V by expressing the element in terms of the base $\mathbf{e}_1, \mathbf{e}_2, \ldots, \mathbf{e}_n$. For let

$$\mathbf{x} = (\xi_1, \xi_2, \ldots, \xi_n) = \xi_1\mathbf{e}_1 + \xi_2\mathbf{e}_2 + \ldots + \xi_n\mathbf{e}_n.$$

Then
$$\theta\mathbf{x} = \sum_{j=1}^{n} \sum_{i=1}^{m} \xi_j a_{ij} \mathbf{f}_i$$
$$= \sum_{i=1}^{m} \left\{ \sum_{j=1}^{n} a_{ij} \xi_j \right\} \mathbf{f}_i$$
$$= (\eta_1, \eta_2, \ldots, \eta_m),$$
where $\eta_i = \sum_{j=1}^{n} a_{ij} \xi_j$.

The mn scalars a_{ij} are usually written in the form

$$\mathbf{A} = \begin{pmatrix} a_{11} & a_{12} & \ldots & a_{1n} \\ a_{21} & a_{22} & \ldots & a_{2n} \\ \ldots & \ldots & \ldots & \ldots \\ a_{m1} & a_{m2} & \ldots & a_{mn} \end{pmatrix}$$

which is called a matrix, the curved brackets being put in to signify that we are thinking of the array in this connection. This matrix has m rows and n columns, and is called an *m by n matrix*. If $\theta\mathbf{x} = \mathbf{y}$, an element of W, we say that $\mathbf{y} = \mathbf{A}\mathbf{x}$. (The double notation $\theta\mathbf{x}$ and $\mathbf{A}\mathbf{x}$ for $\mathbf{y}$ causes no confusion—we use θ when we are thinking in terms of linear transformations and $\mathbf{A}$ when we are manipulating with matrices.)

We have seen that corresponding to any linear transformation of $V \rightarrow W$ there is an m by n matrix: conversely any matrix corresponds to a unique linear transformation. For suppose we have a matrix $\mathbf{A}$ of the form above, sometimes written (a_{ij}) for short. Then we may *define* a linear transformation by its effect on the base $\{\mathbf{e}_j\}$. If $\theta\mathbf{e}_j = \sum_{i=1}^{m} a_{ij}\mathbf{f}_i$ and $\theta\mathbf{x} = \sum_{j=1}^{n} \sum_{i=1}^{m} a_{ij}\mathbf{f}_i$, where $\mathbf{x} = (\xi_1, \xi_2, \ldots, \xi_n)$, θ is seen to be indeed a linear transformation since it obeys the required laws, as the $\mathbf{e}_j$'s are linearly independent (so that their images can be selected as above, not being restricted by any linear combination of them being $\mathbf{0}$) and span V (so that their images define the images of all elements of V).

Thus there is a 1–1 correspondence between all linear transformations of V into W and all m by n matrices with scalar elements. The advantage of using matrices is that these give the images of any element specifically, and in any analytical work (for example in geometrical transformations) such images will be required. Indeed, the study of a particular linear transformation in detail leads inevitably to the consideration of its matrix.

The great disadvantage of matrices lies of course in the fact that their use determines the use of particular bases of V and W: the matrix of a transformation is defined in terms of specified bases, if we choose different bases we obtain different matrices, and the 1–1 correspondence between transformations and matrices presupposes that we have the same bases throughout.

The situation is brought out in geometry. The use of matrices involves the use of particular co-ordinate systems, and this is of course very often convenient, and necessary. But sometimes it is easier to study vectors independently of a co-ordinate system, and in such cases matrices cannot be introduced.

To sum up, if θ corresponds to the matrix (a_{ij}) and if under θ $(\xi_1, \xi_2, \ldots, \xi_n)$ maps into $(\eta_1, \eta_2, \ldots, \eta_m)$ then $\eta_i = \sum_{j=1}^{n} a_{ij}\xi_j$. The base element $\mathbf{e}_j$ maps into $\sum_{i=1}^{m} a_{ij}\mathbf{f}_i$. Notice that the i's are associated with W and the j's with V: this may seem a strange way round, but it makes the formula $\eta_i = \sum_{j=1}^{n} a_{ij}\xi_j$ more convenient, and this is the formula most used in practice, where we are interested in co-ordinate representations. Note that the number of rows of $\mathbf{A}$ is the dimension of W, and the number of columns the dimension of V.

We now proceed to investigate the algebra of matrices, which is of course dependent on the rules of combination of linear transformations. Combination of matrices is defined in terms of the combination of the associated transformations, and thus we ensure that the correspondence between them is isomorphic with respect to any operations introduced.

Equal matrices

Two matrices represent the same linear transformation (with respect to the same bases of course) if and only if all elements are equal: thus $\mathbf{A} = \mathbf{B}$, where $\mathbf{A} = (a_{ij})$ and $\mathbf{B} = (b_{ij})$, if and only if $a_{ij} = b_{ij}$ for all i and j. They must of course both be of the same size m by n (i.e. have m rows and n columns).

Sum of matrices

Bearing in mind the definition of sum of transformations, we see that the sum of **A** and **B** is the matrix $(a_{ij}+b_{ij})$: we add the corresponding elements to obtain the elements of the sum. Again of course this is defined only when **A** and **B** are the same size (we sometimes say in this case that **A** and **B** are *conformable for addition*).

Multiplication of a matrix by a scalar

It is obvious from the definition for linear transformations that the product of λ and **A** is the matrix (λa_{ij}), where every element is multiplied by λ.

Product of matrices

So far the operations have been simple, but the product is more interesting in terms of matrices.

Suppose $\theta: V \to W$ and $\phi: W \to X$ are linear transformations, and let $\dim V = n$, $\dim W = m$, $\dim X = p$. Choose bases $\{\mathbf{e}_j\}$, $\{\mathbf{f}_i\}$ and $\{\mathbf{g}_k\}$ in V, W and X, and let the matrix of θ with respect to the bases $\{\mathbf{e}_j\}$ and $\{\mathbf{f}_i\}$ be $\mathbf{A} = (a_{ij})$, and the matrix of ϕ with respect to the bases $\{\mathbf{f}_i\}$ and $\{\mathbf{g}_k\}$ be $\mathbf{B} = (b_{ki}\}$. Then if

$$\mathbf{x} = (\xi_1, \xi_2, \ldots, \xi_n),\ \mathbf{y} = \theta\mathbf{x} = (\eta_1, \eta_2, \ldots, \eta_m\}$$

and $\mathbf{z} = \phi\mathbf{y} = \phi\theta\mathbf{x} = (\zeta_1, \zeta_2, \ldots, \zeta_p)$ we have $\eta_i = \sum_{j=1}^{n} a_{ij}\xi_j$

and $$\zeta_k = \sum_{i=1}^{m} b_{ki}\eta_i = \sum_{i=1}^{m}\sum_{j=1}^{n} b_{ki}a_{ij}\xi = \sum_{j=1}^{n}\left\{\sum_{i=1}^{m} b_{ki}a_{ij}\right\}\xi_j.$$

Hence if the matrix of $\phi\theta$ with respect to the bases $\{\mathbf{e}_j\}$ and $\{\mathbf{g}_k\}$ is (c_{kj}) we must have $c_{kj} = \sum_{i=1}^{m} b_{ki}a_{ij}$.

We call this matrix **BA** and we have the following rule for multiplying matrices.

To obtain the (kj)th element of the product **BA** consider the kth row of **B** and the jth column of **A**, and multiply the corresponding pairs of elements of these, adding the results.

It is clear that we can do this only when the number of columns of **B** is the same as the number of rows of **A**: if **B** is h by k and **A** is l by m we must have $k = l$: in this case we say that the matrices are *conformable for multiplication*, and **BA** is h by m.

An example will make the process clearer. Let

$$\mathbf{B} = \begin{pmatrix} 1 & 0 & 2 \\ -1 & 2 & 1 \end{pmatrix} \quad \text{and} \quad \mathbf{A} = \begin{pmatrix} 1 & 0 \\ 2 & -2 \\ 3 & 0 \end{pmatrix}.$$

Then

$$\mathbf{BA} = \begin{pmatrix} 1.1+0.2+2.3 & 1.0+0.(-2)+2.0 \\ (-1).1+2.2+1.3 & (-1).0+2.(-2)+1.0 \end{pmatrix}$$
$$= \begin{pmatrix} 7 & 0 \\ 6 & -4 \end{pmatrix}.$$

Notice that, while sum is clearly commutative, product is not in general, and that although **BA** may exist this does not guarantee that **AB** exists. In fact, both products are possible only if **A** is m by n say, and **B** is n by m and, unless $m = n$, **AB** and **BA** are not even of the same order.

Square matrices

If $m = n$, so that V and W have the same dimension, and hence are isomorphic, the matrix of $\theta: V \to W$ is called *square*, and may be considered as the matrix of an endomorphism of V into itself with the same base used in both cases. Two square matrices of the same size are always conformable for multiplication, and the set of all such form a ring, by theorem 9.11.2.

The zero matrix

The matrix of the zero transformation has every element zero.

The identity matrix (or unit matrix)

When $V = W$, the identity transformation sends every element into itself, and so sends $\mathbf{e}_i$ into $\mathbf{e}_i$. Thus its matrix, with respect to the same base taken twice, has unities along the leading diagonal from the top left to bottom right, and zeros everywhere else. In other words, it is (a_{ij}) where $a_{ii} = 1$ and $a_{ij} = 0$ if $i \neq j$. It is called the *identity* or *unit matrix*, and denoted by **I**, or sometimes $\mathbf{I}_n$ if it is n by n. Notice that

$$\lambda\mathbf{I} = \begin{pmatrix} \lambda & 0 & \dots & 0 \\ 0 & \lambda & \dots & 0 \\ . & . & \ddots & . \\ 0 & . & ..0 & \lambda \end{pmatrix}.$$

Vectors as matrices

If $\mathbf{x}$ maps into $\mathbf{y}$ under the transformation whose matrix is $\mathbf{A}$ we say that $\mathbf{y} = \mathbf{Ax}$, and it is convenient to think of $\mathbf{x}$ and $\mathbf{y}$ as being particular examples of matrices.

If $\mathbf{A}$ is m by n, we may think of $\mathbf{x}$ as being an n by 1 matrix, so that $\mathbf{A}$ and $\mathbf{x}$ are indeed conformable for multiplication and $\mathbf{Ax}$ is m by 1. Thus we write the n dimensional vector $\mathbf{x}$ in the form

$$\begin{pmatrix} \xi_1 \\ \xi_2 \\ \vdots \\ \xi_n \end{pmatrix}, \quad \text{and the } m \text{ dimensional } \mathbf{y} \text{ as} \quad \begin{pmatrix} \eta_1 \\ \eta_2 \\ \vdots \\ \eta_m \end{pmatrix}.$$

Vectors written in this way are called *column vectors*, and we may easily check that the equation $\mathbf{y} = \mathbf{Ax}$ does indeed follow the rules for matrix multiplication, since $\eta_i = \sum_{j=1}^{n} a_{ij} \xi_j$.

(A 1 by n matrix is called a *row vector*, but we do not introduce these here.)

Note that the product of a 1 by n matrix with an n by 1 matrix is a 1 by 1 matrix, which is in effect merely a scalar: it corresponds to a mapping of a one-dimensional space into itself, i.e. of the field of scalars into itself, the single element being the image of the unity. The product of an n by 1 matrix with a 1 by n matrix is an n by n matrix.

9.13. The range and null space of a linear transformation

If $\theta: V \to W$ is a linear transformation, where V and W are not necessarily finite dimensional, the image of θ is called the *range* of θ, and the kernel the *null space*. Note that here again we have specialised terminology for vector spaces.

Since the range and null space are the image and kernel of θ considered as a group homomorphism, they are subgroups of W and V respectively. As expected, they are in fact subspaces.

Theorem 9.13.1. *The range R of $\theta: V \to W$ is a vector subspace of W.*

By theorem 2.3.3 the range is a subgroup of W. If $\mathbf{y} \in R$, so that $\mathbf{y} = \theta\mathbf{x}$ for some $\mathbf{x} \in V$, $\theta(\lambda\mathbf{x}) = \lambda\mathbf{y}$ and so $\lambda\mathbf{y} \in R$ for any scalar λ. Hence R is a subspace.

Theorem 9.13.2. *The null space N of $\theta: V \to W$ is a vector subspace of V.*

By theorem 2.3.1 the null space is a subgroup. If $\mathbf{x} \in N$, so that $\theta\mathbf{x} = \mathbf{0}$, then $\theta(\lambda\mathbf{x}) = \lambda\mathbf{0} = \mathbf{0}$, and hence $\lambda\mathbf{x} \in N$. Hence N is a subspace.

The above work is much the same as for any structure. We now consider the case where V and W are finite dimensional, and link the ideas of range and null spaces with the matrix of θ.

If V and W are finite dimensional, and

$$\dim V = n, \quad \dim W = m,$$

we see by theorem 9.9.1 that the range R has finite dimension $\leqslant m$, and the null space N finite dimension $\leqslant n$. The dimension of R is called the *rank* of θ, and that of N is called the *nullity* of θ. These are sometimes denoted by $r(\theta)$ and $\nu(\theta)$ respectively.

Thus $0 \leqslant r(\theta) \leqslant m$ and $0 \leqslant \nu(\theta) \leqslant n$. If θ is the zero transformation, $r(\theta) = 0$ and $\nu(\theta) = n$, whereas if θ is an isomorphism $r(\theta) = n$ and $\nu(\theta) = 0$.

In both the above cases rank plus nullity is n, and this fundamental result is true generally. Intuitively it is fairly clear: if, so to speak, ν dimensions map into the zero we are left with $(n-\nu)$ dimensions which do not become zero, and we expect that the range will be of this dimension. The formal proof is not difficult, but like all such proofs needs a little care and the notation makes it look more complicated than it really is.

Theorem 9.13.3. *If $\theta: V \to W$ is a linear transformation between finite dimensional vector spaces with* $\dim V = n$, *then*

$$r(\theta) + \nu(\theta) = n.$$

The null space N is a subspace of V: choose a base of it, which will have ν elements, which we will denote by

$$\mathbf{e}_{n-\nu+1}, \mathbf{e}_{n-\nu+2}, \ldots, \mathbf{e}_n.$$

By theorem 9.9.2 this is contained in a base of V: say

$$\mathbf{e}_1, \mathbf{e}_2, \ldots, \mathbf{e}_n.$$

We now prove that $\theta\mathbf{e}_1, \theta\mathbf{e}_2, \ldots, \theta\mathbf{e}_{n-\nu}$ form a base of R, so that $r(\theta) = n - \nu(\theta)$ and the result is proved. We show first that these vectors span R. Clearly R is spanned by $\theta\mathbf{e}_1, \theta\mathbf{e}_2, \ldots, \theta\mathbf{e}_n$, and $\theta\mathbf{e}_{n-\nu+1}, \ldots, \theta\mathbf{e}_n$ are zero.

Now to show that $\theta\mathbf{e}_1, \theta\mathbf{e}_2, \ldots, \theta\mathbf{e}_{n-\nu}$ are linearly independent, suppose that

$$\lambda_1\theta\mathbf{e}_1+\lambda_2\theta\mathbf{e}_2+\ldots+\lambda_{n-\nu}\theta\mathbf{e}_{n-\nu} = \mathbf{0}.$$

Then $\theta(\lambda_1\mathbf{e}_1+\lambda_2\mathbf{e}_2+\ldots+\lambda_{n-\nu}\mathbf{e}_{n-\nu}) = \mathbf{0}$ and so $\sum_{i=1}^{n-\nu}\lambda_i\mathbf{e}_i$ is in the null space and hence is of the form $\sum_{j=n-\nu+1}^{n}\mu_j\mathbf{e}_j$. It follows that

$$\sum_{i=1}^{n-\nu}\lambda_i\mathbf{e}_i-\sum_{j=n-\nu+1}^{n}\mu_j\mathbf{e}_j = \mathbf{0}$$

and, since the $\mathbf{e}_i$'s are linearly independent, being a base of V, this means that all the λ_i's are zero, and the result follows.

The rank of a matrix

If A is the matrix of θ with respect to some pair of bases, the rank of A is defined to be the rank of θ. If

$$\mathbf{A} = (a_{ij}), \quad \theta\mathbf{e}_j = \sum_{i=1}^{m} a_{ij}\mathbf{f}_i$$

and is represented by the jth column of $\mathbf{A}$. Thus the rank of $\mathbf{A}$ is the maximum number of linearly independent columns of $\mathbf{A}$ (where these are treated as column vectors).

It turns out that the rank is also given by the maximum number of linearly independent *rows* of $\mathbf{A}$, treated again as vectors, but we do not prove this here. It is also true that the rank is equal to the order of the largest non-zero determinant that can be formed by selecting a number of rows and the same number of columns from $\mathbf{A}$, but again we do not prove this assertion, although this is the way that rank is usually calculated in practice. (For example if

$$\mathbf{A} = \begin{pmatrix} 1 & 2 & 0 & 1 \\ 3 & 5 & -1 & 2 \\ 4 & 7 & -1 & 3 \end{pmatrix},$$

all the third-order determinants, viz.

$$\begin{vmatrix} 1 & 2 & 0 \\ 3 & 5 & -1 \\ 4 & 7 & -1 \end{vmatrix}, \quad \begin{vmatrix} 1 & 2 & 1 \\ 3 & 5 & 2 \\ 4 & 7 & 3 \end{vmatrix}, \quad \begin{vmatrix} 1 & 0 & 1 \\ 3 & -1 & 2 \\ 4 & -1 & 3 \end{vmatrix}, \quad \begin{vmatrix} 2 & 0 & 1 \\ 5 & -1 & 2 \\ 7 & -1 & 3 \end{vmatrix}$$

are zero, but the second-order determinant $\begin{vmatrix} 1 & 2 \\ 3 & 5 \end{vmatrix}$ is not, and so the rank of **A** is 2. A base for the range is given by the first two columns of **A**.)

With respect to different bases a given transformation will have different matrices, and all these will of course have the same rank. Also we must notice that the same matrix will correspond to different transformations if we take various bases, but all these transformations must be of the same rank, namely that of the matrix.

9.14. Inverse transformations

For a linear transformation $\theta: V \to W$ to have an inverse it must be onto W and also 1–1; that is, the range must be W and the null space the zero subspace $\{\mathbf{0}\}$. In this case θ is an isomorphism and V and W are isomorphic spaces.

In the finite dimensional case we must have $\nu(\theta) = 0$ and hence by theorem 9.13.3 $r(\theta) = n$. Thus $m = n$ and V, W have the same dimension.

Inverse of a matrix

If **A** is to have an inverse it must be the matrix of an isomorphism θ and hence must be square (this may be seen also from the fact that if an inverse $\mathbf{A}^{-1}$ exists, **A** and $\mathbf{A}^{-1}$ must be conformable for multiplication with either taken first).

A square matrix then has an inverse if and only if its rank is n (where the matrix is n by n). If this is the case,

$$\mathbf{A}\mathbf{A}^{-1} = \mathbf{A}^{-1}\mathbf{A} = \mathbf{I}_n,$$

since $\mathbf{I}_n$ is the matrix of the identity transformation. It follows at once by the general theory of mappings that if **A** and **B** both have inverses and are of the same order then

$$(\mathbf{AB})^{-1} = \mathbf{B}^{-1}\mathbf{A}^{-1},$$

so that **AB** does have an inverse. It is also immediate that $(\lambda\mathbf{A})^{-1} = (1/\lambda)\mathbf{A}^{-1}$, since $(\lambda\mathbf{A})((1/\lambda)\mathbf{A}^{-1}) = ((1/\lambda)\mathbf{A}^{-1})(\lambda\mathbf{A}) = \mathbf{I}_n$.

A square n by n matrix of rank n is called a *non-singular* matrix. The set of n by n non-singular matrices forms a non-commutative group under product, as may be easily seen.

Calculation of the inverse of a non-singular matrix

In order to perform the important practical problem of calculating the inverse of a matrix, we need to use some standard results on determinants, which we assume the reader already possesses: if not they may be found in any work on this subject.

Let $\mathbf{A} = (a_{ij})$ be a non-singular n by n matrix, and let A_{ij} be the cofactor of a_{ij} in the determinant $|a_{ij}|$. [The cofactor of a_{ij} is the $(n-1)$th order determinant formed by omitting the ith row and jth column of $\mathbf{A}$, multiplied by $(-1)^{i+j}$.] Then we know that $\sum_{i=1}^{n} a_{ki}A_{ki} = |\mathbf{A}|$ and $\sum_{i=1}^{n} a_{ki}A_{ji} = 0$ if $k \neq j$. Now consider the matrix $\mathbf{X} = (\alpha_{ij})$ where α_{ij} is defined to be A_{ji} (note the interchange of suffices—the rows of the cofactors correspond to the columns of the matrix (α_{ij})). Then the (k,j)th element of $\mathbf{XA}$ is given by $\sum_{i=1}^{n} a_{ki}A_{ji}$, and we see from the above that this is zero if $k \neq j$ and $|\mathbf{A}|$ if $k = j$. Hence $\mathbf{XA} = |\mathbf{A}|\mathbf{I}_n$. Similarly $\sum_{i=1}^{n} a_{ik}A_{ik} = |\mathbf{A}|$ and $\sum_{i=1}^{n} a_{ik}\mathbf{A}_{ij} = 0$ if $k \neq j$, and so $\mathbf{AX} = |\mathbf{A}|\mathbf{I}_n$.

Thus the inverse of $\mathbf{A}$ (unique by general theory) is given by $\mathbf{X}/|\mathbf{A}|$. $\mathbf{X}$ is sometimes called the *adjugate* matrix of $\mathbf{A}$, and is formed, we repeat, by forming the matrix of cofactors of the elements of $\mathbf{A}$ in the determinant $|\mathbf{A}|$, and then interchanging rows and columns.

Example. If

$$\mathbf{A} = \begin{pmatrix} 1 & 0 & 1 \\ 2 & 1 & 0 \\ 3 & 1 & 4 \end{pmatrix}, \text{ the matrix of cofactors is } \begin{pmatrix} 4 & -8 & -1 \\ -4 & 1 & -1 \\ -1 & 2 & 1 \end{pmatrix}$$

and hence the adjugate is $\begin{pmatrix} 4 & -4 & -1 \\ -8 & 1 & 2 \\ -1 & -1 & 1 \end{pmatrix}$. But $|\mathbf{A}| = 3$ and

so the inverse is
$$\begin{pmatrix} \frac{4}{3} & \frac{-4}{3} & \frac{-1}{3} \\[4pt] \frac{-8}{3} & \frac{1}{3} & \frac{2}{3} \\[4pt] \frac{-1}{3} & \frac{-1}{3} & \frac{1}{3} \end{pmatrix}.$$

The above process breaks down if $|\mathbf{A}| = 0$. But from our discussion on rank we see that $\mathbf{A}$ is non-singular if and only if it possesses an n by n determinant which is not zero, and this can only be $|\mathbf{A}|$ itself.

9.15. Solution of simultaneous equations

An important practical application of the theory of vector spaces and matrices is in the solving of linear simultaneous equations.

The general problem is to find the n unknowns $x_1, x_2, \ldots, x_n$ which satisfy the m linear equations

$$\begin{aligned} a_{11}x_1 + a_{12}x_2 + \ldots + a_{1n}x_n &= b_1, \\ a_{21}x_1 + a_{22}x_2 + \ldots + a_{2n}x_n &= b_2, \\ &\ldots\ldots \\ a_{m1}x_1 + a_{m2}x_2 + \ldots + a_{mn}x_n &= b_m, \end{aligned}$$

where the a_{ij}'s and the b_i's are elements of some field (of scalars) F.

In matrix form these equations become merely $\mathbf{Ax} = \mathbf{b}$, where $\mathbf{A} = (a_{ij})$, an m by n matrix, $\mathbf{x}$ is the n-dimensional column vector

$$\begin{pmatrix} x_1 \\ x_2 \\ \vdots \\ x_n \end{pmatrix} \quad \text{and } \mathbf{b} \text{ is the } m\text{-dimensional column vector} \quad \begin{pmatrix} b_1 \\ b_2 \\ \vdots \\ b_m \end{pmatrix}.$$

With respect to some choice of bases, which we imagine given, $\mathbf{A}$ represents a linear transformation of V into W, where V has dimension n and W has dimension m. For simplicity we will also use $\mathbf{A}$ to denote this transformation, which should cause no confusion in this section. $\mathbf{x}$ is an element of V and $\mathbf{b}$ an element of W, and the problem is to find which elements of V map into a given $\mathbf{b}$ in W. Three cases may arise:

(*a*) $\mathbf{b}$ is not in the range of $\mathbf{A}$: in which case there is no solution.

(*b*) $\mathbf{b}$ is in the range of $\mathbf{A}$ and the nullity of $\mathbf{A}$ is zero, so that there is a unique solution $\mathbf{x}$.

(*c*) **b** is in the range of **A** and the nullity of **A** is $\geqslant 1$, so that there are many solutions for **x**.

We will deal briefly with these cases.

Case (*b*)

This is the simplest case, and the most important in practice.

Since $\nu(\mathbf{A}) = 0$, we must have $r(\mathbf{A}) = n$ and hence $m \geqslant n$: there must be at least as many equations as unknowns.

The usual case will be when $m = n$, so that **A** is square and, since $r(\mathbf{A}) = n$, is also non-singular with $|\mathbf{A}| \neq 0$. The unique solution is then given by $\mathbf{x} = \mathbf{A}^{-1}\mathbf{b}$, and may be calculated, although in simple cases it is often best to solve by elementary methods rather than by calculation of $\mathbf{A}^{-1}$.

Example

$$\begin{aligned} x_1+2x_2+x_3 &= 1, \\ 3x_1-x_2+2x_3 &= 2, \\ 2x_1+x_2-x_3 &= 0. \end{aligned}$$

Here $|\mathbf{A}| = \begin{vmatrix} 1 & 2 & 1 \\ 3 & -1 & 2 \\ 2 & 1 & -1 \end{vmatrix} = 18$ and hence there is a unique solution.

The adjugate matrix of **A** is $\begin{pmatrix} -1 & 3 & 5 \\ 7 & -3 & 1 \\ 5 & 3 & -7 \end{pmatrix}$ and hence

$$\begin{pmatrix} x_1 \\ x_2 \\ x_3 \end{pmatrix} = \frac{1}{18}\begin{pmatrix} -1 & 3 & 5 \\ 7 & -3 & 1 \\ 5 & 3 & -7 \end{pmatrix}\begin{pmatrix} 1 \\ 2 \\ 0 \end{pmatrix}$$

so that the solution is $x_1 = \frac{5}{18}$, $x_2 = \frac{1}{18}$, $x_3 = \frac{11}{18}$.

If $m > n$, with $r(\mathbf{A}) = n$, for **b** to be in the range of **A** the vector **b** must be in the subspace of W spanned by the columns of **A**, since these columns are the images of the base of V and span the range. Hence the rank of the matrix $\mathbf{A}|\mathbf{b}$ formed by adding the column **b** to **A**, thus:

$$\begin{pmatrix} a_{11} & a_{12} & \dots & a_{1n} & b_1 \\ \vdots & \vdots & & \vdots & \vdots \\ a_{m1} & a_{m2} & \dots & a_{mn} & b_m \end{pmatrix},$$

must also be n. If this is the case we may choose n linearly independent rows of $\mathbf{A}$ and thus form a new square matrix $\tilde{\mathbf{A}}$, which will be non-singular, and the solution will be given by $(\tilde{\mathbf{A}})^{-1}\tilde{\mathbf{b}}$, where $\tilde{\mathbf{b}}$ corresponds to $\tilde{\mathbf{A}}$. An example should make this clear.

Example.

$$\begin{aligned} 2x_1+3x_2 &= 5, \\ x_1-2x_2 &= -1, \\ 3x_1+x_2 &= 4. \end{aligned}$$

Here $r(\mathbf{A}) = 2$, since there is a non-zero 2×2 determinant in $\mathbf{A}$, viz.

$$\begin{vmatrix} 2 & 3 \\ 1 & -2 \end{vmatrix},$$

and $r(\mathbf{A}|\mathbf{b})$ also is 2, since

$$|\mathbf{A}|\mathbf{b}| = \begin{vmatrix} 2 & 3 & 5 \\ 1 & -2 & -1 \\ 3 & 1 & 4 \end{vmatrix} = 0.$$

Thus there is a unique solution (if $r(\mathbf{A}|\mathbf{b})$ did equal 3 there would be no solution, since $\mathbf{b}$ would not be in the range of $\mathbf{A}$—given three equations in only two unknowns this latter case would be the expected one; in this example we are fortunate in that $\mathbf{b}$ *is* in the range). The first two rows of $\mathbf{A}$ are linearly independent, and we therefore use the first two equations, obtaining $\tilde{\mathbf{A}}\mathbf{x} = \tilde{\mathbf{b}}$, where

$$\tilde{\mathbf{A}} = \begin{pmatrix} 2 & 3 \\ 1 & -2 \end{pmatrix},$$

$$\tilde{\mathbf{b}} = \begin{pmatrix} 5 \\ -1 \end{pmatrix}.$$

$$(\tilde{\mathbf{A}})^{-1} = -\frac{1}{7}\begin{pmatrix} -2 & -3 \\ -1 & 2 \end{pmatrix},$$

and so $$\mathbf{x} = -\frac{1}{7}\begin{pmatrix} -2 & -3 \\ -1 & 2 \end{pmatrix}\begin{pmatrix} 5 \\ -1 \end{pmatrix} = \begin{pmatrix} -\frac{1}{7} \\ -\frac{1}{7} \end{pmatrix}.$$

Case (*a*)

If the matrix $\mathbf{A}|\mathbf{b}$ has rank greater than the rank of $\mathbf{A}$ (1 greater in fact), so that $\mathbf{b}$ is not in the subspace of W spanned by the columns of **A**, which subspace is the range of **A**, then the equations have no solution and are inconsistent.

If $m > n$ we expect this to happen: other cases, as for instance that in the work on case (*b*) above, are exceptional.

Example.

$$2x_1+3x_2 = 5,$$
$$x_1-2x_2 = -1,$$
$$3x_1+x_2 = 3.$$

Here, as before, $r(\mathbf{A}) = 2$, but $r(\mathbf{A}|\mathbf{b}) = 3$ since $|\mathbf{A}|\mathbf{b}| \neq 0$.

If $m < n$ this case can occur only if $r(\mathbf{A}) <$ its maximum possible value m, and not always then, of course. Such examples are fairly uncommon: we expect in this case that case (*c*) will apply and that there will be more than one solution.

Example.

$$x_1+2x_2-3x_3 = 4,$$
$$2x_1+4x_2-6x_3 = 9.$$

Here $r(\mathbf{A}) = 1$ and $r(\mathbf{A}|\mathbf{b}) = 2$, so that there are no solutions. This can be seen at once in this example, since the left-hand side of the second equation is twice that of the first, and doubling the first and subtracting the second gives the contradiction $0 = -1$.

If $m = n$ there are no solutions if $r(\mathbf{A}) < m$, i.e. if $|\mathbf{A}| = 0$, and if $r(\mathbf{A}|\mathbf{b}) > r(\mathbf{A})$.

Example.

$$x_1+2x_2+3x_3 = 2,$$
$$-x_1+x_2+2x_3 = 1,$$
$$x_1+5x_2+8x_3 = 6.$$

Here

$$|\mathbf{A}| = \begin{vmatrix} 1 & 2 & 3 \\ -1 & 1 & 2 \\ 1 & 5 & 8 \end{vmatrix} = 0, \text{ but since } \begin{vmatrix} 2 & 3 & 2 \\ 1 & 2 & 1 \\ 5 & 8 & 6 \end{vmatrix} \neq 0,$$

$$r(\mathbf{A}|\mathbf{b}) = \operatorname{rank}\begin{pmatrix} 1 & 2 & 3 & 2 \\ -1 & 1 & 2 & 1 \\ 1 & 5 & 8 & 6 \end{pmatrix} = 3,$$

and the equations have no solution.

Case (*c*)

If the null space N of $\mathbf{A}$ has dimension $\geqslant 1$ (i.e. is not the trivial subspace $\{\mathbf{0}\}$), and if $\mathbf{b}$ is in the range of $\mathbf{A}$, there will be more than one solution of the set of equations $\mathbf{Ax} = \mathbf{b}$. In fact, by theorem 2.3.2, the set of solutions will form a coset (or linear manifold) $\mathbf{x}+N$, where $\mathbf{x}$ is any one solution.

If $m > n$ this case is very exceptional: for it to occur we must have $r(\mathbf{A}) < n$, and $r(\mathbf{A}|\mathbf{b}) = r(\mathbf{A})$.

Example.

$$2x_1+3x_2 = 5,$$

$$4x_1+6x_2 = 10,$$

$$6x_1+9x_2 = 15.$$

Here all second-order determinants in both $\mathbf{A}$ and $\mathbf{A}|\mathbf{b}$ are zero and so both matrices have rank 1. $\nu(\mathbf{A}) = 1$, and the three equations are not independent, all being in fact the same. We thus find the general solution of the first (in the general case we take r independent equations from the set, where r is rank of $\mathbf{A}$) and obtain $x_2 = -\frac{2}{3}x_1+\frac{5}{3}$ as the general solution. This may of course be given in many forms: a form involving the coset is $x_1 = 1+\frac{1}{2}\lambda$, $x_2 = 1-\frac{1}{3}\lambda$, where the null space N is the set $\{(\frac{1}{2}\lambda, -\frac{1}{3}\lambda)\}$.

If $m < n$, so that there are more unknowns than equations, case (*c*) is the usual one. Here $r(\mathbf{A}) \leqslant m < n$, and so $\nu(\mathbf{A})$ must be > 0: if $\mathbf{b}$ is in the range then there is more than one solution: we choose r independent equations to find them.

Example.

$$x_1+2x_2+3x_3 = 1,$$

$$2x_1+x_2-x_3 = 1.$$

Here $r(\mathbf{A}) = r(\mathbf{A}|\mathbf{b}) = 2$ and $\nu(\mathbf{A}) = 1$. The general solution may be given in many forms: for example, x_3 may be chosen arbitrarily and then $x_1 = \frac{5}{3}x_3+\frac{1}{3}$, $x_2 = -\frac{7}{3}x_3+\frac{1}{3}$, as may be found by simply solving the two equations by considering them as simultaneous equations in the two unknowns x_1 and x_2.

If $m = n$ and $|\mathbf{A}| = 0$, so that $r(\mathbf{A}) < n$, then if $\mathbf{b}$ is in the range of $\mathbf{A}$ (i.e. if $r(\mathbf{A}|\mathbf{b}) = r(\mathbf{A})$), there will be many solutions, found as in the previous example.

Example.

$$x_1+2x_2+3x_3 = 1,$$
$$2x_1+x_2-x_3 = 1,$$
$$3x_1+3x_2+2x_3 = 2.$$

Here $|\mathbf{A}| = 0$ and it is easily seen that $r(\mathbf{A}) = r(\mathbf{A}|\mathbf{b}) = 2$. The first two equations are independent and we use these to find the general solution, as in the last example.

To sum up, if we are presented with a set of simultaneous linear equations and wish to find the nature of their solution, we find $r(\mathbf{A})$ and $r(\mathbf{A}|\mathbf{b})$, by finding the order of the largest non-zero determinant in each. If $r(\mathbf{A}|\mathbf{b}) > r(\mathbf{A})$ there is no solution. If the ranks are equal ($r(\mathbf{A}|\mathbf{b})$ cannot be less than $r(\mathbf{A})$) then a solution exists, unique if $r(\mathbf{A}) = n$ and with ν independent parameters if $r(\mathbf{A}) = n-\nu$.

The most important case is when $m = n$. Here there is a unique solution if and only if $|\mathbf{A}| \neq 0$.

If $\mathbf{b} = \mathbf{0}$ we have a set of homogeneous linear equations. In this case $\mathbf{b}$ must be in the range, since this is a subspace and contains the zero. If $r(\mathbf{A}) = n$ the only solution is $\mathbf{x} = \mathbf{0}$: if $r(A) < n$ there will also be non-zero solutions, the set of which will form the null space of $\mathbf{A}$.

9.16. Fields as vector spaces

Suppose a field F^* has a subfield F. Then we may consider F^* as a vector space over F by defining sum as the ordinary field sum and scalar product as the field product of elements of F^* with those of F. We obviously obtain a vector space.

This idea is sometimes useful, particularly in connection with linear dependence and bases: we may think of the dimension of F^* over F for example. But it does not use all the field properties of F^*: any two elements of F^* may be multiplied, which is not the case with a general vector space.

Examples.

If F is the trivial subfield F^*, F^* forms a vector space of dimension 1 over itself.

The real field is a vector space over the field of rationals: it has infinite dimension, as may be seen by considering the elements $\sqrt{2}$, $\sqrt{3}$, $\sqrt{5}$, $\sqrt{7}$, ..., $\sqrt{p}$, ..., where p is the general prime. It is fairly easily seen that none of these is a linear combination of the preceding ones (with the rationals as coefficients), and thus there are more than n linearly independent elements for any finite n: thus there cannot be a finite base.

The complex field is a vector space of dimension 2 over the real field.

The set $\{a+b\sqrt{2}\}$, with a and b rational, is a vector space of dimension 2 over the rationals, a base being given by 1 and $\sqrt{2}$, for example. The set $\{a+b\sqrt[3]{3}+c\sqrt[3]{9}\}$ is a vector space of dimension 3 over the rationals.

Transcendental extensions as a vector space

A simple transcendental extension of F is a vector space over F, and has infinite dimension, since the elements

$$1, x, x^2, x^3, \ldots, x^n, \ldots$$

are such that any finite set of them is linearly independent, and hence there cannot be a base with a finite number of elements (by theorem 9.6.3).

Algebraic extensions as a vector space

We recollect that a simple algebraic extension of F is formed by taking an irreducible polynomial $f(x)$ with leading coefficient 1, and forming the quotient field $F[x]/[f(x)]$. If the degree of the extension is n, so that $f(x)$ has degree n, the n elements $1, x, x^2, \ldots, x^{n-1}$ are linearly independent over F, and span the extension field, since any polynomial in $F[x]$ of degree $\geqslant n$ is reduced to one of degree $\leqslant n-1$ by working modulo $f(x)$. Hence $1, x, x^2, \ldots, x^{n-1}$ form a base and the extension thus has dimension n.

Galois fields as vector spaces

$GF[p^n]$ has as subfield the field of residues modulo p, and clearly its dimension over this subfield is n. We are now in fact in a position to prove theorem 4.8.2, whereas in chapter 4 this was merely outlined. For, referring to that theorem, F has a subfield J_p and F is finite: hence it has finite dimension n say over J_p and hence contains p^n elements.

9.17. Vector spaces over rings

If the set of scalars is a commutative ring R with unity, we may still define a vector space V over R in precisely the same ways as we did over a field F. The elementary results of §9.3 are still valid except for theorem 9.3.4 and its corollaries, but this is of course a significant defect of such spaces.

We may still define linear dependence, but theorem 9.4.1, the proof of which involved dividing by λ^{k+1}, is not valid. Bases may similarly be defined, but we do not have a unique expression for $\mathbf{x}$ in terms of a base, since this property depended on theorem 9.3.4. Neither do we have the concept of dimension, since the fundamental theorem 9.6.1 uses theorem 9.4.1, which is not valid.

Worked exercises

1. Prove that $\mathbf{x}_1 = (3, 1, 2)$, $\mathbf{x}_2 = (0, -1, 5)$ and $\mathbf{x}_3 = (1, 2, -2)$ are linearly independent.

Suppose $\lambda_1\mathbf{x}_1+\lambda_2\mathbf{x}_2+\lambda_3\mathbf{x}_3 = \mathbf{0}$. Then we have the three scalar equations

$$\begin{aligned} 3\lambda_1 \qquad + \lambda_3 &= 0, \\ \lambda_1 - \lambda_2 + 2\lambda_3 &= 0, \\ 2\lambda_1 + 5\lambda_2 - 2\lambda_3 &= 0 \end{aligned}$$

and we wish to show that this implies $\lambda_1, \lambda_2, \lambda_3$ are all zero, i.e. that the system of equations has no non-zero solution.

The quickest way of doing this is to show that the determinant of coefficients is non-zero. In this case the determinant is

$$\begin{vmatrix} 3 & 0 & 1 \\ 1 & -1 & 2 \\ 2 & 5 & -2 \end{vmatrix} = -17 \neq 0,$$

and so the vectors are linearly independent.

If we wish to avoid using the general theory of simultaneous equations we may proceed directly as follows.

From the three equations above, eliminating λ_2, we obtain

$$7\lambda_1 + 8\lambda_3 = 0$$

and, eliminating λ_3 from this and the first equation, $17\lambda_1 = 0$. Hence $\lambda_1 = 0$, so $\lambda_3 = 0$ from the first equation, and $\lambda_2 = 0$ from the second.

2. Prove that there exists an infinite set $\mathbf{x}_1, \mathbf{x}_2, \ldots$, such that $\mathbf{x}_1, \ldots, \mathbf{x}_k$ are linearly independent for all k, if and only if V is infinite dimensional.

Suppose first that such a set exists. Then if V were finite dimensional, of dimension n, by theorem 9.6.3 $\mathbf{x}_1, \mathbf{x}_2, \ldots, \mathbf{x}_{n+1}$ cannot be linearly independent, giving a contradiction. Hence V is infinite dimensional.

Conversely suppose V is infinite dimensional, so that no finite set of linearly independent vectors can span V. We proceed to construct the required set of vectors by induction.

Suppose we have found the linearly independent set $\mathbf{x}_1, \mathbf{x}_2, \ldots, \mathbf{x}_k$. These cannot span V and hence we can find $\mathbf{x}_{k+1}$ not in the subspace spanned by $\mathbf{x}_1, \mathbf{x}_2, \ldots, \mathbf{x}_k$. Then if $\mathbf{x}_1, \mathbf{x}_2, \ldots, \mathbf{x}_{k+1}$ were linearly dependent, by theorem 9.4.1 some $\mathbf{x}_i$ is a linear combination of its predecessors. But since $\mathbf{x}_1, \mathbf{x}_2, \ldots, \mathbf{x}_k$ are linearly independent this cannot be true for $\mathbf{x}_1, \mathbf{x}_2, \ldots$, or $\mathbf{x}_k$; and $\mathbf{x}_{k+1}$ cannot be a linear combination of $\mathbf{x}_1, \mathbf{x}_2, \ldots, \mathbf{x}_k$ since it is not in the subspace spanned by them. Hence $\mathbf{x}_1, \mathbf{x}_2, \ldots, \mathbf{x}_{k+1}$ are linearly independent. We start the process by taking an arbitrary non-zero $\mathbf{x}_1$, and can continue indefinitely by induction.

3. If θ has rank r show that there exist bases $\mathbf{e}_1, \ldots, \mathbf{e}_n$ of V and $\mathbf{f}_1, \ldots, \mathbf{f}_m$ of W with respect to which the matrix of θ is

$$\begin{pmatrix} 1\ldots0 & 0\ldots0 \\ \vdots\ \ddots\ \vdots & \vdots\quad\vdots \\ 0\ldots1 & 0\ldots0 \\ 0\ldots0 & 0\ldots0 \\ \vdots\quad\vdots & \vdots\quad\vdots \\ 0\ldots0 & 0\ldots0 \end{pmatrix} \begin{matrix} r \text{ rows} \\ \\ (m-r) \text{ rows} \end{matrix}$$

$$r \text{ columns} \quad (n-r) \text{ columns}$$

where the first r elements of the leading diagonal are 1 and all other elements are zero.

As in the proof of theorem 9.13.3 choose a base $\mathbf{e}_{r+1}, \mathbf{e}_{r+2}, \ldots, \mathbf{e}_n$ of the null space of θ and extend it to form a base $\mathbf{e}_1, \mathbf{e}_2, \ldots, \mathbf{e}_n$ of V. (We assume the result of theorem 9.13.3 that $r = n - \nu(\theta)$.) Then we proved in theorem 9.13.3 that $\theta\mathbf{e}_1, \theta\mathbf{e}_2, \ldots, \theta\mathbf{e}_r$ is a base of the range of θ. Let

$$\theta\mathbf{e}_i = \mathbf{f}_i, \quad i = 1, \ldots, r.$$

Since the range is a subspace of W, the set $\mathbf{f}_1, \ldots, \mathbf{f}_r$ can be extended, by theorem 9.9.2, to give a base $\mathbf{f}_1, \mathbf{f}_2, \ldots, \mathbf{f}_m$ of W.

Then $\theta\mathbf{e}_i = \mathbf{f}_i$, $i = 1 \ldots r$ and $\theta\mathbf{e}_i = \mathbf{0}$, $i = (r+1), \ldots, n$, and so the matrix of θ with respect to these bases is as required.

4. If M and N are subspaces of a finite dimensional vector space V, prove that

$$\dim M + \dim N = \dim (M \cap N) + \dim (M+N).$$

Let $\dim M = \mu$, $\dim N = \nu$, $\dim (M \cap N) = \rho$, all finite by theorem 9.9.1. Take a base $\mathbf{z}_1, \ldots, \mathbf{z}_\rho$ of $M \cap N$. Since $M \cap N$ is a subspace of both M

and N, this can be extended by theorem 9.9.2 to a base

$$\mathbf{z}_1, \ldots, \mathbf{z}_\rho, \quad \mathbf{x}_1, \ldots, \mathbf{x}_{\mu-\rho} \quad \text{of} \quad M$$

and a base

$$\mathbf{z}_1, \ldots, \mathbf{z}_\rho, \quad \mathbf{y}_1, \ldots, \mathbf{y}_{\nu-\rho} \quad \text{of} \quad N.$$

We now prove that

$$\mathbf{z}_1, \ldots, \mathbf{z}_\rho, \quad \mathbf{x}_1, \ldots, \mathbf{x}_{\mu-\rho}, \quad \mathbf{y}_1, \ldots, \mathbf{y}_{\nu-\rho}$$

is a base of $M+N$.

Any element of M is a linear combination of $\mathbf{z}_1, \ldots, \mathbf{z}_\rho, \mathbf{x}_1, \ldots, \mathbf{x}_{\mu-\rho}$ and any element of N is a linear combination of $\mathbf{z}_1, \ldots, \mathbf{z}_\rho, \mathbf{y}_1, \ldots, \mathbf{y}_{\nu-\rho}$. Thus any element of $M+N$ is a linear combination of the $\mathbf{z}_i$'s, $\mathbf{x}_j$'s and $\mathbf{y}_k$'s, and so this set spans $M+N$.

Now suppose $\Sigma\alpha_i\mathbf{z}_i+\Sigma\beta_j\mathbf{x}_j+\Sigma\gamma_k\mathbf{y}_k = \mathbf{0}$. Then

$$\Sigma\gamma_k\mathbf{y}_k = -\Sigma\alpha_i\mathbf{z}_i-\Sigma\beta_j\mathbf{x}_j \in M,$$

and also is in N, and so in $M \cap N$. Thus $\Sigma\gamma_k\mathbf{y}_k = \Sigma\delta_i\mathbf{z}_i$, since $\mathbf{z}_1, \ldots, \mathbf{z}_\rho$ are a base of $M \cap N$. But the $\mathbf{y}_k$'s and $\mathbf{z}_i$'s together form a linearly independent set, being a base of N, and so all γ_k's are zero. Similarly all β_j's are zero, and so $\Sigma\alpha_i\mathbf{z}_i = \mathbf{0}$ and all α_i's are zero because $\mathbf{z}_1, \ldots, \mathbf{z}_\rho$ are linearly independent.

Hence $\mathbf{z}_1, \ldots, \mathbf{z}_\rho, \mathbf{x}_1, \ldots, \mathbf{x}_{\mu-\rho}, \mathbf{y}_1, \ldots, \mathbf{y}_{\nu-\rho}$ are linearly independent and hence form a base of $M+N$. Thus

$$\dim(M+N) = \rho+\mu-\rho+\nu-\rho = \mu+\nu-\rho.$$

5. Find the conditions which must be satisfied by u, v, w in order that the equations

$$x+y+z = u,$$
$$(b+c)\,x+(c+a)\,y+(a+b)\,z = v,$$
$$bcx+cay+abz = w,$$

may be consistent (i) when a, b, c are all different, (ii) when two of a, b, c are equal and different from the third, (iii) when $a = b = c$.

If $u = 1$, $v = a$, $w = 0$, and $a \neq 0$, solve the equations completely in those cases in which they are consistent. (Mathematical Tripos.)

Write the equations in the form

$$\begin{pmatrix} 1 & 1 & 1 \\ b+c & c+a & a+b \\ bc & ca & ab \end{pmatrix} \begin{pmatrix} x \\ y \\ z \end{pmatrix} = \begin{pmatrix} u \\ v \\ w \end{pmatrix}$$

or $\mathbf{AX} = \mathbf{U}$, say.

(i) $$|\mathbf{A}| = -\Sigma bc(c-b), \quad \text{expanding by the third row,}$$
$$= -(b-c)(c-a)(a-b).$$

Hence if a, b, c are all different $|\mathbf{A}| \neq 0$ and so $\mathbf{A}$ has rank 3 and the equations are consistent and have a unique solution, whatever the values of u, v and w. To solve the equations we could find the inverse of $\mathbf{A}$ and use the equation $\mathbf{X} = \mathbf{A}^{-1}\mathbf{U}$, but it is simpler to proceed by elementary methods.

When $u = 1$, $v = a$ and $w = 0$ the equations become

$$x+y+z = 1 \qquad (\alpha),$$
$$(b+c)\,x+(c+a)\,y+(a+b)\,z = a \qquad (\beta),$$
$$bcx+cay+abz = 0 \qquad (\gamma).$$

$(\beta)-a.(\alpha)$ gives $(b+c-a)\,x+cy+bz = 0;$
$(\gamma)\div a$ $(a \neq 0)$ gives $(bc/a)\,x+cy+bz = 0.$

Hence
$$[(b+c-a)-(bc/a)]\,x = 0,$$
or
$$-\frac{(a-b)\,(a-c)}{a}\,x = 0.$$

Thus, since $a-b \neq 0$ and $a-c \neq 0$, $x = 0$.

(α) gives $y+z = 1$ and (γ) gives $cy+bz = 0$ $(a \neq 0)$.

Hence $(b-c)\,y = b$, $y = b/(b-c)$ and $z = 1-y = -c/(b-c)$.

(ii) Suppose $c = b$ and $a \neq b$.
The equations become

$$x+y+z = u, \quad 2bx+(a+b)\,(y+z) = v,$$
$$b^2x+ab(y+z) = w.$$

$r(\mathbf{A}) = 2$ in this case, since $|\mathbf{A}| = 0$ and there is a non-zero second order determinant, viz.

$$\begin{vmatrix} 1 & 1 \\ 2b & a+b \end{vmatrix} = a-b, \quad \text{in} \quad \mathbf{A}.$$

Hence the condition for consistency is that $r(\mathbf{A}|\mathbf{U}) = 2$, i.e. that $\mathbf{A}|\mathbf{U}$ has no non-zero third-order determinants. This condition is that

$$\begin{vmatrix} 1 & 1 & u \\ 2b & a+b & v \\ b^2 & ab & w \end{vmatrix} = 0,$$

which on being expanded becomes

$$ub^2(a-b) \;-vb(a-b)+w(a-b) = 0.$$

But $a-b \neq 0$ and so the condition is $ub^2-vb+w = 0$.

If $u = 1, v = a$ and $w = 0$ this condition is $b^2-ab = 0$, i.e. since $a \neq b$, $b = 0$.

The equations become

$$x+y+z = 1, \quad a(y+z) = a \quad \text{or} \quad y+z = 1,$$

and the third equation disappears. Thus $x = 0$, $y = 1-z$ and z arbitrary is the general solution.

(iii) If $a = b = c$ the equations become

$$x+y+z = u, \quad 2a(x+y+z) = v, \quad a^2(x+y+z) = w.$$

For consistency we must have $v = 2au$ and $w = a^2u$, and these are clearly necessary and sufficient conditions. In the case where $u = 1$, $v = a$ and $w = 0$ the consistency conditions are $a = 2a$ and $0 = a^2$, and are not satisfied if $a \neq 0$.

Exercises 9A

Prove **1–5** from the axioms.

1. $\lambda(\mu\mathbf{x}) = \mu(\lambda\mathbf{x})$.

2. $(\lambda+\mu)(\mathbf{x}+\mathbf{y}) = \lambda\mathbf{x}+\lambda\mathbf{y}+\mu\mathbf{x}+\mu\mathbf{y}$.

3. $(\lambda+\mu+\nu)\mathbf{x} = \lambda\mathbf{x}+\mu\mathbf{x}+\nu\mathbf{x}$.

4. $(-\lambda)\mathbf{x} = -(\lambda\mathbf{x}) = \lambda(-\mathbf{x})$.

5. $\lambda\mathbf{x}-\mu\mathbf{x} = (\lambda-\mu)\mathbf{x}$.

6. Prove that all the solutions of $\sum_{i=1}^{n} p_i(x)\dfrac{d^i y}{dx^i} = 0$ form a vector space.

7. If $\mathbf{x}_1 = \mathbf{0}$ prove that $\mathbf{x}_1, \ldots, \mathbf{x}_n$ are linearly dependent.

8. If $\mathbf{x}_1 = \mathbf{x}_2$ prove that $\mathbf{x}_1, \ldots, \mathbf{x}_n$ are linearly dependent.

9. Prove that a single non-zero vector forms by itself a linearly independent set.

10. If $\mathbf{x}_1$ and $\mathbf{x}_2$ form a linearly dependent set and both are non-zero, prove that $\mathbf{x}_2 = k\mathbf{x}_1$ for some k.

11. Prove that a non-empty subset of a linearly independent set is linearly independent.

12. Prove that if $\mathbf{x}_1, \ldots, \mathbf{x}_n$ are linearly dependent, so are $\mathbf{x}_1, \ldots, \mathbf{x}_m$ for $m > n$.

Find $\lambda_1, \lambda_2, \lambda_3$ so that $\sum_{i=1}^{3} \lambda_i \mathbf{x}_i = \mathbf{0}$ in **13–15**.

13. $\mathbf{x}_1 = (1, 0, 0)$, $\mathbf{x}_2 = (0, 0, 1)$, $\mathbf{x}_3 = (3, 0, 5)$.

14. $\mathbf{x}_1 = (1, 1, 0)$, $\mathbf{x}_2 = (1, 3, 2)$, $\mathbf{x}_3 = (2, 0, -2)$.

15. $\mathbf{x}_1 = (-1, 2, 1)$, $\mathbf{x}_2 = (-2, 4, -4)$, $\mathbf{x}_3 = (-2, 4, -1)$.

16. In **15** above express $\mathbf{x}_3$ as a linear combination of $\mathbf{x}_1$ and $\mathbf{x}_2$.

Prove that $\mathbf{x}_1, \mathbf{x}_2, \mathbf{x}_3$ in **17** and **18** are linearly independent.

17. $\mathbf{x}_1 = (1, 1, 0)$, $\mathbf{x}_2 = (1, 0, 1)$, $\mathbf{x}_3 = (0, 1, 1)$.

18. $\mathbf{x}_1 = (2, 1, 0)$, $\mathbf{x}_2 = (-1, 1, 2)$, $\mathbf{x}_3 = (3, -1, 1)$.

19. If $\mathbf{x}_1 = (1, 1, 0)$, $\mathbf{x}_2 = (1, 3, 2)$, $\mathbf{x}_3 = (2, 0, -2)$, $\mathbf{x}_4 = (3, 1, -1)$ prove that

(i) $\mathbf{x}_1, \mathbf{x}_2, \mathbf{x}_3$ are linearly dependent (see **14** above),
(ii) $\mathbf{x}_1, \mathbf{x}_2, \mathbf{x}_4$ are linearly independent,
(iii) $\mathbf{x}_1, \mathbf{x}_2, \mathbf{x}_3, \mathbf{x}_4$ are linearly dependent,
(iv) $\mathbf{x}_4$ cannot be expressed as a linear combination of $\mathbf{x}_1, \mathbf{x}_2, \mathbf{x}_3$,
(v) $\mathbf{x}_3$ *can* be expressed as a linear combination of $\mathbf{x}_1$ and $\mathbf{x}_2$.

20. If $\mathbf{x}_1, \ldots, \mathbf{x}_p$ are linearly independent and $\mathbf{x}_1, \ldots, \mathbf{x}_{p+1}$ are linearly dependent prove that $\mathbf{x}_{p+1}$ is a linear combination of $\mathbf{x}_1, \ldots, \mathbf{x}_p$.

21. If $\mathbf{x}_1, \ldots, \mathbf{x}_m$ are linearly independent and $\mathbf{x}_1, \ldots, \mathbf{x}_m, \mathbf{x}_{m+1}, \ldots, \mathbf{x}_p$ span V, prove that there exists a base of V containing $\mathbf{x}_1, \ldots, \mathbf{x}_m$ and contained in $\mathbf{x}_1, \ldots, \mathbf{x}_p$.

22. Prove that if F is a subring of a commutative ring R with unity and F is a field, then R may be considered as a vector space over F. Give an example of such a vector space.

(*Hint.* Consider spaces of functions.)

23. Prove in detail that the set (ξ_1, ξ_2) for which $\lambda_1\xi_1+\lambda_2\xi_2 = 0$ (for some fixed λ_1 and λ_2 not both zero) forms a subspace of the space of 2-dimensional vectors.

24. Prove that the set (ξ_1, ξ_2, ξ_3) for which

$$\lambda_1\xi_1+\lambda_2\xi_2+\lambda_3\xi_3 = 0 \quad \text{and} \quad \mu_1\xi_1+\mu_2\xi_2+\mu_3\xi_3 = 0$$

(for some fixed λ_i's and μ_j's, at least one λ_i and one μ_j not zero) form a subspace M of the space of three-dimensional vectors. Find the dimension of M (i) if

$$\frac{\lambda_1}{\mu_1} = \frac{\lambda_2}{\mu_2} = \frac{\lambda_3}{\mu_3},$$

(ii) if these ratios are not all equal.

25. Prove that M is a subspace of V if and only if

$$\mathbf{x} \in M \text{ and } \mathbf{y} \in M \Rightarrow \lambda\mathbf{x}+\mu\mathbf{y} \in M \ \forall \ \lambda, \mu.$$

26. Give examples to show that if V is infinite dimensional and M is a subspace of V, then M may have finite or infinite dimension.

27. If V is the vector space of all polynomials $\mathbf{x}(t)$ with coefficients in F, let M be the set of elements of V with the property that $\mathbf{x}(-t) = \mathbf{x}(t)$ and N be the set with the property that $\mathbf{x}(-t) = -\mathbf{x}(t)$. Prove that (i) M and N are subspaces, (ii) $M \cap N = \{\mathbf{0}\}$ and $M+N = V$, and deduce that

$$V \cong M \oplus N.$$

28. If V and W are vector spaces over the field of residues modulo p and θ is a group-homomorphism of the group structure of V into that of W, prove that if $\theta\mathbf{x} = \mathbf{y}$, $\theta(r\mathbf{x}) = r\mathbf{y}$ and deduce that θ is a linear transformation of V into W.

29. Prove that θ is a linear transformation if and only if

$$\theta\left(\sum_{i=1}^{k} \lambda_i \mathbf{x}_i\right) = \sum_{i=1}^{k} \lambda_i . \theta(\mathbf{x}_i) \quad (\forall\ k \text{ and } \forall\ \mathbf{x}_1, \ldots, \mathbf{x}_k).$$

30. If V is the set of real valued functions defined for all real t, show that $\theta: V \to V$ defined by $\theta(\mathbf{x}(t)) = \mathbf{x}(t+1)$ is a linear transformation.

31. Show that $\theta: V \to V$ defined by $\theta(\mathbf{x}) = \mathbf{x}+\mathbf{a}$, $\mathbf{a} \neq \mathbf{0}$, is not a linear transformation.

32. Show that $\theta: V \to V$ defined by $\theta(\mathbf{x}) = \lambda\mathbf{x}$, $\lambda \neq 0$, is a linear transformation.

33. If P is the point with co-ordinates (ξ_1, ξ_2, ξ_3) (over the reals) prove that

$$OP^2 = (\xi_1, \xi_2, \xi_3) \begin{pmatrix} \xi_1 \\ \xi_2 \\ \xi_3 \end{pmatrix}.$$

34. If the n by n matrix $\mathbf{A} = (a_{ij})$, and $a_{ij} = 0 \;\forall\; i \neq j$, we call $\mathbf{A}$ a *diagonal* matrix. Prove that if $\mathbf{A}$ and $\mathbf{B}$ are n by n diagonal matrices, $\mathbf{AB}$ is diagonal.

35. Prove that all m by n matrices form a vector space. What is the dimension of this space?

36. Prove that all n by n matrices form a ring, non-commutative if $n > 1$, and with a unity. Is the ring an integral domain for $n > 1$?

37. Do the set of n by n non-singular matrices form a field?

38. Show that an n by 1 matrix represents a linear transformation of $F \to V$, where V has dimension n.

39. If F is finite and has r elements, prove that a vector space V of dimension n over F has r^n elements.

40. If V has dimension n over F show that the group structure of V is isomorphic to $F \times F \times \ldots \times F$, with n F's.

41. List the non-singular 2 by 2 matrices over the finite field of residues modulo 2 and identify the group of such matrices under the matrix product.

42. If $\theta: V \to W$, where V and W are finite dimensional, and if M is a subspace of V with dimension p, and if $\dim (M \cap N) = q$ where N is the null space of θ, show that $\theta(M)$ is a subspace of W and that $\dim \theta(M) = p-q$.

Find the ranks of the matrices in **43–47.**

43. $\begin{pmatrix} 1 & 2 \\ 0 & 0 \end{pmatrix}$.

44. $\begin{pmatrix} 1 & 2 \\ 4 & 6 \end{pmatrix}$.

45. $\begin{pmatrix} 1 & 0 & 1 \\ 3 & 2 & 1 \\ 7 & 4 & 2 \end{pmatrix}$.

46. $\begin{pmatrix} 1 & 2 & 3 \\ 2 & 4 & 6 \\ 3 & 6 & 9 \end{pmatrix}$.

47. $\begin{pmatrix} 1 & 0 & -1 & 2 \\ 2 & 1 & 3 & 0 \\ 3 & 1 & 2 & 2 \\ 9 & 3 & 6 & 6 \end{pmatrix}$.

Find the inverses of the matrices in **48** and **49.**

48. $\begin{pmatrix} 1 & 3 \\ 2 & 5 \end{pmatrix}$.

49. $\begin{pmatrix} 1 & 2 & 3 \\ -1 & 3 & 0 \\ 0 & 4 & 3 \end{pmatrix}$.

50. Prove that $\mathbf{I}^{-1} = \mathbf{I}$.

Solve completely the simultaneous equations in **51–56.**

51. $x+y+z = 1, \quad x+2y+3z = 0, \quad 3x+4y+4z = 3.$

52. $x+y+2z = 1, \quad 2x-y-z = -1, \quad x-2y-3z = -2.$

53. $x+y+2z = 1, \quad 2x-y-z = -1, \quad x-2y-3z = -3.$

54. $x-y+2z = 1, \quad 2x+y+z = 2.$

55. $x+y = 1, \quad 2x+3y = 2, \quad 5x+6y = 5.$

56. $x+y = 1, \quad 2x+3y = 2, \quad 4x+6y = 5.$

57. Prove that the equations $\mathbf{Ax} = \mathbf{y}$, where $\mathbf{A}$ is m by n, have a solution for every $\mathbf{y}$ if and only if $r(\mathbf{A}) = m$.

58. If $\mathbf{a}$ is a solution of $\mathbf{Ax} = \mathbf{y}$ prove that the general solution is

$$\mathbf{a} + \sum_{i=1}^{\nu(\mathbf{A})} \lambda_i \mathbf{b}_i \quad \text{where} \quad \mathbf{b}_1, \ldots, \mathbf{b}_{\nu(\mathbf{A})}$$

are linearly independent solutions of $\mathbf{Ax} = \mathbf{0}$.

Solve completely the simultaneous equations in **59** and **60.**

59. $x+2y-z = 0, \quad 2x-5y+z = 0, \quad x-7y+2z = 0.$

60. $2x-y+z = 0, \quad 4x-2y+2z = 0, \quad -6x+3y-3z = 0.$

Exercises 9B

1. Prove that the set of vectors $(\xi_1, \ldots, \xi_n)$ whose co-ordinates satisfy the m equations $\sum_{j=1}^{n} \alpha_{ij}\xi_j = 0, i = 1, \ldots, m$, forms a subspace M of the space of n-dimensional vectors. What is the dimension of M? Interpret M geometrically in the cases

(i) $n = 2, m = 1$; (ii) $n = 3, m = 1$;

(iii) $n = 3, m = 2$; (iv) $n = 3, m = 3$

and explain how the various possible dimensions of M occur.

2. Prove that if M and N are subspaces of V, $V \cong M \oplus N$ if and only if any $\mathbf{v} \in V$ is expressible in the form $\mathbf{v} = \mathbf{m}+\mathbf{n}$, $\mathbf{m} \in M$, $\mathbf{n} \in N$, *uniquely*.

3. Prove that if V is finite dimensional and M is a subspace of V,

$$\dim (V/M) = \dim V - \dim M.$$

(*Hint.* Consider the range and null space of the natural homomorphism $V \to V/M$.)

4. Prove that the result of exercise **9**A, no. **29** is true for vector spaces over the rational field.

5. If $\mathbf{x}_1, \ldots, \mathbf{x}_p$ are not all zero prove that they form a linearly dependent set if and only if there is a proper subset of $\mathbf{x}_1, \ldots, \mathbf{x}_p$ spanning the same subspace.

6. If V has dimension n over F show that the set of linear transformations of V into F is a vector space over F of dimension n. (This is called the *dual space*, or *adjoint space*, of V.)

7. Let $\mathbf{a}_1, \mathbf{a}_2, \ldots, \mathbf{a}_n$ be linearly independent vectors of a vector space V, and let $\mathbf{b}_1, \ldots, \mathbf{b}_n$ span the vector space W. Let ρ be a permutation of $1, 2, \ldots, n$. In each of the following cases state whether the conclusion is always true or can be false, and give either a proof or a counter-example.

(i) There is a unique linear transformation of $V \to W$ sending

$$\mathbf{a}_i \quad \text{to} \quad \mathbf{b}_i \quad (i = 1, 2, \ldots, n).$$

(ii) There is a unique linear transformation of $W \to V$ sending

$$\mathbf{b}_i \quad \text{to} \quad \mathbf{a}_i \quad (i = 1, 2, \ldots, n).$$

(iii) There is a linear transformation of $V \to V$ sending

$$\mathbf{a}_i \quad \text{to} \quad \mathbf{a}_{\rho(i)} \quad (i = 1, 2, \ldots, n).$$

(iv) There is a linear transformation of $W \to W$ sending

$\mathbf{b}_i$ to $\mathbf{b}_{\rho(i)}$ $(i = 1, 2, \ldots, n)$. (Cambridge Preliminary Examination.)

8. If M, N, N' are subspaces of V and if

(i) $M \cap N = M \cap N'$; (ii) $M+N = M+N'$; (iii) $N \subseteq N'$,

prove that $N = N'$.

9. If $\mathbf{A}$ is m by n prove that an n by m matrix $\mathbf{X}$ exists with $\mathbf{XA} = \mathbf{I}_n$ if and only if $\nu(\mathbf{A}) = 0$. Prove also that an n by m matrix $\mathbf{Y}$ exists with

$$\mathbf{AY} = \mathbf{I}_m \quad \text{if and only if} \quad r(\mathbf{A}) = m.$$

10. If $\theta: V \to W$ and $\phi: W \to X$ with $\dim W = n$, prove that

(i) $r(\phi\theta) \leqslant r(\phi)$; (ii) $r(\phi\theta) \leqslant r(\theta)$;

(iii) $r(\phi\theta) \geqslant r(\phi)+r(\theta)-n$; (iv) $\nu(\phi\theta) \leqslant \nu(\theta)+\nu(\phi)$.

11. If θ and ϕ are both linear transformations from $V \to W$, prove that $r(\theta+\phi) \leqslant r(\theta)+r(\phi)$.

(*Hint*. Use worked exercise no. 4.)

12. Find conditions that the simultaneous equations

$$ax+by+c = 0, \quad a'x+b'y+c' = 0$$

have (i) a unique solution, (ii) no solutions, (iii) many solutions. Solve completely in the consistent cases.

13. Find all the solutions of the equations

$$Bx+7y+8z = A, \quad 4x-2y-3z+9 = 0, \quad x+y+z = 11$$

for all possible values of the constants A and B. (Cambridge Open Schol.)

14. If $\mathbf{e}_1, \ldots, \mathbf{e}_n$ form a base of V, and $\mathbf{e}_i^* = \sum_{j=1}^{n} p_{ji}\mathbf{e}_j$ show that $\mathbf{e}_1^*, \ldots, \mathbf{e}_n^*$ form a base if and only if the matrix $\mathbf{P} = (p_{ij})$ is non-singular (i.e. has rank n). In this case if a given vector of V has co-ordinate representation

$$\mathbf{x} = \begin{pmatrix} \xi_1 \\ \vdots \\ \xi_n \end{pmatrix} \quad \text{with respect to the base } \{\mathbf{e}_i\}$$

and co-ordinate representation

$$\mathbf{x}^* = \begin{pmatrix} \xi_1^* \\ \vdots \\ \xi_n^* \end{pmatrix} \quad \text{with respect to the base } \{\mathbf{e}_i^*\},$$

prove that $\mathbf{x} = \mathbf{P}\mathbf{x}^*$.

15. Suppose $\theta: V \to W$, $\{\mathbf{e}_j\}$, $\{\mathbf{e}_j^*\}$ are bases of V and $\{\mathbf{f}_i\}$, $\{\mathbf{f}_i^*\}$ are bases of W, and suppose

$$\mathbf{e}_j^* = \sum_{k=1}^{n} p_{kj}\mathbf{e}_k,$$

$$\mathbf{f}_i^* = \sum_{k=1}^{m} q_{ki}\mathbf{f}_k.$$

Then if $\mathbf{A}$ is the matrix of θ with respect to $\{\mathbf{e}_j\}$ and $\{\mathbf{f}_i\}$ and $\mathbf{B}$ is the matrix of θ with respect to $\{\mathbf{e}_j^*\}$ and $\{\mathbf{f}_i^*\}$, prove that $\mathbf{B} = \mathbf{Q}^{-1}\mathbf{A}\mathbf{P}$.

($\mathbf{A}$ and $\mathbf{B}$ are called *equivalent* matrices.)

16. Prove that in the set of m by n matrices, the relation of equivalence as defined in **15** above is an equivalence relation, and prove that $\mathbf{A}$ and $\mathbf{B}$ are in the same equivalence class if and only if $r(\mathbf{A}) = r(\mathbf{B})$.

$\left(\textit{Hint}.\text{ Show that } \mathbf{A} \text{ is equivalent to } \begin{pmatrix} \mathbf{I}_r & \mathbf{0} \\ \mathbf{0} & \mathbf{0} \end{pmatrix}.\right)$

10

GEOMETRICAL APPLICATIONS

10.1. Euclidean geometry

There are several ways of placing the various types of geometry on a rigorous axiomatic footing. One way is by means of purely geometrical axioms: thus projective geometry may be based on axioms of incidence, together with one or two others, such as that postulating the idea of cross-ratio. One very satisfying method is by means of algebra: basing geometry on the idea of a vector space. The connection between a vector space and geometry based on a co-ordinate system is obvious, and since it is easy to define the former abstractly, as in the previous chapter, it is comparatively straightforward to give a rigorous development of a geometry by this means. Such a treatment loses something in beauty—analytical geometry is never as fascinating as pure geometry—but it gains tremendously in the ease with which firm foundations may be laid. In this section we will indicate briefly the formulation of Euclidean geometry by this means, and in the following sections we will apply the method to some other geometries.

Euclidean geometry is a complicated structure, much more so than is, for example, projective geometry. Even if we define lines, planes etc., in terms of linear manifolds in the obvious ways the ideas that we have already introduced about vector spaces are not sufficient for our present purpose. The idea that distinguishes Euclidean geometry, and which we have not so far introduced, is that of distance, and its consequence, the notion of angle.

To make the work simpler we will deal in this chapter only with finite dimensional spaces, although in fact a great deal of the work applies also to the infinite dimensional case. It is also convenient in this section, when dealing with Euclidean geometry, to restrict our base field, the scalars or co-ordinates, to be the real field. Other fields raise difficulties, although the theory is perfectly well developed for the complex field in particular.

Thus a Euclidean space is basically a vector space plus some idea of distance. It may be thought that this could be given very simply in terms of co-ordinates: thus the distance from O to $P(\xi_1, \xi_2, \ldots, \xi_n)$ might be defined simply as $\sqrt{(\xi_1^2+\xi_2^2+\ldots+\xi_n^2)}$ in line with the familiar Pythagorean result. But such a definition depends on a choice of base, and, speaking intuitively, there is no reason why a base chosen need be composed of mutually perpendicular vectors; i.e. the axes need not be rectangular and the ordinary sum of squares formula need not in fact be valid. Neither need the base vectors be of unit length (again speaking intuitively from the point of view of a given Euclidean space). Thus the straightforward definition is not adequate. In fact we need to start at the other end, to postulate a distance function which obeys certain axioms and hence possesses certain properties and then to define rectangular axes and unit base vectors in terms of this, and so to arrive at the usual distance formula relative to such a base.

In chapter 8 of volume 1 we defined a scalar or inner product of two vectors, and obtained expressions for distance and angle in terms of this. This is the approach that we will follow in the abstract case. The concept will be referred to here as *inner product*. The axioms we lay down will clearly be satisfied by the intuitive 'scalar product' and are sufficient for the theory to be developed. We proceed to give the definition.

Definition. *If V is a finite dimensional vector space over the real field of scalars, an* inner product *on V is a real valued function, defined on any ordered pair of vectors* $\mathbf{x}$ *and* $\mathbf{y}$ *in V and denoted by* $(\mathbf{x}, \mathbf{y})$, *which satisfies the following axioms.*

(1) $(\mathbf{x}, \mathbf{y}) = (\mathbf{y}, \mathbf{x})$;

(2) $(\mathbf{x}_1+\mathbf{x}_2, \mathbf{y}) = (\mathbf{x}_1, \mathbf{y})+(\mathbf{x}_2, \mathbf{y})$;

(3) $(\lambda\mathbf{x}, \mathbf{y}) = \lambda(\mathbf{x}, \mathbf{y})$ where λ is a scalar;

(4) $(\mathbf{x}, \mathbf{x}) \geqslant 0 \; \forall \; \mathbf{x} \in V$ and $(\mathbf{x}, \mathbf{x}) = 0 \Leftrightarrow \mathbf{x} = \mathbf{0}$.

V together with such an inner product is called a *Euclidean vector space*, or merely a *Euclidean space*.

That $(\mathbf{0}, \mathbf{0}) = 0$ is immediate by 3. 4 states that for $\mathbf{x} \neq \mathbf{0}$, $(\mathbf{x}, \mathbf{x})$ is greater than zero. It is also immediately seen using 1, 2 and 3 that

$$(\sum_i \lambda_i \mathbf{x}_i, \mathbf{y}) = \sum_i \lambda_i(\mathbf{x}_i, \mathbf{y}) \quad \text{and} \quad (\mathbf{x}, \sum_i \mu_i \mathbf{y}_i) = \sum_i \mu_i(\mathbf{x}, \mathbf{y}_i).$$

We now define the *length* of a vector $\mathbf{x}$ to be $\sqrt{(\mathbf{x}, \mathbf{x})}$, taking the positive square root. By axiom 4 above, this is always real. We write this as $|\mathbf{x}|$.

Theorem 10.1.1. (i) $|\mathbf{x}| \geqslant 0$ *and* $|\mathbf{x}| = 0 \Leftrightarrow \mathbf{x} = \mathbf{0}$;
(ii) $|\lambda\mathbf{x}| = |\lambda||\mathbf{x}|$;
(iii) $|-\mathbf{x}| = |\mathbf{x}|$.

(i) follows at once from axiom 4. By axioms 1 and 3

$$(\lambda\mathbf{x}, \lambda\mathbf{x}) = \lambda^2(\mathbf{x}, \mathbf{x}),$$

and hence (ii) follows. (iii) is merely a special case of (ii).

The *distance* between any two vectors $\mathbf{x}$ and $\mathbf{y}$ is now defined simply to be the length of $\mathbf{x}-\mathbf{y}$, and is written, of course, $|\mathbf{x}-\mathbf{y}|$. Notice that this is defined in terms of the *vectors*, but in geometrical terms is thought of in terms of the points at the end of these vectors: this identification of a vector **OP** and the 'point' P occurs throughout the work.

The basic properties of distance are proved in the following theorem.

Theorem 10.1.2. (i) $|\mathbf{x}-\mathbf{y}| \geqslant 0$ *and* $|\mathbf{x}-\mathbf{y}| = 0 \Leftrightarrow \mathbf{x} = \mathbf{y}$;
(ii) $|\mathbf{x}-\mathbf{y}| = |\mathbf{y}-\mathbf{x}|$.

(i) follows at once from theorem 10.1.1 (i) and (ii) from theorem 10.1.1 (iii).

There is one other important property we wish to establish, the so-called triangle inequality. But in order to do this we need first to prove an important inequality in connection with lengths, which is also fundamental to the introduction of angle. This is Schwarz's inequality, and is by no means trivial to prove, though not over difficult.

Theorem 10.1.3. *Schwarz's inequality.*

For any $\mathbf{x}$ *and* $\mathbf{y} \in V$, $|(\mathbf{x}, \mathbf{y})| \leqslant |\mathbf{x}|.|\mathbf{y}|$.

Note that the modulus sign on the LHS has its everyday meaning as the numerical value of the real number $(\mathbf{x}, \mathbf{y})$, whereas those on the RHS denote the lengths of $\mathbf{x}$ and $\mathbf{y}$. This double use of the sign need cause no confusion.

Let $\lambda = |\mathbf{y}|^2$ and $\mu = (\mathbf{x}, \mathbf{y})$.

Then $0 \leqslant (\lambda\mathbf{x}-\mu\mathbf{y}, \lambda\mathbf{x}-\mu\mathbf{y})$ by axiom 4

$$= \lambda^2(\mathbf{x}, \mathbf{x}) - 2\lambda\mu(\mathbf{x}, \mathbf{y}) + \mu^2(\mathbf{y}, \mathbf{y})$$

by an obvious extension of axioms 2 and 3 and using axiom 1,

$$= |\mathbf{y}|^4|\mathbf{x}|^2 - 2|\mathbf{y}|^2(\mathbf{x}, \mathbf{y})^2 + (\mathbf{x}, \mathbf{y})^2|\mathbf{y}|^2$$

by definition of λ and μ,

$$= |\mathbf{y}|^2[|\mathbf{y}|^2|\mathbf{x}|^2 - (\mathbf{x}, \mathbf{y})^2].$$

Hence since $|\mathbf{y}|^2 \geqslant 0$, we have $(\mathbf{x}, \mathbf{y})^2 \leqslant |\mathbf{x}|^2|\mathbf{y}|^2$ and the result follows on taking positive square roots.

Theorem 10.1.4. *The triangle inequality.*

For any $\mathbf{x}$, $\mathbf{y}$ *and* $\mathbf{z}$, $|\mathbf{x}-\mathbf{y}| \leqslant |\mathbf{x}-\mathbf{z}| + |\mathbf{z}-\mathbf{y}|$.

Write $\mathbf{u} = \mathbf{x}-\mathbf{z}$ and $\mathbf{v} = \mathbf{z}-\mathbf{y}$ so that $\mathbf{x}-\mathbf{y} = \mathbf{u}+\mathbf{v}$. Then we wish to prove that $|\mathbf{u}+\mathbf{v}| \leqslant |\mathbf{u}| + |\mathbf{v}|$. But

$$\begin{aligned} |\mathbf{u}+\mathbf{v}|^2 &= (\mathbf{u}+\mathbf{v}, \mathbf{u}+\mathbf{v}) \\ &= (\mathbf{u}, \mathbf{u}) + 2(\mathbf{u}, \mathbf{v}) + (\mathbf{v}, \mathbf{v}) \\ &= |\mathbf{u}|^2 + 2(\mathbf{u}, \mathbf{v}) + |\mathbf{v}|^2 \\ &\leqslant |\mathbf{u}|^2 + 2|\mathbf{u}||\mathbf{v}| + |\mathbf{v}|^2 \quad \text{by Schwarz's inequality,} \\ &= (|\mathbf{u}| + |\mathbf{v}|)^2 \end{aligned}$$

and the result follows on taking positive square roots.

This theorem states the well-known fact that the sum of the lengths of any two sides of a triangle is greater than the length of the third side.

Having defined distance we now proceed to define the angle between any two vectors. We recollect that in both two and three-dimensional Cartesian spaces the scalar product $\mathbf{x}.\mathbf{y} = xy\cos\theta$

where θ is the angle between **x** and **y**, and we define angle in the general case according to this idea.

Definition. *The angle between* **x** *and* **y** *is the angle between* 0° *and* 180° *whose cosine is*

$$\frac{(\mathbf{x}, \mathbf{y})}{|\mathbf{x}|.|\mathbf{y}|}.$$

Note that by Schwarz's inequality the quantity

$$\frac{(\mathbf{x}, \mathbf{y})}{|\mathbf{x}|.|\mathbf{y}|}$$

lies between -1 and $+1$ and so such an angle exists: the Schwarz inequality is vital to the definition being meaningful.

The most important case is when $(\mathbf{x}, \mathbf{y}) = 0$ and the angle is 90°. In this case **x** and **y** are said to be *orthogonal*.

We have now defined distance and angle. It remains to consider these, and the inner product, relative to given bases, in particular relative to 'rectangular' bases. But such bases must be defined in terms of a given postulated inner product. If we have a base $(\mathbf{e}_1, \mathbf{e}_2, ..., \mathbf{e}_n)$ we require that any pair of the base vectors are orthogonal and that each has length 1, in symbols that $(\mathbf{e}_i, \mathbf{e}_j) = \delta_{ij}$, where δ_{ij} is the so-called 'Kronecker delta' whose value is 0 if $i \neq j$ and 1 if $i = j$.

Definition. *A set of vectors* $\{\mathbf{x}_1, \mathbf{x}_2, ..., \mathbf{x}_r\}$ *in a Euclidean vector space* V *is called an* orthonormal *set if* $(\mathbf{x}_i, \mathbf{x}_j) = \delta_{ij}$.

(The word *orthonormal* is used because any pair of distinct vectors is orthogonal, and each vector is *normal* in the sense of having unit length.)

A base that forms an orthonormal set is naturally called an *orthonormal base*, and we now wish to show that such bases exist. We do this by starting with any base and 'orthogonalising' it by a process known as the 'Gram–Schmidt orthogonalisation process'. Since ordinary bases exist, so then do orthonormal bases, although distinct finite bases will not necessarily give distinct orthonormal ones. We first prove a subsidiary theorem.

Theorem 10.1.5. Any orthonormal set is linearly independent.

Suppose $\{\mathbf{x}_1, \mathbf{x}_2, \ldots, \mathbf{x}_r\}$ is an orthonormal set and $\sum_{i=1}^{r} \lambda_i \mathbf{x}_i = \mathbf{0}$. Then

$$0 = (\mathbf{0}, \mathbf{x}_j) = (\Sigma\lambda_i \mathbf{x}_i, \mathbf{x}_j)$$
$$= \Sigma\lambda_i(\mathbf{x}_i, \mathbf{x}_j)$$
$$= \lambda_j,$$

for any j, since by the definition of orthonormality $(\mathbf{x}_i, \mathbf{x}_j) = 0$ $i \neq j$ and $(\mathbf{x}_j, \mathbf{x}_j) = 1$.

Hence $\Sigma\lambda_i \mathbf{x}_i = \mathbf{0} \Rightarrow \lambda_i = 0$ for all i and so the set is linearly independent.

Theorem 10.1.6. The Gram–Schmidt orthogonalisation process.

If $(\mathbf{e}_1, \mathbf{e}_2, \ldots, \mathbf{e}_n)$ *is a base in a Euclidean vector space* V, *we may construct an orthonormal base* $(\mathbf{f}_1, \mathbf{f}_2, \ldots, \mathbf{f}_n)$ *such that for all* j, $\mathbf{f}_j$ *is a linear combination of* $\mathbf{e}_1, \mathbf{e}_2, \ldots, \mathbf{e}_j$.

We proceed by induction. The induction starts by putting $\mathbf{f}_1 = \mathbf{e}_1/|\mathbf{e}_1|$, since of course $\mathbf{e}_1 \neq \mathbf{0}$: then $\mathbf{f}_1$ has length 1. Now suppose that $\mathbf{f}_1, \mathbf{f}_2, \ldots, \mathbf{f}_r$ have been found, being an orthonormal set and with each $\mathbf{f}_j$ being a linear combination of $\mathbf{e}_1, \mathbf{e}_2, \ldots, \mathbf{e}_j$. Put $\mathbf{g} = \mathbf{e}_{r+1} - \sum_{i=1}^{r} (\mathbf{e}_{r+1}, \mathbf{f}_i)\, \mathbf{f}_i$, so that $\mathbf{g}$ is certainly a linear combination of $\mathbf{e}_1, \mathbf{e}_2, \ldots, \mathbf{e}_{r+1}$. Then

$$(\mathbf{g}, \mathbf{f}_j) = (\mathbf{e}_{r+1}, \mathbf{f}_j) - \sum_{i=1}^{r} (\mathbf{e}_{r+1}, \mathbf{f}_i)\,(\mathbf{f}_i, \mathbf{f}_j)$$
$$= (\mathbf{e}_{r+1}, \mathbf{f}_j) - (\mathbf{e}_{r+1}, \mathbf{f}_j), \quad \text{since } \mathbf{f}_1, \mathbf{f}_2, \ldots, \mathbf{f}_r \text{ are orthonormal,}$$
$$= 0, \quad \text{and this is true for } j = 1, \ldots, r.$$

We now note that $\mathbf{g} \neq \mathbf{0}$, since if this were not so it would imply that $\mathbf{e}_1, \mathbf{e}_2, \ldots, \mathbf{e}_{r+1}$ were linearly dependent. Hence we may put $\mathbf{f}_{r+1} = \mathbf{g}/|\mathbf{g}|$, so that

$$|\mathbf{f}_{r+1}| = 1 \quad \text{and} \quad (\mathbf{f}_{r+1}, \mathbf{f}_j) = \frac{1}{|\mathbf{g}|}(\mathbf{g}, \mathbf{f}_j) = 0 \quad \text{by the above.}$$

Hence $\mathbf{f}_1, \mathbf{f}_2, \ldots, \mathbf{f}_{r+1}$ form an orthonormal set as required and by induction we obtain finally an orthonormal set $(\mathbf{f}_1, \mathbf{f}_2, \ldots, \mathbf{f}_n)$. But these are a linearly independent set by theorem 10.1.5, and so, containing n elements, must be a base of V, thus proving the theorem.

Having shown the existence of orthonormal bases it is now easy to express the inner product in terms of them.

Theorem 10.1.7. *If* $\mathbf{x} = (\xi_1, \xi_2, \ldots. \xi_n)$ *and* $\mathbf{y} = (\eta_1, \eta_2, \ldots, \eta_n)$ *relative to an orthonormal base, then*

$$(\mathbf{x}, \mathbf{y}) = \xi_1\eta_1 + \xi_2\eta_2 + \ldots + \xi_n\eta_n.$$

If the base be $(\mathbf{e}_i)$,

$$\mathbf{x} = \sum_i \xi_i \mathbf{e}_i \quad \text{and} \quad \mathbf{y} = \sum_j \eta_j \mathbf{e}_j,$$

so that

$$(\mathbf{x}, \mathbf{y}) = (\sum_i \xi_i \mathbf{e}_i, \sum_j \eta_j \mathbf{e}_j)$$

$$= \sum_i \sum_j \xi_i \eta_j (\mathbf{e}_i, \mathbf{e}_j)$$

by the inner product axioms,

$$= \sum_i \xi_i \eta_i$$

by the orthonormality of the base.

Corollary. *The length of* $\mathbf{x}(\xi_1, \xi_2, \ldots, \xi_n)$ *is* $\sqrt{(\xi_1^2 + \xi_2^2 + \ldots + \xi_n^2)}$.

Thus we obtain the usual expression for length and similarly, of course, for distance, in terms of co-ordinates. But note that these co-ordinates must be relative to an orthonormal base, which must be defined in terms of the inner product. To fix the ideas, and to dispel any illusion that we are using a circular argument, let us summarise the work so far.

For V to be a Euclidean space it must possess an inner product which satisfies the axioms we gave. We then define length, distance and angle in terms of this. The definition of angle leads to the idea of orthogonality and orthonormal bases, which we show exist (they are of course dependent on the inner product that V possesses). We then show that relative to any such base the inner product, which has been present all the time but has not hitherto been expressed in terms of co-ordinates, must take the familiar form $\sum_i \xi_i \eta_i$; it follows at once that length is given by the usual Pythagorean formula. (Conversely, given any base of V we may define $((\xi_i), (\eta_i))$ to be $\sum_i \xi_i \eta_i$: this is easily shown to be an inner product and the given base is orthonormal with respect to this inner product.)

All the metrical properties of Euclidean geometry follow in the usual way, and we will not attempt to develop the theory any further, our purpose being to show how the foundations may be made firm by the use of algebraic ideas. Note once again that this work applies for any finite dimension, but only to spaces over the real field.

We have not yet mentioned the basic geometrical notions of lines, planes, etc. These ideas are easy to define and are common to many types of geometry, not merely Euclidean: it is distance that characterises Euclidean geometry and that causes the main trouble in its definition. Our definition of line, plane and, in general, linear subspace, will not involve the idea of inner product.

An r-dimensional linear subspace in a finite dimensional vector space V is merely a linear manifold $\mathbf{x}+M$, where M is an r-dimensional vector subspace of V (we recollect theorem 9.9.1, that any non-trivial vector subspace has finite dimension less than the dimension of V). If M is the trivial subspace $\{\mathbf{0}\}$ then $\mathbf{x}+M$ is merely a point (identified of course with the vector $\mathbf{x}$) which may be considered as a 0-dimensional linear subspace, and if $M = V$ we obtain the n-dimensional linear subspace consisting of the whole space.

Thus if V is two-dimensional we have lines, which are the linear manifolds $\mathbf{x}+M$ where M is a one-dimensional vector subspace consisting of all scalar multiples of a vector $\mathbf{y}$. If V is three-dimensional we obtain lines, given by one-dimensional subspaces, and planes given by two-dimensional ones. In general one-dimensional linear subspaces are called lines, two-dimensional are planes and three-dimensional are called solids.

The ordinary definition and properties of incidence follow. We say, for instance, that a point $\mathbf{y}$ is on a line $\mathbf{x}+M$ if the vector $\mathbf{y}$ is a member of the coset $\mathbf{x}+M$. Two linear subspaces of the same dimension are *parallel* if and only if they are cosets of the same vector subspace: a linear subspace R of dimension r is parallel to S, one of dimension s, where $s > r$, if and only if there is a linear subspace of dimension s parallel to S and containing R as a subset.

In terms of co-ordinates let M have a base $\{\mathbf{f}_1, \mathbf{f}_2, \ldots, \mathbf{f}_r\}$. Then the linear subspace $\mathbf{a}+M$ consists of all points $\mathbf{x}$ where $\mathbf{x} = \mathbf{a}+\sum_1^r \lambda_i \mathbf{f}_i$. By writing this as n scalar equations and eliminating the λ_i's, we see that an $(n-1)$-dimensional linear subspace is given by a linear equation in the co-ordinates $(\xi_1, \xi_2, \ldots, \xi_n)$. An $(n-2)$-dimensional linear subspace is given as the intersection of two such equations, and so on. For example, to find the two equations for an $(n-2)$-dimensional linear subspace we eliminate the $(n-2)$ λ_i's from $(n-1)$ of the scalar equations above and then repeat for another set of $(n-1)$ of the equations.

The general intersection theory of linear subspaces is a difficult subject, with many special cases arising. We will discuss the two- and three-dimensional cases only.

If V is two-dimensional consider the two lines $\mathbf{a}+M$, $\mathbf{b}+N$, where M and N are of course one-dimensional subspaces of V. If $\mathbf{x}$ is a common point, $\mathbf{x} = \mathbf{a}+\mathbf{m} = \mathbf{b}+\mathbf{n}$ for some $\mathbf{m} \in M$ and $\mathbf{n} \in N$. Hence $\mathbf{a}-\mathbf{b} = \mathbf{n}-\mathbf{m} \in M+N$, the linear sum of M and N, and each different expression of $\mathbf{a}-\mathbf{b}$ in this form leads to a common point of the lines. Using the result of worked exercise 4 of chapter 9, that

$$\dim(M+N)+\dim(M \cap N) = \dim M+\dim N,$$

we see that if M and N intersect only at O then $\dim(M+N) = 2$ and so $M+N$ is the whole space V. Hence $\mathbf{a}-\mathbf{b} \in M+N$ and the expression $\mathbf{a}-\mathbf{b} = \mathbf{n}-\mathbf{m}$ is unique since $M \cap N = \{\mathbf{0}\}$: thus in this case the lines intersect in a unique point. On the other hand, if M and N intersect in more than the origin, so that $M = N$, the lines $\mathbf{a}+M$ and $\mathbf{b}+N$ either do not intersect (if $\mathbf{a}-\mathbf{b}$ is not in $M+N$, i.e. not in $M\,(= N)$), or intersect in a line (i.e. are coincident) if $\mathbf{a}-\mathbf{b} \in M$.

If V is three-dimensional the above analysis applies in a similar manner. If M and N have dimension two, so that $\mathbf{a}+M$ and $\mathbf{b}+N$ are planes, in general $\dim(M+N) = 3$, the dimension of V, and $\dim(M \cap N)$ therefore is 1. Hence $\mathbf{a}-\mathbf{b}$ is expressible in the form $\mathbf{n}-\mathbf{m}$ in a singly infinite number of ways and the planes meet in a line. But special cases may occur—if

$$\dim(M+N) = 2, \quad \text{so that} \quad M = N,$$

the planes either do not intersect or intersect in a plane (being identical).

Similar analysis leads to the result that in three dimensions a plane and a line in general intersect in a point, but if they are parallel they either do not intersect or intersect in a line. Similarly two lines in three dimensions generally do not intersect, but may in special cases do so in a point or a line.

For higher dimensions the results are even more complicated, but may be discussed in a similar way.

10.2. Affine geometry

The main theoretical difficulty in the definition of Euclidean geometry as given in the last section was in connection with the ideas of distance and angle. If we do not define these we obtain a much simpler geometry, called affine geometry. An *affine space* is a vector space considered in connection with the definitions of lines, planes and linear subspaces as given in the last section, and affine geometry is the study of the properties that arise from this.

We will again restrict ourselves to the finite dimensional case, and in this section we will consider vector spaces over any field that does not have characteristic 2. (It turns out that fields with characteristic 2 lead to anomalies in affine geometry and it is simpler to exclude them.)

Thus in such a vector space we define an r-dimensional *linear (affine) subspace* to be a linear manifold of dimension r and obtain the same theory of incidence and intersection as previously. Parallelism is defined as before.

Length and distance are not defined, but affine geometry differs from projective geometry in two important respects: the notion of parallelism leads to special cases in intersection theory—for example two lines in two dimensions may not intersect—and we may speak in a meaningful way about ratio of lengths on the same or parallel lines.

Consider the line segment **AB** on the line $\mathbf{c}+M$, where $\dim M = 1$. Then $\mathbf{a} = \mathbf{c}+\mathbf{m}_a$ and $\mathbf{b} = \mathbf{c}+\mathbf{m}_b$ for some $\mathbf{m}_a$ and $\mathbf{m}_b$ in M. Then $\mathbf{AB} = \mathbf{b}-\mathbf{a} = \mathbf{m}_b-\mathbf{m}_a$. Now let $\mathbf{x}$ be a fixed vector in M, so that as M is one-dimensional, $\mathbf{m}_a = \lambda_a \mathbf{x}$ and $\mathbf{m}_b = \lambda_b \mathbf{x}$ for some λ_a and λ_b. Thus $\mathbf{AB} = (\lambda_b - \lambda_a)\,\mathbf{x}$.

Now if $\mathbf{CD}$ is any segment on a line parallel to AB, or on AB itself, $\mathbf{CD} = (\lambda_d - \lambda_c)\,\mathbf{x}$, with the same $\mathbf{x}$ as before. We now define the ratio AB/CD to be $(\lambda_b - \lambda_a)/(\lambda_d - \lambda_c)$, and this is independent of the particular choice of $\mathbf{x}$, for any other choice would be $\mathbf{y} = \mu\mathbf{x}$ and the μ's would cancel.

Hence ratios of lengths on the same or parallel lines may be defined, but not of course on non-parallel lines, since then we would depend on a particular choice of the basic vector $\mathbf{x}$. Nor, of course, can length be defined.

Affine geometry is more general than Euclidean, but does not have the same beauty and economy as projective geometry. Many elementary geometrical theorems are really affine theorems, and usually lend themselves easily to vector treatment. Examples are the centroid theorems, the theorems of Ceva and Menelaus, and the theory of conjugate diameters of an ellipse (which is equivalent in affine geometry to a circle).

10.3. Projective geometry

In both Euclidean and affine geometry special cases arise because of the presence of parallel elements—for example in two dimensions two distinct lines do not always meet. In projective geometry these special cases disappear and we obtain a geometry of great simplicity and beauty, which nevertheless admits of a great deal of geometrical work still being applicable.

We still restrict ourselves to the finite dimensional case but now admit any field of scalars.

An n-dimensional *projective space* is an $(n+1)$-dimensional vector space V over any field F, where an r-dimensional linear projective subspace is defined to be an $(r+1)$-dimensional vector subspace of V (*not* a general linear manifold). A point is a one-dimensional vector subspace, a line a two-dimensional vector subspace, and so on.

A linear subspace defined by the vector subspace M is said to be in that defined by N if M is a subspace of N. As we have said, properties of incidence are much simpler than for the other geometries. Thus in two dimensions two lines (given by two-dimensional subspaces) are either coincident or meet in a point (if the subspaces are M and N and $\dim(M+N) = 3$, $\dim(M \cap N) = 1$ and $M \cap N$ represents a point). Similar arguments in three dimensions tell us that two non-coincident planes always meet in a line, a plane meets a line which is not in it in a point, and two distinct lines either do not intersect or meet in a point (but cannot be coplanar unless they meet in a point or are coincident).

In terms of co-ordinates a point in an n-dimensional projective space is given by all $(n+1)$-dimensional vectors of the form $(c\xi_1, c\xi_2, \ldots, c\xi_{n+1})$ where c is arbitrary: thus only the ratios matter and all equations must be homogeneous in the sense that the sum of the powers of the variables is the same for every term. The zero vector may represent any point and its use is excluded. An $(n-1)$-dimensional linear subspace is given by an n-dimensional subspace and its co-ordinates satisfy the single linear equation $l_1\xi_1 + l_2\xi_2 + \ldots + l_{n+1}\xi_{n+1} = 0$. An $(n-2)$-dimensional linear subspace is given by the intersection of two such linear equations, and so on.

In projective geometry there is no meaningful definition of length, nor of ratio of lengths as there was in affine geometry. The corresponding concept is that of the *cross-ratio* of four points on a line, which may be defined as follows (there are many other possible definitions, but all are equivalent).

Let a line be given by a two-dimensionsl subspace M, and let $\{\mathbf{e}_1, \mathbf{e}_2\}$ be a base for M. A point A of the line will be the set of vectors of the form $\mu(\mathbf{e}_1 + \lambda_A \mathbf{e}_2)$ for a fixed λ_A and variable μ (let a vector representing the point have co-ordinates (α, β) and write $\lambda_A = \beta/\alpha$). Then the cross-ratio of the four points (AB, CD) on the line, written in pairs in that order, is defined as

$$\frac{\lambda_C - \lambda_A}{\lambda_B - \lambda_C} \cdot \frac{\lambda_B - \lambda_D}{\lambda_D - \lambda_A}.$$

The important thing about this definition is that it is independent of the choice of base for M, and hence the concept is meaningful. To prove this let any other base be $\{\mathbf{f}_1, \mathbf{f}_2\}$, where $\mathbf{f}_1 = a\mathbf{e}_1 + b\mathbf{e}_2$ and $\mathbf{f}_2 = c\mathbf{e}_1 + d\mathbf{e}_2$ and $ad \neq bc$ (for otherwise

$\mathbf{f}_1$ and $\mathbf{f}_2$ would not be linearly independent). Let

$$\mu(\mathbf{e}_1+\lambda\mathbf{e}_2) = \mu'(\mathbf{f}_1+\lambda'\mathbf{f}_2).$$

Then $$\mu(\mathbf{e}_1+\lambda\mathbf{e}_2) = \mu'(a\mathbf{e}_1+b\mathbf{e}_2+\lambda'(c\mathbf{e}_1+d\mathbf{e}_2)),$$

and so $$\mu = \mu'a+\mu'\lambda'c, \quad \mu\lambda = \mu'b+\mu'\lambda'd.$$

Hence $$\lambda = \frac{b+\lambda'd}{a+\lambda'c}.$$

Thus

$$\lambda_C-\lambda_A = \frac{(b+\lambda'_C d)(a+\lambda'_A c)-(b+\lambda'_A d)(a+\lambda'_C c)}{(a+\lambda'_A c)\,(a+\lambda'_C c)}$$

$$= \frac{(\lambda'_C-\lambda'_A)\,(ad-bc)}{(a+\lambda'_A c)(a+\lambda'_C c)},$$

and it follows that

$$\frac{\lambda_C-\lambda_A}{\lambda_B-\lambda_C}\cdot\frac{\lambda_B-\lambda_D}{\lambda_D-\lambda_A} = \frac{\lambda'_C-\lambda'_A}{\lambda'_B-\lambda'_C}\cdot\frac{\lambda'_B-\lambda'_D}{\lambda'_D-\lambda'_A}$$

and the cross-ratio is unchanged.

10.4. Transformations in geometry

When dealing with geometrical spaces we are interested in those transformations that preserve the geometrical structure of the space. The transformations in question will depend on the particular geometry under consideration, and in this section we consider the problem for the various geometries we have introduced.

In every case we wish to preserve the geometrical properties of the whole space, and hence we are restricted to transformations of the space V *onto itself* (or onto an isomorphic space). Furthermore, any *linear* transformations considered will be automorphisms; often called *non-singular* since they correspond to non-singular matrices. But note that we will be including transformations that are *not* linear, i.e. not homomorphisms of the vector space V.

We will find that in all cases the set of transformations forms a group: we wish for the product $\phi\theta$ to be included if θ and ϕ are, we include the identity and we require inverses so that we can transform back. The group will not generally be commutative of course.

We are approaching the subject from the geometrical point of view, having already considered some of the various types of geometry and wishing to discover the groups of transformations that will preserve their geometrical properties. The question may be thought of the other way round: given a particular group of transformations of a vector space, a geometry may be considered to be the study of the properties that are preserved by these transformations (such properties are said to be 'invariant'). The choice of groups is of course helped by a knowledge of which geometries are likely to be fruitful. This way of looking at geometry (a very precise definition of a particular geometry, in fact) was first put forward by Felix Klein in his 'Erlangen Program' of 1872.

The projective group

Projective transformations form the simplest group and will be considered first.

Let the n-dimensional projective space be given by the $(n+1)$-dimensional vector space V, and let $\theta: V \to V$ be an automorphism of V (a non-singular linear transformation of V into itself). If M is a vector subspace of V, the image $\theta(M)$ can easily be shown to be a subspace having the same dimension as M (the images of a base of M are clearly linearly independent if θ is non-singular, and span $\theta(M)$). Hence θ preserves points, lines and other projective subspaces. It also preserves incidence properties, for if $\mathbf{x} \in M \cap N$, $\theta(\mathbf{x}) \in \theta(M) \cap \theta(N)$. That cross-ratio is preserved follows by defining it in terms of a particular base $\{\mathbf{e}_1, \mathbf{e}_2)$ for the line in V, and for the base $\{\theta\mathbf{e}_1, \theta\mathbf{e}_2\}$ for the image line, and using the property of independence with respect to base proved in the last section.

Any non-singular linear transformation $\theta: V \to V$ thus gives a transformation which leaves the basic projective properties of V invariant, but all such are not distinct. θ and ϕ are the same, considered from the geometrical point of view, if and only if they map points into the same images, that is if and only if they map one-dimensional subspaces into the same images.

Theorem 10.4.1. *θ and ϕ give the same projective transformation if and only if $\phi = \lambda\theta$ for any non-zero scalar λ.*

If $\phi = \lambda\theta$ then for any vector $\mathbf{x}$, $\phi\mathbf{x} = \lambda(\theta\mathbf{x})$ and these represent the same point. Now suppose that θ and ϕ give the same projective transformation, so that $\theta^{-1}\phi$ is the identity projective transformation. Let $\{\mathbf{e}_1, \mathbf{e}_2, \ldots, \mathbf{e}_{n+1}\}$ be a base of V. Then $\theta^{-1}\phi(\mathbf{e}_i) = \lambda_i\mathbf{e}_i$ since $\mathbf{e}_i$ and its image must represent the same point. Now consider $\mathbf{e}_1+\mathbf{e}_2+\ldots+\mathbf{e}_{n+1}$. The image is $\Sigma\lambda_i\mathbf{e}_i$ and this must be a multiple of $\Sigma\mathbf{e}_i$, so that all the λ_i's must be equal, equalling λ say. Thus $\theta^{-1}\phi(\Sigma\xi_i\mathbf{e}_i) = \lambda\Sigma\xi_i\mathbf{e}_i$ and so $\theta^{-1}\phi = \lambda\mathbf{I}$, where $\mathbf{I}$ is the identity linear transformation. Thus $\phi = \lambda\theta$.

The group of all non-singular linear transformations of V is called the *full linear group* in $(n+1)$ dimensions, and the group of all projective transformations is called the *projective group* in n dimensions (that these are indeed groups is obvious). It is clear from the above theorem that the projective group in n dimensions is the quotient group of the full linear group in $(n+1)$ dimensions relative to the subgroup of non-zero scalar multiples of the identity.

In terms of co-ordinates, let θ map $(\xi_1, \xi_2, \ldots, \xi_{n+1})$ into $(\eta_1, \eta_2, \ldots, \eta_{n+1})$ relative to some fixed base. Then

$$\begin{aligned}
\eta_1 &= a_{11}\xi_1+a_{12}\xi_2+\ldots+a_{1,\,n+1}\xi_{n+1},\\
\eta_2 &= a_{21}\xi_1+a_{22}\xi_2+\ldots+a_{2,\,n+1}\xi_{n+1},\\
&\ldots\ldots\ldots\ldots\ldots\ldots\ldots\ldots\ldots\ldots\ldots\\
\eta_{n+1} &= a_{n+1,\,1}\xi_1+a_{n+1,\,2}\xi_2+\ldots+a_{n+1,\,n+1}\xi_{n+1},
\end{aligned}$$

where the matrix $\mathbf{A} = (a_{ij})$ is non-singular (that is, the determinant $|\mathbf{A}|$ is non-zero), and the matrices $\mathbf{A}$ and $\lambda\mathbf{A}$ give the same projective transformation.

The Affine group

If V is an n-dimensional affine space, any non-singular linear transformation $\theta: V \to V$ preserves the geometry. For under θ the affine subspace $\mathbf{a}+M$ transforms into $\theta(\mathbf{a})+\theta(M)$, which is a subspace of the same dimension. Parallel subspaces remain parallel and the properties of incidence are unchanged, as is the ratio of two lengths on the same or parallel lines (since this is independent of the choice of unit on the lines).

Thus the affine group contains the full linear group. In this case however the origin O (corresponding to the unit vector) does not have the special role that it possesses in projective geometry, since affine subspaces are linear manifolds and not vector subspaces. Hence we are not restricted to linear transformations.

Consider the transformation (called a 'translation') T defined by $T(\mathbf{x}) = \mathbf{x}+\mathbf{k}$, where $\mathbf{k}$ is a fixed vector. Under this the image of the affine subspace $\mathbf{a}+M$ is the parallel subspace $\mathbf{a}+\mathbf{k}+M$, so that parallelism and incidence are again invariant, as is also the ratio of lengths on parallel lines. Thus translations preserve the affine geometry of the space and are included in the affine group.

If we combine a non-singular linear transformation and a translation we still obtain a transformation which preserves the affine properties, and we will prove below that all such transformations form a group. This is the *affine group*, and it is the study of the properties invariant under this that forms the subject matter of affine geometry.

Theorem 10.4.2. *The set of all affine transformations, that is all transformations of the form* $\mathbf{x} \to \theta(\mathbf{x})+\mathbf{k}$, *where* θ *is a non-singular linear transformation and* $\mathbf{k}$ *a fixed vector, forms a group.*

If two such transformations are $\mathbf{x} \to \theta(\mathbf{x})+\mathbf{k}$ and

$$\mathbf{x} \to \phi(\mathbf{x})+\mathbf{h},$$

their product is given by

$$\mathbf{x} \to \phi(\theta(\mathbf{x})+\mathbf{k})+\mathbf{h} = \phi\theta(\mathbf{x})+(\phi(\mathbf{k})+\mathbf{h})$$

and this is affine since $\phi\theta$ is non-singular and $\phi(\mathbf{k})+\mathbf{h}$ is fixed. The identity is affine and the inverse of $\mathbf{x} \to \theta(\mathbf{x})+\mathbf{k}$ is given by $x \to \theta^{-1}(\mathbf{x}-\mathbf{k}) = \theta^{-1}(\mathbf{x})-\theta^{-1}(\mathbf{k})$, which is also affine. Product is easily seen to be associative.

Note. A translation followed by a non-singular linear transformation is an affine transformation, since $\mathbf{x} \to \theta(\mathbf{x}+\mathbf{k})$ is of the form $\mathbf{x} \to \theta(\mathbf{x})+\theta(\mathbf{k})$.

In terms of co-ordinates, let $\mathbf{A}$ be the matrix of θ with respect to some base and let $\mathbf{k}$ have co-ordinates $(k_1, k_2, \ldots, k_n)$. Then if $(\xi_1, \xi_2, \ldots, \xi_n)$ transforms to $(\eta_1, \eta_2, \ldots, \eta_n)$ under the affine transformation we see at once that

$$\begin{aligned} \eta_1 &= a_{11}\xi_1 + a_{12}\xi_2 + \ldots + a_{1n}\xi_n + k_1, \\ \eta_2 &= a_{21}\xi_1 + a_{22}\xi_2 + \ldots + a_{2n}\xi_n + k_2, \\ &\ldots\ldots\ldots\ldots\ldots\ldots\ldots\ldots\ldots \\ \eta_n &= a_{n1}\xi_1 + a_{n2}\xi_2 + \ldots + a_{nn}\xi_n + k_n. \end{aligned}$$

The Euclidean group

If V is a Euclidean vector space we wish to consider those transformations that preserve the Euclidean geometrical properties. Clearly these will be affine transformations. Let us consider first the effect of a translation.

The translation $\mathbf{x} \to \mathbf{x}+\mathbf{k}$ does not alter the difference between two vectors: $\mathbf{x}-\mathbf{y} \to (\mathbf{x}+\mathbf{k})-(\mathbf{y}+\mathbf{k}) = \mathbf{x}-\mathbf{y}$. Hence distance is invariant, as is the angle $Y\hat{X}Z$, defined as the angle between the vectors $\mathbf{y}-\mathbf{x}$ and $\mathbf{z}-\mathbf{x}$. Thus translations preserve the essential concepts of distance and angle, but not of course the length of a vector nor the angle between two vectors, where in these cases the origin plays an essential role in being the 'starting point' of the vectors concerned: the origin is changed by a translation.

We now pass to the much harder question of discovering which non-singular linear transformations leave the essential ideas invariant. We may define such a transformation by the invariancy of either inner product or length, and the following theorem shows that these are equivalent.

Theorem 10.4.3. *A linear transformation preserves length if and only if it preserves inner product, and in either case it preserves distance, angle and orthogonality. Such a transformation is called* orthogonal.

If the transformation preserves inner product then it clearly preserves length and the other concepts listed, since these are defined in terms of inner product.

To show that invariance of length implies that of inner product, notice that

$$\begin{aligned} |\mathbf{x}+\mathbf{y}|^2 &= (\mathbf{x}+\mathbf{y}, \mathbf{x}+\mathbf{y}) = (\mathbf{x}, \mathbf{x}) + 2(\mathbf{x}, \mathbf{y}) + (\mathbf{y}, \mathbf{y}) \\ &= |\mathbf{x}|^2 + 2(\mathbf{x}, \mathbf{y}) + |\mathbf{y}|^2. \end{aligned}$$

Hence $2(\mathbf{x}, \mathbf{y}) = |\mathbf{x}+\mathbf{y}|^2 - |\mathbf{x}|^2 - |\mathbf{y}|^2$ and so inner product may be defined in terms of lengths, and the invariance of the latter implies that of the former.

The set of all orthogonal linear transformations forms a group called the *orthogonal group*, and all transformations of the form $\mathbf{x} \to \theta(\mathbf{x}) + \mathbf{k}$, where θ is orthogonal, also form a group (a subgroup of the affine group) called the *Euclidean group*. Euclidean geometry is the study of properties invariant under this group. We have seen already that distance and angle are invariant (but not the origin or zero vector): it follows that the image of any figure is congruent to the object. Such transformations are often called *isometries*.

Let us now consider the matrix $\mathbf{A}$ of an orthogonal linear transformation θ relative to some fixed orthonormal base $(\mathbf{e}_1, \mathbf{e}_2, \ldots, \mathbf{e}_n)$. Since θ is orthogonal the images $\theta(\mathbf{e}_i)$ form an orthonormal set. But these images, in terms of co-ordinates relative to $(\mathbf{e}_i)$, are merely the columns of $\mathbf{A}$, and the fact that these form an orthonormal set is given in terms of the elements of $\mathbf{A}$ by the equations

$$\sum_{i=1}^{n} a_{ij} a_{ij} = 1, \quad \sum_{i=1}^{n} a_{ij} a_{ik} = 0 \quad \text{for} \quad j \neq k,$$

or, expressed as one equation,

$$\sum_{i=1}^{n} a_{ij} a_{ik} = \delta_{jk},$$

where δ_{jk} is the Kronecker delta.

The converse is also true: if the columns of $\mathbf{A}$ form an orthonormal set then θ is orthogonal (assuming the base is orthonormal of course). For if $(\xi_1, \xi_2, \ldots, \xi_n) \to (\eta_1, \eta_2, \ldots, \eta_n)$ we know that

$$\eta_i = \sum_{j=1}^{n} a_{ij} \xi_j$$

and so
$$\sum_{i=1}^{n} \eta_i^2 = \sum_{i=1}^{n} \sum_{j=1}^{n} \sum_{k=1}^{n} a_{ij} a_{ik} \xi_j \xi_k = \sum_{j} \sum_{k} \delta_{jk} \xi_j \xi_k$$
$$= \sum_{j} \xi_j^2$$

showing that length is preserved.

The above criterion may be expressed more simply. Write $\mathbf{A}'$ for the matrix (a_{ji}), interchanging rows and columns in $\mathbf{A}$.

Then the (j, k)th element of $\mathbf{A}'\mathbf{A}$ is $\sum_{i=1}^{n} a'_{ji}\, a_{ik}$ (where a'_{ji} is the (j, i)th element of $\mathbf{A}'$)

$$= \sum_{i=1}^{n} a_{ij} a_{ik} = \delta_{ik}.$$

Hence $\mathbf{A}'\mathbf{A}$ has 1 in the leading diagonal and 0 everywhere else and so is $\mathbf{I}$, the identity. Conversely if $\mathbf{A}\mathbf{A}' = \mathbf{I}$, $\sum_{i=1}^{n} a_{ij} a_{ik} = \delta_{jk}$. We have thus proved the following theorem.

Theorem 10.4.4. *Relative to an orthonormal base a matrix* $\mathbf{A}$ *represents an orthogonal linear transformation if and only if* $\mathbf{A}'\mathbf{A} = \mathbf{I}$.

Such a matrix is said, naturally enough, to be an *orthogonal matrix*.

It is immediate that for an orthogonal matrix (which is of course non-singular) $\mathbf{A}' = \mathbf{A}^{-1}$, and that $\mathbf{A}\mathbf{A}' = \mathbf{I}$, so that the *rows* of $\mathbf{A}$, considered as vectors, also form an orthonormal set. It also follows that since all orthogonal linear transformations form a group, so do all orthogonal matrices (this may be proved directly from the matrix definition).

We finally mention the obvious fact that a general Euclidean transformation in terms of co-ordinates is given by

$$(\xi_1, \xi_2, \ldots, \xi_n) \to (\eta_1, \eta_2, \ldots, \eta_n),$$

where

$$\begin{aligned}
\eta_1 &= a_{11}\xi_1 + a_{12}\xi_2 + \ldots + a_{1n}\xi_n + k_1, \\
\eta_2 &= a_{21}\xi_1 + a_{22}\xi_2 + \ldots + a_{2n}\xi_n + k_2, \\
&\ldots\ldots\ldots\ldots\ldots\ldots\ldots\ldots\ldots\ldots \\
\eta_n &= a_{n1}\xi_1 + a_{n2}\xi_2 + \ldots + a_{nn}\xi_n + k_n
\end{aligned}$$

and the matrix (a_{ij}) is orthogonal.

10.5. Simultaneous equations and geometry

We may interpret the solution of linear simultaneous equations (considered in §9.15) geometrically in terms of the intersection theory of linear subspaces. The intersection theory for

Euclidean and affine geometry is identical and we first consider this. We will refer to the examples and notation of §9.15.

The general problem given in that section is, in geometrical terms, that of finding the intersection in n-dimensional space of m linear subspaces of dimension $(n-1)$. (Linear subspaces of lesser dimension may be considered as being given by the intersection of two or more linear subspaces of dimension $(n-1)$, that is by two or more linear equations.) We will call $(n-1)$-dimensional linear subspaces *hyperplanes*: this is a convenient and common usage.

Case (b)

The common non-degenerate case when $m = n$ and $\mathbf{A}$ is non-singular gives us n hyperplanes with one common point. If $m > n$ we have the exceptional case of more than n hyperplanes having a common point, whereas if $m < n$ the case cannot arise—the hyperplanes must intersect in at least a line of points.

Case (a)

This is the case where we have no solution, that is no common point of the m hyperplanes. It is the usual case when $m > n$. If $m = n$ we must have $|\mathbf{A}| = 0$ and $r(\mathbf{A}|\mathbf{b}) > r(\mathbf{A})$. This means that one of the given set of equations, E say, has its LHS obtained as a linear combination of the LHS's of the preceding equations, but the RHS is not the same linear combination of the RHS's of these: thus the intersection space of these preceding equations lies in a hyperplane parallel to E and so is itself parallel to E according to the definition of parallelism given on p. 280. To see what this means we refer to the example given in §9.15:

$$\begin{aligned} x_1+2x_2+3x_3 &= 2,\\ -x_1+x_2+2x_3 &= 1,\\ x_1+5x_2+8x_3 &= 6. \end{aligned}$$

Here we have three planes in three dimensions. The line of intersection of the first two lies in the plane (a hyperplane in this case of course) $x_1+5x_2+8x_3 = 5$, obtained as twice the first equation plus the second. This plane is parallel to the third (E) and so the line of intersection of the first two is parallel to E, and there are of course no finite solutions.

If $m < n$ we have a similar situation: the example given in §9.15 is of two parallel planes in three dimensions.

Case (c)

The null space N of $\mathbf{A}$ is the intersection of parallel hyperplanes through O, and we see that the intersection of the given hyperplanes is a coset of this, i.e. is a linear subspace parallel to it. If $m < n$ this is the expected case: the example in §9.15 is of two planes in three dimensions which meet in a line. If $m = n$ the equations are linearly dependent: in three dimensions we obtain three planes with a common line of intersection (or in exceptional cases a common plane of intersection, when all three planes are coincident). If $m > n$ the case is very exceptional: in the example of §9.15 we have three coincident lines in two dimensions: in three dimensions we could have four planes all through the same line and so on.

Projective geometry

The intersection theory is much simpler than for affine and Euclidean geometry. Geometrically the solutions of m homogeneous linear equations in n unknowns give us the common points of m hyperplanes in $(n-1)$-*dimensional* projective space. If $r(\mathbf{A}) = n$ the only solution is $\mathbf{x} = \mathbf{0}$ and there is no point of intersection, whereas if $r(\mathbf{A}) < n$ there are non-zero solutions forming the null space of $\mathbf{A}$, corresponding geometrically to a linear subspace of dimension one less than that of the algebraic dimension of the null space. Note that if $m < n$ then the latter case must arise: thus $(n-1)$ hyperplanes always have common points of intersection in $(n-1)$-dimensional projective space—there is no concept of parallelism to give rise to exceptional cases. We may of course have an intersection space of higher dimension than expected—three planes in three-dimensional space must intersect in at least a point, but may in special cases intersect in a line, or even in a plane.

10.6. The field of scalars in a geometry

We have so far restricted our field of scalars to be the real field for Euclidean geometry, but for projective and affine geometry the field has been unrestricted (save only that we do not admit a field of characteristic 2 for affine space). The fundamental properties in the latter cases are independent of the choice of field of scalars: for example the cross-ratio is well defined, being itself a scalar. In particular the intersection theory of linear subspaces that we have sketched is the same for all choices of scalars.

Many of the less fundamental properties however do depend on a particular choice or choices of scalars. The most important is the intersection theory for subspaces which are non-linear. Taking the familiar case of the intersection of two curves in a plane, a curve of degree m (i.e. one represented by a polynomial equation of degree m) will intersect one of degree n in not more than mn points in general, some of which may be repeated, and making no allowance for the many exceptional cases that may arise. But over a field that is not algebraically closed (say the reals, or the rationals) these curves will often intersect in *less* than mn points. To give our geometry its full generality and beauty we need a field such as the complex field that is algebraically closed, and then indeed the curves will in general intersect in precisely mn points: for example any line will meet any circle in two points, which may be real, complex or coincident.

Complex Euclidean geometry

In §10.1 we restricted our field of scalars to be the real field, and the definition of inner product that we gave depended on this being so. It is useful to be able to construct a similar geometry over the complex field, and we indicate very briefly how this may be done.

An inner product is defined to be a complex valued function satisfying axioms which are identical with those in §10.1 except that the first is replaced by $(\mathbf{x}, \mathbf{y}) = \overline{(\mathbf{y}, \mathbf{x})}$, where $\overline{(\mathbf{y}, \mathbf{x})}$ is the complex conjugate of $(\mathbf{y}, \mathbf{x})$. It follows then that $(\mathbf{x}, \mathbf{x})$ must be real, since it equals its conjugate, and the fourth axiom is meaningful. Such a space is called a *unitary space*.

Length and distance are defined as in the real case and are of course real. Orthogonality is still defined by $(\mathbf{x}, \mathbf{y}) = 0$, but is not now linked with angle. Orthonormal bases exist as before and the Gram–Schmidt orthogonalisation process goes through in exactly the same way.

By the new first axiom it follows easily that

$$(\mathbf{x}, \mathbf{y}_1+\mathbf{y}_2) = (\mathbf{x}, \mathbf{y}_1)+(\mathbf{x}, \mathbf{y}_2) \quad \text{and} \quad (\mathbf{x}, \lambda\mathbf{y}) = \bar{\lambda}(\mathbf{x}, \mathbf{y}),$$

where $\bar{\lambda}$ is of course the complex conjugate of λ. It is easily seen then that relative to an orthonormal base if $\mathbf{x} = (\xi_i)$ and $\mathbf{y} = (\eta_i)$ the inner product is given by

$$(\mathbf{x}, \mathbf{y}) = \xi_1\bar{\eta}_1+\xi_2\bar{\eta}_2+\ldots+\xi_n\bar{\eta}_n,$$

and that the norm or length of $\mathbf{x}$ is $\sqrt{(\xi_1\bar{\xi}_1+\xi_2\bar{\xi}_2+\ldots+\xi_n\bar{\xi}_n)}$. (Note how this links up with the ordinary definition of modulus of the complex number $\xi = \alpha+i\beta$, where

$$\sqrt{(\xi\bar{\xi})} = \sqrt{[(\alpha+i\beta)(\alpha-i\beta)]} = \sqrt{(\alpha^2+\beta^2)} = |\xi|.)$$

Linear transformations of a unitary space which preserve the inner product are called *unitary transformations*, and the set of such is the *unitary group*: these transformations combined with the translations form the group of isometries of complex Euclidean geometry, preserving the essential property of distance.

If $\mathbf{A}$ is the matrix of a unitary transformation relative to an orthonormal base, a similar analysis to that in § 10.4 shows us that $\bar{\mathbf{A}}'\mathbf{A} = \mathbf{I}$, where $\bar{\mathbf{A}}'$ is obtained by interchanging the rows and columns of $\mathbf{A}$ *and* taking the complex conjugates of all elements. This condition is also sufficient for $\mathbf{A}$ to represent a unitary transformation, and such a matrix is of course called a *unitary matrix*.

10.7. Finite geometries

If the field of scalars is a finite field, that is a Galois field $GF[p^n]$ for prime p, then we obtain a geometry with only a finite number of points (assuming we still restrict ourselves to the finite dimensional case).

It is difficult to define a satisfactory inner product in the finite case, since there is no concept of order and the fourth axiom becomes meaningless: we will therefore not concern ourselves with finite Euclidean geometries.

We may construct finite affine geometries, but these are on the whole not very interesting. Figure 20 shows a representation of a two-dimensional affine space over the finite field of residues modulo 3.

× $F(0, 2)$	× $G(1, 2)$	× $H(2,2)$
× $C(0, 1)$	× $D(1, 1)$	× $E(2, 1)$
× $O(0, 0)$	× $A(1, 0)$	× $B(2, 0)$

Fig. 20

There are nine points as shown, and twelve lines, in four sets each containing three parallel lines thus: *OAB*, *CDE*, *FGH*; *OCF*, *ADG*, *BEH*; *ODH*, *AEF*, *CGB*; *OEG*, *BDF*, *ACH*.

Any two points are joined by one and only one line and any two non-parallel lines meet in a unique point. Clearly an affine geometry of dimension n over a field of k elements has k^n points.

Finite projective geometries are rather more interesting. Over the Galois field $GF[p^n]$ we obtain a projective geometry in each dimension, and these exhibit the usual properties of projective geometry, except that if $p = 2$ there are only three points on each line and there is no useful concept of cross-ratio.

Let us consider the finite plane projective geometries. If the field has k elements there are k^3-1 possible triads of co-ordinates (excepting $(0, 0, 0)$). But each point has $k-1$ different representations, and so there are $(k^3-1)/(k-1) = k^2+k+1$ points, and the same number of lines (since the lines are given by equations of the form $l_1\xi_1+l_2\xi_2+l_3\xi_3 = 0$ and are specified by the ratios of the elements l_1, l_2, l_3 which thus act in an analogous manner to the co-ordinates of the points). There are $k+1$ points on each line (as may be shown similarly, since

$$(k^2-1)/(k-1) = (k+1)$$

and $k+1$ lines through a point. We must not expect our lines to look like Euclidean lines, and must bear in mind that they consist of the $k+1$ points *only*, and are not continuous.

If $k = 2$, so that the scalars form the field of residues modulo 2, we have seven lines each containing three points: the resulting geometry is shown in figure 21, the lines being *BDC*, *CEA*, *AFB*, *AGD*, *BGE*, *CGF* and *DEF*. Their equations are given in terms of co-ordinates (x, y, z).

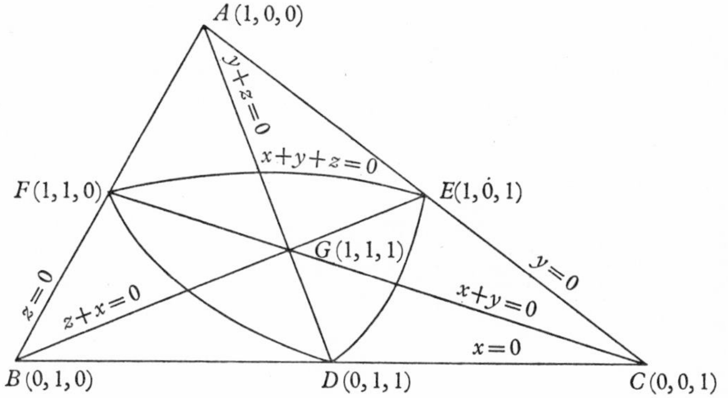

Fig. 21

If $k = 3$ we obtain the geometry over the field of residues modulo 3 and this contains thirteen lines and thirteen points, with four points on each line. Figure 22 shows this geometry, with the points and lines labelled. Bear in mind that *HJX* and *A* are collinear, as are *JGY* and *B*, *GHZ* and *C* and *XYZ* and *U*.

10.8. Algebraic varieties and ideals

Consider a three-dimensional geometrical space whose points are given by the triads of co-ordinates (x, y, z). In this section the coordinates may be members of any field, and it is immaterial whether we consider our geometry as Euclidean or affine or, indeed, as either: the work applies to all algebraic geometries over a field.

A curve in three dimensions is represented algebraically by the simultaneous vanishing of two polynomial equations in x, y and z. For example the equations $x^2+y^2-1 = 0$ and $z-1 = 0$ both vanish on the circle in the plane $z-1 = 0$ whose centre is on the z-axis and whose radius is 1. Let us call this circle C. The particular equations we chose to give C are

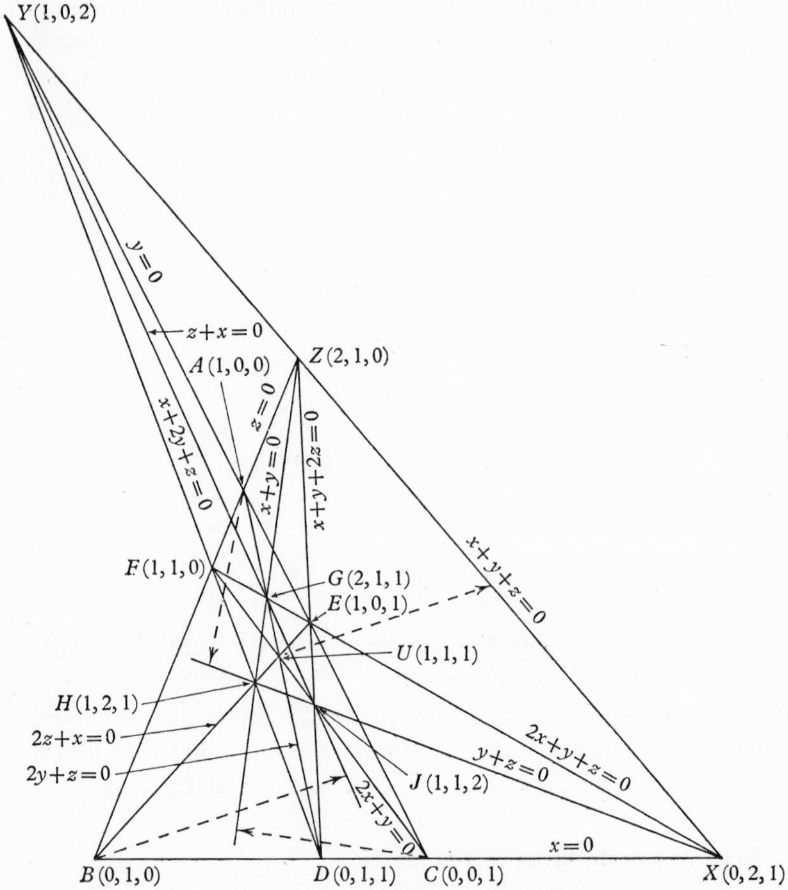

Fig. 22

not unique: another possible pair is $x^2+y^2+z^2-2 = 0$ and $z-1 = 0$, for example. We may in fact take any two of the surfaces through C and give C as their intersection (or part of their intersection in exceptional cases). The algebraic description of C is best given by *all* the polynomials which vanish at all points of it.

Consider then the ring of polynomials $F[x, y, z]$ in the three indeterminates x, y and z over the coordinate field F. If P and Q are polynomials which have zeros at all points of the circle C, then clearly $P-Q$ and XP for any polynomial X also vanish on

C. Thus the set of polynomials that vanish at all points of C is an ideal I_c in $F[x, y, z]$, and I_c determines C as the intersection of the zeros of all its members. I_c is generated by the two polynomials x^2+y^2-1 and $z-1$. It consists of all polynomials of the form $(x^2+y^2-1)P+(z-1)Q$, and C may be given as the intersection of the zeros of any two independent generators of I_c. But it is the ideal itself rather than any particular choice of generators that gives the best definition of C.

We may easily generalise the above work to any finite number of dimensions. In n-dimensional algebraic geometry, where a point is given by n co-ordinates $(x_1, x_2, \ldots, x_n)$, an *algebraic variety* is the set of points which satisfy a finite set of polynomial equations. Thus if $n = 2$, algebraic varieties are either curves or finite sets of points (the latter given by two equations), if $n = 3$ they are surfaces, curves or finite sets of points, and so on. Note that we allow only *polynomial* equations of the form

$$P(x_1, x_2, \ldots, x_n) = 0:$$

equations such as $\sqrt{x_1}+\sqrt{x_2} = 1$, $x_1-\sin x_2 = 0$ do not give algebraic varieties.

Theorem 10.8.1. *The set of polynomials which vanish at all points of an algebraic variety V is an ideal I_v in*

$$F[x_1, x_2, \ldots, x_n].$$

If P and Q are two such polynomials, so that $P = 0$ and $Q = 0$ at any point of V, then $P-Q = 0$ at any point of V and so the set of such polynomials is a subgroup. Also

$$\forall\, X \in F[x_1, x_2, \ldots, x_n], \quad XP = PX = 0,$$

at all points of V and so the set is an ideal.

Conversely any ideal gives us a set of points which are zeros of all polynomials in the ideal. It can be proved (although we will not do so here) that the ideal has a finite set of generators and it follows that the point set is an algebraic variety: thus any ideal corresponds to an algebraic variety.

Irreducible varieties and prime ideals

In two-dimensional geometry the equations of two or more curves may be combined to give us an equation which is that of all those curves: thus $(x^2+y^2-1)(x^2+y^2-4) = 0$ represents two circles, and $(x-y)(x+y+1)y = 0$ represents three straight lines. There is an obvious distinction between such curves and those given by an irreducible equation, and an obvious notation is to call such curves *reducible* curves as distinct from *irreducible* curves such as $x^2+y^2-1 = 0$.

We obtain similar cases in more than 2 dimensions. Thus in three dimensions we may consider two distinct circles together as forming a variety (an example is given by the simultaneous vanishing of $x^2+y^2-1 = 0$ and $z^2-1 = 0$, given equally well by $x^2+y^2-1 = 0$ and $x^2+y^2+z^2-2 = 0$) and we see that here there is no necessity for either of the polynomials to be reducible. We thus require a general definition of a reducible variety, and this we proceed to give.

Definition. *An algebraic variety is* reducible *if it can be expressed as the union of two sets of points V_1 and V_2 such that there exist polynomials P_1 and P_2 where P_1 vanishes on V_1 and not on V_2, and P_2 vanishes on V_2 and not on V_1.*

(When we say a polynomial vanishes *on* a set we mean that it must vanish at *all* points of the set.)

Example. In the case of the two circles given above we may take $P_1 \equiv z-1$ and $P_2 \equiv z+1$.

A variety that is not reducible is of course called an *irreducible variety*, and it is to such varieties that we often wish to restrict our work.

The question now arises as to which ideals correspond to irreducible varieties. With our definition above it is quite easy to see that such ideals are the prime ideals as defined in §7.9. We recall that I is prime if and only if $xy \in I \Rightarrow x \in I$ or $y \in I$.

Theorem 10.8.2. *If I_V is the ideal corresponding to a variety V, V is irreducible if and only if I_V is prime.*

Suppose first that V is reducible, so that V is the union $V_1 \cup V_2$ where polynomials P_1 and P_2 exist vanishing on V_1 and V_2

respectively but not on both. Thus neither P_1 nor $P_2 \in I_V$. But P_1P_2 vanishes on V_1 and on V_2 and so on V and hence $P_1P_2 \in I_V$. Hence I_V is not prime.

Conversely suppose that I_V is not prime, so that P_1 and P_2 exist where $P_1P_2 \in I_V$ but neither P_1 nor P_2 is. Denote the set of points of V at which $P_1 = 0$ by V_1 and the set at which $P_2 = 0$ by V_2. Then since $P_1P_2 \in I_V$, $P_1P_2 = 0$ at all points of V and so either P_1 or $P_2 = 0$ at all points of V: hence $V = V_1 \cup V_2$. But $P_1 \notin I_V$ and so does not vanish at every point of V. Hence P_1 does not vanish on V_2 (using the fact that $V = V_1 \cup V_2$) and similarly P_2 does not vanish on V_1. Hence V is reducible.

(Note that P_1, P_2 and P_1P_2 may vanish at points not in V and also, of course, that P_1 may vanish at *some* points of V_2 and P_2 at some of V_1.)

Worked exercises

1. In an affine space prove that there is a unique point P dividing the line segment AB in the ratio $t:(1-t)$, and P is given by $\mathbf{p} = (1-t)\mathbf{a}+t\mathbf{b}$. Deduce that there is a unique point Q dividing AB in the ratio $u:v$ given by $q = \dfrac{v\mathbf{a}+u\mathbf{b}}{v+u}$ provided $u+v \neq 0$.

Suppose P is a point on AB such that $AP:PB = t:(1-t)$, so that $AP:AB = t:1$. Then $(\mathbf{p}-\mathbf{a}) = t(\mathbf{b}-\mathbf{a})$ so that $\mathbf{p} = \mathbf{a}+(\mathbf{b}-\mathbf{a})t$ and the result follows. Conversely, since the argument is reversible, the point given by the vector $(1-t)\mathbf{a}+t\mathbf{b}$ *does* divide AB in the ratio $t:(1-t)$ (and does of course lie on AB since it is $\mathbf{a}+t\mathbf{AB}$). (The equation $\mathbf{x} = (1-t)\mathbf{a}+t\mathbf{b}$ gives the vector equation of the line AB in terms of the parameter t.)

Now put $t = \dfrac{u}{v+u}$, so that $1-t = \dfrac{v}{v+u}$. Then the second result follows at once. If $v+u = 0$ this is of course not possible: in this case we see that there is no finite point dividing AB in the ratio $1:-1$.

2. In figure 23 ABC and $A'B'C'$ are any two triangles in a projective plane (i.e. two-dimensional projective space) such that AA', BB', CC' are concurrent in O. BC and $B'C'$ meet in L, CA and $C'A'$ meet in M and AB and $A'B'$ meet in N. Prove that LMN are collinear. (This is Desargue's theorem, a fundamental projective theorem.)

We first show that we can choose a co-ordinate system in which A, B and C are $(1, 0, 0)$, $(0, 1, 0)$ and $(0, 0, 1)$ respectively, and O is $(1, 1, 1)$.

Bearing in mind that the plane is given by a three-dimensional vector space with one-dimensional subspaces representing points, we assume that A, B and C are not collinear and hence that $\mathbf{a}$, $\mathbf{b}$ and $\mathbf{c}$ (any vectors representing A, B and C) are linearly independent. Hence we may take $\mathbf{a}$, $\mathbf{b}$

and **c** as the vectors of a base, and so A, B and C have the required co-ordinates. But **a**, **b** and **c** are arbitrary with respect to scalar multiples, and we may choose them so that O is $(1, 1, 1)$: the coordinates of A, B and C remain as required, since it is only the ratios of the co-ordinates that matter—the point $(a, 0, 0)$ is the same as $(1, 0, 0)$ for instance. (ABC is often called the *triangle of reference* and O the *unit point*: any four points, no three of which are collinear, may be taken as the vertices of the triangle of reference and the unit point in a similar way.)

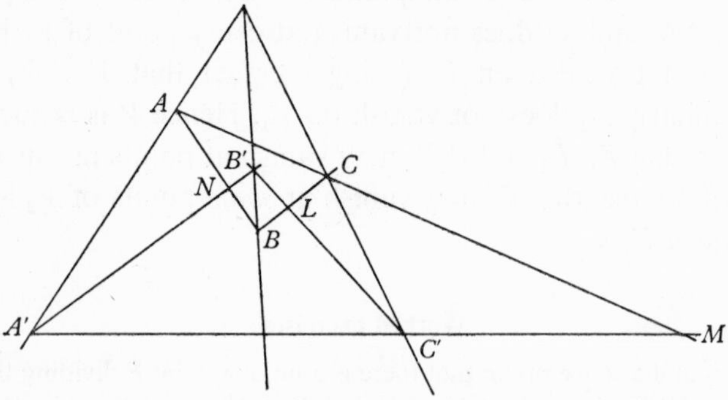

Fig. 23

A' is on OA and hence may be taken as $(f, 1, 1)$, since it must be a linear combination of $(1, 1, 1)$ and $(1, 0, 0)$ and only ratios of co-ordinates matter. Similarly B' is $(1, g, 1)$ and C' is $(1, 1, h)$. The point L is on $B'C'$ and hence its vector is a linear combination of $(1, g, 1)$ and $(1, 1, h)$, but it is also on BC and hence its first co-ordinate is zero. Thus it must be $(0, g-1, 1-h)$, subtracting the vectors of B' and C'.

Calling the co-ordinates x, y and z, we see that L lies on the line

$$\frac{x}{f-1}+\frac{y}{g-1}+\frac{z}{h-1} = 0,$$

and this is symmetrical in x, y and z, so that M and N also lie on it. Hence LMN are collinear.

(Note the economy and symmetry of the above work: such beauty is typical of the theorems in projective geometry.)

3. Show that

$$\mathbf{A} = \begin{pmatrix} -1 & 0 & 0 \\ 0 & -1 & 0 \\ 0 & 0 & 1 \end{pmatrix}$$

is orthogonal. Of what transformation of three-dimensional Euclidean space is **A** the matrix?

By ordinary matrix multiplication

$$\mathbf{A'A} = \begin{pmatrix} -1 & 0 & 0 \\ 0 & -1 & 0 \\ 0 & 0 & 1 \end{pmatrix}\begin{pmatrix} -1 & 0 & 0 \\ 0 & -1 & 0 \\ 0 & 0 & 1 \end{pmatrix} = \begin{pmatrix} 1 & 0 & 0 \\ 0 & 1 & 0 \\ 0 & 0 & 1 \end{pmatrix} = \mathbf{I}.$$

Now suppose A transforms the point (x, y, z) into (x', y', z'). Then

$$\begin{pmatrix} x' \\ y' \\ z' \end{pmatrix} = \begin{pmatrix} -1 & 0 & 0 \\ 0 & -1 & 0 \\ 0 & 0 & 1 \end{pmatrix} \begin{pmatrix} x \\ y \\ z \end{pmatrix}$$

so that $x' = -x$, $y' = -y$, $z' = z$. It is clear that A represents a rotation of 180° about the z-axis.

4. Seven dinners are to be arranged for 7 men, 3 of the men to be present at each dinner, in such a way that no two men eat together more than once. Show how this might be done.

Call the men A, B, C, D, E, F and G. Think of these in geometrical terms as points and a dinner as a line, and interpret the presence of a given man at a given dinner as meaning that the point lies on the line. Then the problem becomes that of constructing 7 lines each through 3 of the given 7 points such that any two points lie on at most one line.

Notice first that each man must attend exactly 3 dinners. For if not there would be some man attending more than 3, which is impossible since he needs two different companions at each and there are only 6 such companions to choose from. Hence there are 3 lines through each point. Also each man, since he attends 3 dinners, must meet all the other 6, so that there *is* a line through any two points.

The solution now is clearly given by the finite projective geometry of figure 21, and we see that the 7 dinners are for the following triads of men: *BDC*, *CEA*, *AFB*, *AGD*, *BGE*, *CGF* and *DEF*. Note that any pair of dinners must have one man in common.

With the specification of 3 men at each dinner, less than 7 dinners for the same number of men could not be arranged, whereas it would be easy to fulfil the conditions for more than 7 men and the corresponding number of dinners, provided we did not insist on each man attending the same number (3) of dinners. If each man must attend 3 dinners, a solution for 9 men is given by the figure for Pappus' theorem (figure 25 in exercises 10B).

Figure 22 solves the same problem for 4-man dinners for the relevant number (13) of men.

Exercises 10A

Note. Some of these exercises are difficult in comparison with the A exercises in other chapters.

Prove **1–3** for the inner product $(\mathbf{x}, \mathbf{y})$ in a Euclidean vector space:

1. $(\mathbf{x}, \mathbf{y}_1+\mathbf{y}_2) = (\mathbf{x}, \mathbf{y}_1)+(\mathbf{x}, \mathbf{y}_2)$.

2. $(\mathbf{x}, \mu\mathbf{y}) = \mu(\mathbf{x}, \mathbf{y})$.

3. $(\lambda_1\mathbf{x}_1+\lambda_2\mathbf{x}_2, \mu_1\mathbf{y}_1+\mu_2\mathbf{y}_2)$

$$=\lambda_1\mu_1(\mathbf{x}_1, \mathbf{y}_1)+\lambda_1\mu_2(\mathbf{x}_1, \mathbf{y}_2)+\lambda_2\mu_1(\mathbf{x}_2, \mathbf{y}_1)+\lambda_2\mu_2(\mathbf{x}_2, \mathbf{y}_2).$$

4. Prove that, in a Euclidean vector space, (i) $|\mathbf{x}+\mathbf{y}| \leqslant |\mathbf{x}|+|\mathbf{y}|$, (ii) $|\mathbf{x}+\mathbf{y}| \geqslant |\mathbf{x}|-|\mathbf{y}|$, (iii) $|\mathbf{x}-\mathbf{y}| \geqslant |\mathbf{x}|-|\mathbf{y}|$.

5. Prove that the equality sign in Schwarz's inequality (theorem 10.1.3) applies if and only if $\mathbf{x}$ and $\mathbf{y}$ are linearly dependent.

6. Prove that the equality sign in the triangle inequality (theorem 10.1.4) applies if and only if $(\mathbf{x}-\mathbf{z})$ and $(\mathbf{y}-\mathbf{z})$ are linearly dependent, and show that this is true if and only if the points representing the vectors $\mathbf{x}$, $\mathbf{y}$ and $\mathbf{z}$ are collinear.

7. Prove the extension of theorem 10.1.4 that

$$|\mathbf{x}_1-\mathbf{x}_n| \leqslant |\mathbf{x}_1-\mathbf{x}_2|+|\mathbf{x}_2-\mathbf{x}_3|+\ldots+|\mathbf{x}_{n-1}-\mathbf{x}_n|.$$

8. Using co-ordinates relative to a given orthonormal base, so that the usual formula for inner product applies, use the Gram–Schmidt orthogonalisation process (theorem 10.1.6) to form an orthonormal base from the base

$$\mathbf{e}_1 = (1, 2, 2), \quad \mathbf{e}_2 = (3, 0, 4), \quad \mathbf{e}_3 = (6, 2, 3)$$

in three-dimensional Euclidean space.

9. In three-dimensional Euclidean space, using co-ordinates relative to a given orthonormal base, determine which of the following are orthonormal bases:

(i) $\mathbf{e}_1 = (0, 1, 1), \quad \mathbf{e}_2 = (0, 1, -1), \quad \mathbf{e}_3 = (1, 0, 0);$

(ii) $\mathbf{e}_1 = \left(0, \frac{1}{\sqrt{2}}, \frac{1}{\sqrt{2}}\right), \quad \mathbf{e}_2 = \left(0, -\frac{1}{\sqrt{2}}, \frac{1}{\sqrt{2}}\right), \quad \mathbf{e}_3 = (1, 0, 0);$

(iii) $\mathbf{e}_1 = \left(\frac{1}{\sqrt{3}}, \frac{1}{\sqrt{3}}, \frac{1}{\sqrt{3}}\right), \quad \mathbf{e}_2 = \left(0, -\frac{1}{\sqrt{2}}, \frac{1}{\sqrt{2}}\right), \quad \mathbf{e}_3 = (1, 0, 0).$

10. Show that Schwarz's inequality, in terms of co-ordinates relative to an orthonormal base, becomes

$$\sum_{i<j} (\xi_i\eta_j-\xi_j\eta_i)^2 \geqslant 0.$$

11. If $\{\mathbf{e}_1, \ldots, \mathbf{e}_n\}$ is an orthonormal base of V, prove that

$$\forall\, \mathbf{x}, \mathbf{y} \in V, \quad (\mathbf{x}, \mathbf{y}) = \sum_i (\mathbf{x}, \mathbf{e}_i)\,(\mathbf{e}_i, \mathbf{y})$$

and deduce that $|\mathbf{x}|^2 = \sum_i (\mathbf{x}, \mathbf{e}_i)^2$.

12. Prove that $|\mathbf{x}+\mathbf{y}|^2+|\mathbf{x}-\mathbf{y}|^2 = 2|\mathbf{x}|^2+2|\mathbf{y}|^2$, where $|\mathbf{x}|$ is the length of $\mathbf{x}$ in a Euclidean vector space.

13. Show that if $\mathbf{x}$ is orthogonal to each of $\mathbf{y}_1, \mathbf{y}_2, \ldots, \mathbf{y}_m$ then it is orthogonal to any vector in the subspace spanned by $\mathbf{y}_1, \mathbf{y}_2, \ldots, \mathbf{y}_m$.

14. In an affine space show that the mid-point of AB is $\frac{1}{2}(\mathbf{a}+\mathbf{b})$.

15. In an affine space prove that the three medians of a triangle are concurrent.

16. Prove that the hyperplanes

$$a_1x_1+\ldots+a_nx_n+a_{n+1} = 0, \quad b_1x_1+\ldots+b_nx_n+b_{n+1} = 0$$

are parallel (or coincident) if and only if

$$\frac{a_1}{b_1} = \frac{a_2}{b_2} = \ldots = \frac{a_n}{b_n}.$$

When are they coincident?

17. In four-dimensional affine space let M be the two-dimensional vector subspace spanned by (1, 0, 0, 0) and (0, 1, 0, 0). Give a pair of equations which determine the plane $\mathbf{a}+M$ where $\mathbf{a}$ is (a_1, a_2, a_3, a_4).

18. Prove that two parallel linear affine subspaces in an affine space, whether or not they have the same dimension, cannot intersect. Is the converse true?

19. Prove that two distinct points in an affine space lie on one and only one line.

20. Prove that the relation of being parallel is an equivalence relation on the set of linear subspaces of a given dimension in an affine space.

21. Prove that in a projective plane two distinct lines always meet in a unique point and two distinct points always lie on a unique line.

22. In projective space of n dimensions prove that two linear projective subspaces of dimensions r and s intersect in a linear subspace of dimension $\geqslant r+s-n$ (if this $\geqslant 0$).

23. In n-dimensional projective space prove that two distinct hyperplanes intersect in a subspace of dimension exactly $n-2$.

24. Figure 24 shows points in a projective plane. Find the co-ordinates of D, E, F, X, Y, Z.

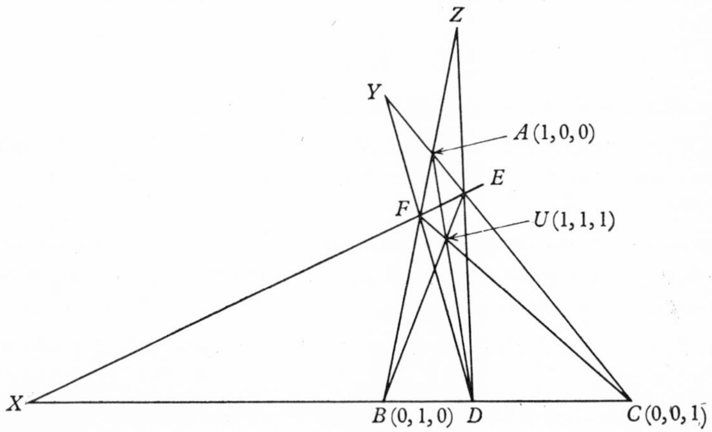

Fig. 24

25. In figure 24 find the equations of the lines *AD*, *BE*, *CF* and *EF*, *FD*, *DE*.

26. In figure 24 shows that *XYZ* are collinear and find the equation of this line.

27. In figure 24 show that the cross-ratio (*BC*, *XD*) is -1. What is (*CA*, *EY*)?

28. Prove that there is a unique projective transformation of a projective line (i.e. a one-dimensional projective space) which transforms three given distinct points to any given three distinct images.

29. Prove that the translations of an affine space *V* form a group isomorphic to *V*.

30. Defining an orthogonal square matrix by the property $\mathbf{A}'\mathbf{A} = \mathbf{I}$, prove that all $n \times n$ orthogonal matrices form a group. (Note that

$$(\mathbf{AB})' = \mathbf{B}'\mathbf{A}'$$

by the matrix multiplication rule.)

31. Assuming the standard result that $|\mathbf{AB}| = |\mathbf{A}|\,.\,|\mathbf{B}|$ prove that if $\mathbf{A}$ is orthogonal, $|\mathbf{A}| = \pm 1$. Give an example to show that the converse is untrue.

32. Show that
$$\begin{pmatrix} \cos\theta & -\sin\theta \\ \sin\theta & \cos\theta \end{pmatrix}$$
is orthogonal. Of what transformation of two-dimensional Euclidean space is this the matrix?

33. Show that
$$\begin{pmatrix} 1 & 0 \\ 0 & -1 \end{pmatrix} \quad \text{and} \quad \begin{pmatrix} -1 & 0 \\ 0 & 1 \end{pmatrix}$$
are orthogonal. Of what transformations are these the matrices?

34. Of what transformation is
$$\begin{pmatrix} k & 0 \\ 0 & k \end{pmatrix}$$
the matrix, and for which values of k is it orthogonal?

35. Prove directly that the set of all Euclidean transformations (i.e. all transformations of the form $\mathbf{x} \rightarrow \theta(\mathbf{x}) + \mathbf{k}$ where θ is orthogonal) is a group.

36. Prove that the group of translations of an affine space is an invariant subgroup of the affine group, and that the quotient group is isomorphic to the full linear group.

37. Prove that the full linear group is not an *invariant* subgroup of the affine group.

38. Interpret geometrically the solutions of the simultaneous equations in exercises 9A, nos. **52–7**.

39. Find the points of intersection in complex two-dimensional Euclidean geometry of the circle $x^2+y^2 = 1$ and the line $x = 2$.

40. Find the points of intersection in complex two-dimensional Euclidean geometry of the circle $x^2+y^2 = 1$ and the ellipse $\frac{1}{9}x^2+\frac{1}{4}y^2 = 1$.

41. Prove that relative to an orthonormal base in a unitary space

$$(\mathbf{x}, \mathbf{y}) = \xi_1 \bar{\eta}_1+\ldots+\xi_n \bar{\eta}_n.$$

42. If $\mathbf{A}$ is unitary and $|\mathbf{A}| = \alpha+i\beta$ where α and β are real prove that $\alpha^2+\beta^2 = 1$ (i.e. the modulus of the determinant is 1).

43. How many points and lines are there in a two-dimensional affine space over the finite field of residues modulo p? How many lines pass through each point and how many points lie on each line?

44. In figure 24 show that if the field of scalars has characteristic 2 (i.e. is $GF[2^k]$ for some k) then X and D, Y and E, Z and F coincide.

45. In figure 22 find the cross-ratios (BC, DX), (CB, DX), (DC, BX).

Exercises 10B

1. Use the result of worked exercise 1 to prove Ceva's theorem in an affine space: if ABC is a triangle and U any point in the plane of ABC and if AU, BU, CU meet BC, CA, AB respectively at P, Q, R, then

$$\frac{BP}{PC}\cdot\frac{CQ}{QA}\cdot\frac{AR}{RB} = +1.$$

2. Prove that in an affine space of dimension n, if U and V are linear subspaces of dimensions r and s, U and V intersect in general in a linear subspace of dimension $r+s-n$ if this is $\geqslant 0$, and do not intersect in general if $r+s-n < 0$. Deduce that if W is a linear subspace of dimension t the intersection of U, V and W is in general a linear subspace of dimension $r+s+t-2n$, if this $\geqslant 0$.

3. Prove that if a linear affine subspace contains two points then it also contains any point on the line joining them.

4. If $A_1, \ldots, A_n$ are n points in an affine space show that the linear subspace of least dimension containing all of these has dimension $\leqslant n$, and that there is only one such subspace of this dimension.

(*Hint.* Consider the intersection of all subspaces containing $A_1, \ldots, A_n$.),

5. Figure 25 shows points in a projective plane. Prove Pappus' theorem, that X_1, X_2, X_3 are collinear.

6. Prove Desargue's theorem (see worked exercise 2) when the figure lies in three-dimensional projective space (so that the triangles ABC, $A'B'C'$ are not necessarily coplanar).

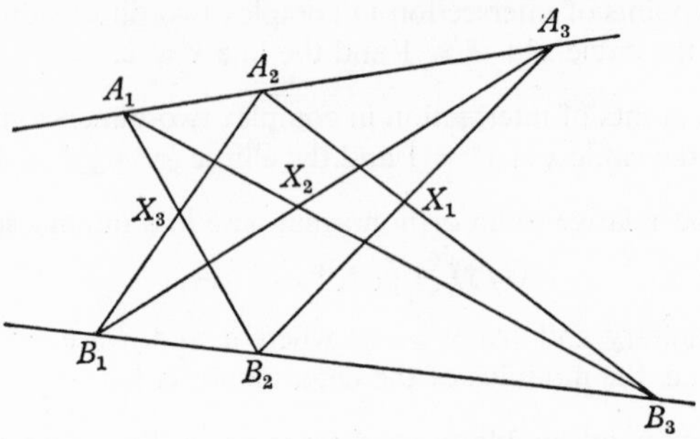

Fig. 25

7. Show that a two-dimensional projective space (i.e. a projective plane) may be represented in three-dimensional Euclidean space as follows.

Take a unit sphere centre O and consider the hemisphere for which $z \geqslant 0$. Points in the projective plane are given by points on this hemisphere, but points at the opposite ends of a diameter in $z = 0$ (i.e. a diameter of the 'open' face of the hemisphere) are identified to give the same point in the projective plane. The Euclidean co-ordinates (x, y, z) of a point on the hemisphere are taken as a possible triad of homogeneous co-ordinates for the corresponding projective point.

What are the lines in the projective plane represented by in this representation?

(This is known as *stereographic projection* and may be generalised to give a representation of an n-dimensional projective space in an $(n+1)$-dimensional Euclidean space.)

8. If the cross-ratio of 4 points (AB, CD) in that order is k, prove that $(BA, DC) = (CD, AB) = (DC, BA) = k$. Prove that the 24 orders of writing the points give 6 cross-ratios, the other 5 being

$$\frac{1}{k}, \quad 1-k, \quad \frac{1}{1-k}, \quad \frac{k}{k-1}, \quad \frac{k-1}{k}.$$

9. Prove that the affine group of n-dimensional affine space is isomorphic to the group of non-singular $(n+1)\times(n+1)$ matrices whose $(n+1)$th row is $(0, 0, \ldots, 0, 1)$, the transformation $\mathbf{x} \to \theta(\mathbf{x})+\mathbf{k}$ corresponding to the matrix

$$\begin{pmatrix} a_{11} & \cdots & a_{1n} & k_1 \\ \vdots & \ddots & \vdots & \vdots \\ a_{n1} & \cdots & a_{nn} & k_n \\ 0 & \cdots & 0 & 1 \end{pmatrix},$$

where (a_{ij}) is the matrix of the linear transformation θ.

10. Let θ be a non-singular linear transformation of n-dimensional Euclidean space which alters all lengths in the same ratio k $(\neq 0)$, so that $|\theta(\mathbf{x})| = k|\mathbf{x}|$.

(i) Prove that $(\theta(\mathbf{x}), \theta(\mathbf{y})) = k^2(\mathbf{x}, \mathbf{y})$, so that θ alters inner products in the ratio k^2.

(ii) Prove that θ leaves angle and orthogonality invariant, but alters distances in the ratio k.

(iii) Prove that all such linear transformations, for variable non-zero k, form a group.

(iv) Prove that all transformations of the form $\mathbf{x} \to \theta(\mathbf{x})+\mathbf{k}$, where θ is as defined above, form a group. (This is called the *similarity group*: it preserves shape but not size.)

11. If $\{\mathbf{e}_1, \ldots, \mathbf{e}_n\}$ is an orthonormal base of a Euclidean space V, and if $\mathbf{e}_i^* = \sum_{j=1}^{n} p_{ij}\mathbf{e}_j$, show that $\{\mathbf{e}_1^*, \ldots, \mathbf{e}_n^*\}$ is an orthonormal base if and only if (p_{ij}) is orthogonal.

12. In an n-dimensional projective space over a finite field with k elements show that there are $k^n+k^{n-1}+\ldots+k+1$ points. If $n = 3$ find the number of lines.

13. Identify the full linear group of the two-dimensional vector space over the field of residues modulo 2.

14. Prove that if θ is a linear transformation of a Euclidean vector space V into itself then $(\theta(\mathbf{x}), \mathbf{y}) = 0 \;\forall\; \mathbf{x}, \mathbf{y} \in V$ if and only if θ is the zero transformation.

15. Let V be the infinite dimensional vector space of the polynomials over the reals. Define (P, Q) to be $\int_0^1 P(t)\,Q(t)\,dt$ for any two polynomials P and Q in V. Show that this definition satisfies the four axioms for an inner product given in §10.1. (Thus we have here an example of an infinite dimensional Euclidean vector space.)

Give suitable definitions of length, distance, angle and orthogonality for this space.

16. Figure 26 shows a figure in a projective plane, where A, B, C, D are any four points no three being collinear, joined by six lines which meet in the three other points X, Y and Z. Prove that XYZ are collinear if and only

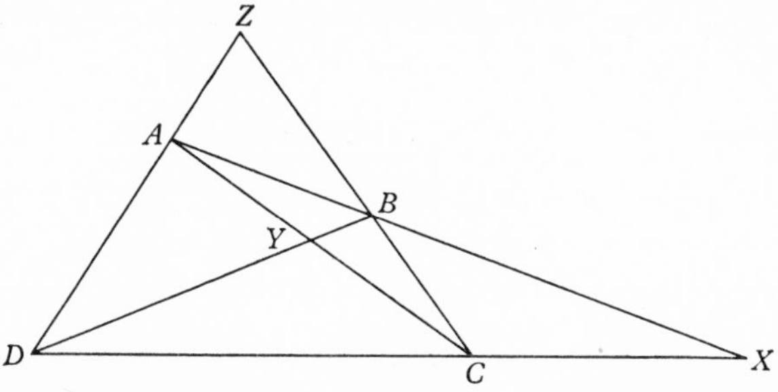

Fig. 26

if the field of co-ordinates has characteristic 2. (That XYZ are *not* collinear is often known as Fano's axiom: its postulation for a projective geometry removes the possibility of the scalar field having characteristic 2. The figure is known as a *complete quadrangle*, the triangle XYZ being the *diagonal point triangle*.)

17. Two teams play a series of tennis matches, each match being a singles match between opponents, one from either team. No pair of opponents play together more than once and not all pairs play together. Any two members of the same team have one and only one common opponent. Show how the matches might be arranged if there are n^2+n+1 members of each team (for some positive integer n). In how many matches does each player take part?

(*Hint*. Represent the players of one team by points and those of the other by lines, and let a point lie on a line if and only if the two players play one another.)

11

BOOLEAN ALGEBRA

11.1. Introduction—the algebra of sets and the algebra of statements

In 1854 George Boole published *The Laws of Thought*, in which he showed that many of the basic ideas of logic could be expressed in symbolical terms and that laws similar to the ordinary rules of algebra could be applied to these symbols. This book was one of the most important and far reaching in the history of mathematics: for the first time it was seen that algebra, hitherto applied almost exclusively to numbers or similar mathematical entities, was in fact of much wider application. Abstract algebra could be said to have begun with Boole.

The algebra that Boole used to express logical ideas, when formalised into an abstract system in the modern manner, is now known as *Boolean algebra*. Besides having local applications, it is the natural algebra for work in the theory of sets, and other applications have also been discovered.

In this chapter we define in §11.2 a Boolean algebra as an abstract system and consider some simple consequences of the laws in the following section. In §11.4 we investigate the idea of a *Boolean function*, continuing this study in §11.5. In §11.6 we consider some applications to logic, and in §11.7 indicate that any finite Boolean algebra may be represented as an algebra of sets. In the final sections we investigate briefly some other applications and a related idea, that of a *lattice*. First we must say a little about set algebra and statement algebra (the basic logical application) in general.

We studied the elementary theory of sets in volume 1 (chapter 3) and we assume that the reader is familiar with the ideas, which are not difficult. Briefly, the elements are the subsets of a given universal set I, and are combined in the following ways. $A \cup B$ (the *union* of A and B) is the set of elements in either A

or B or both, and $A \cap B$ (the *intersection* of A and B) is the set of elements in both A and B. The *complement* of A (written $\bar{A}$ in volume 1, but now written in the more usual form A') is the set of elements not in A, and we write $A \subseteq B$ if all elements of A are in B, and $A \subset B$ if $A \subseteq B$ and $A \neq B$. The empty set is denoted by $\varnothing$.

The above rules of combination satisfy various laws, which will all be given in the next section. Examples are the Commutative Laws that $A \cup B = B \cup A$ and $A \cap B = B \cap A$, the Absorption Laws that $A \cup (A \cap B) = A \cap (A \cup B) = A$, the Laws that $A \cup I = I$ and $A \cap \varnothing = \varnothing$, and so on. These can all be easily proved by considering elements of the sets, or by the use of Venn diagrams (see volume 1 or any standard work on set theory for details).

The algebra of statements was discussed briefly also in volume 1 (§3.6). A *statement*, or *proposition*, is a declaratory sentence which is either true or false, but not both. We do not attempt to define the terms 'true' and 'false', nor do we attempt any analysis of the meaning of words or construction of sentences (beyond a few simple constructions). We are not interested, as far as our logic is concerned, in the practical sense of our statements: for example a statement such as 'monkeys are insects and have fifteen legs' may appear nonsensical, but is perfectly valid as a statement. We repeat that a sentence is a statement if it is either true or false, but not both.

Examples of statements, to be used in the following paragraphs, are (denoted by capital letters):

A: 'all monkeys are stupid',
B: 'apples are good this year',
C: 'I like cats',
D: 'all mammals are stupid'.

Two statements are *equal* if they are logically equivalent, that is if they are either both true or both false in all cases. For example the statements 'all monkeys are stupid and I like cats' and 'I like cats and all monkeys are stupid' are logically equivalent. To take a more complicated example (an example of the De Morgan Laws to be discussed later), the following statements are logically equivalent: 'that all monkeys are stupid *and* I like cats is not true' and 'either not all monkeys are stupid or I do not like cats'. Note that the statements 'apples are good this year' and 'I like cats', even if they both happen to be true for a particular year and a particular person, are not *logically* equivalent and therefore not equal statements.

We combine two statements in two ways as follows. The *disjunction* of A and B is the statement 'either A or B or both'. This is written $A \cup B$ in line with the notation for set theory, and is usually known as *union* for convenience. (In volume 1 we used the notation $\vee$ but now revert to the more general form $\cup$.) Thus with our examples above, $A \cup B$ is the statement 'either all monkeys are stupid or apples are good this year'. $A \cup B$ is true if either A or B or both are true, and false if both A and B are false.

The *conjunction* of A and B is the statement 'both A and B': it is written $A \cap B$ and known as *intersection* (in volume 1 we used $A \wedge B$). In the example $A \cap B$ is the statement 'all monkeys are stupid and apples are good this year'. $A \cap B$ is true if both A and B are true and false otherwise.

The *negation*, or *complement*, of A, written A', is the statement 'it is false that A', or briefly 'not A'. Thus A' in our example is 'it is false that all monkeys are stupid' or, in other words, 'not all monkeys are stupid' or 'there are some monkeys that are not stupid'. A' is false when A is true and true when A is false.

If the truth of A logically implies that of B then we say that $A \subseteq B$, and if $A \subseteq B$ and $A \neq B$ then $A \subset B$. Thus in the above example (assuming that we are given that all monkeys are mammals) we can say that $D \subseteq A$.

Finally we introduce the symbols I to stand for any statement that is always true, and $\varnothing$ for any statement that is always false. (We will replace this notation later by 1 and 0 for both set theory and statement algebra and in fact for any Boolean algebra.)

The rules of combination we have given satisfy the same laws as they do in set theory (this is of course the reason for using the same notation). The reader may easily check this in the case of those laws given earlier in this section, and we will show it generally in §11.3.

11.2 Boolean algebra: its definition and basic laws

The algebra of sets and the algebra of statements satisfy the same basic laws and are both examples of the abstract structure known as a Boolean algebra. We now proceed to define this and to prove the fundamental laws and properties. As with other abstract structures, the reason for its importance and for the choice of particular axioms lies in the fact that it forms the common basis to both the types of algebra examined in § 11.1, and to other practical systems also.

We must first say a word about notation. As in most branches of mathematics there is no generally accepted usage. We choose to use the notation and language of set theory—we will use $\cup$, $\cap$ and $'$ and talk of *union*, *intersection* and *complement*, although it is sometimes convenient in oral work to shorten the first two to 'cup' and 'cap'. Another common notation is that of sum and product, writing $A+B$ and AB for $A \cup B$ and $A \cap B$ respectively, but we prefer not to use this, partly because it obscures the duality between the two processes and leads to strange looking equations such as

$$A+BC = (A+B)(A+C),$$

the second Distributive Law. We are applying the set theory notation to general Boolean algebras, but the reader must not assume without proof that we can apply all the methods and ideas of set theory to the general case: thus we will not prove general theorems by Venn diagrams, although these are very useful in giving checks and indications to our work.

Instead of the notation I and $\varnothing$ we will use 1 and 0, and call these the *unity* and *zero*, in line with notation for rings and fields.

There are many fundamental laws for a Boolean algebra (we will give 21 in fact, mainly in pairs), but not all are independent. We could define the structure by giving all these as axioms, but this would be needlessly artificial. At the other extreme, we could select the least number of independent ones and prove the others from them. One such set, given by Huntington in 1904, consists of 8 laws grouped in 4 pairs, from which all others can be proved. The proofs of most of these others are simple, but that of the Associative Laws is tedious (though not difficult) and so we will find it convenient to assume these laws as our 9th and 10th axioms. We will thus take 10 axioms, in 5 pairs, and prove the remaining laws from them. (The reader who is interested in the proof of the Associative Laws from the others may see J. E. Whitesitt: 'Boolean Algebra and its Applications', §2.3.) Our axiomatic definition is then as follows.

Definition of a Boolean algebra

A Boolean algebra is a set S of elements $A, B, \ldots$ such that $\forall A$ and $B \in S$ there exist unique elements $A \cup B$ and $A \cap B$ in S satisfying the following axioms:

Axiom 1. *The Commutative Laws*

$$(a)\ A \cup B = B \cup A;$$

$$(b)\ A \cap B = B \cap A.$$

Axiom 2. *The Associative Laws*

$$(a)\ A \cup (B \cup C) = (A \cup B) \cup C;$$

$$(b)\ A \cap (B \cap C) = (A \cap B) \cap C.$$

Axiom 3. *The Distributive Laws*

$$(a)\ A \cup (B \cap C) = (A \cup B) \cap (A \cup C);$$

$$(b)\ A \cap (B \cup C) = (A \cap B) \cup (A \cap C).$$

Axiom 4. *There exist elements* 0 *and* 1 *in S such that*

$$(a)\ A \cup 0 = A \text{ and}$$

$$(b)\ A \cap 1 = A \text{ for all } A \in S.$$

Axiom 5. *For any* $A \in S$ *there exists an element* $A' \in S$ *such that* $(a)\ A \cup A' = 1$ *and* $(b)\ A \cap A' = 0$.

These laws exhibit obvious similarities with the laws for groups, rings and fields. We will leave the reader to make his own observations about these similarities and the differences: he should be mature enough to do so at this stage, and some of these observations are to be found in chapter 3 of volume 1. Similarly, the reader should by now be able to follow abstract proofs readily, and so the theorems in this and later sections will be proved without too much verbal explanation.

It can be seen at once that the axioms above are consistent (assuming that set theory is), since they are all satisfied in the particular case where the elements of S are the subsets of a given set.

Duality. The axioms are in pairs: each of any pair is obtained from the other by interchanging $\cup$ and $\cap$, and also 0 and 1. Hence from any theorem we may obtain another by a similar interchange, and in fact in the following pairs of theorems we will prove only one in each case and obtain the other by applying this principle, the so-called *principle of duality*. The operations $\cup$ and $\cap$ are in fact identical in their properties (unlike $+$ and $\times$ in ordinary algebra and in fields) and it is immaterial which is called '$\cup$' and which '$\cap$': but once a notation is chosen it must of course be adhered to in any work involving the given Boolean algebra.

We now proceed to prove the other basic laws from the axioms.

Theorem 11.2.1. *The Idempotent Laws.*

For any $A \in S$, (*a*) $A \cup A = A$ *and* (*b*) $A \cap A = A$.

$$\begin{aligned} A \cup A &= (A \cup A) \cap 1 && \text{by axiom } 4(b),\\ &= (A \cup A) \cap (A \cup A') && \text{by } 5(a),\\ &= A \cup (A \cap A') && \text{by } 3(a),\\ &= A \cup 0 && \text{by } 5(b),\\ &= A && \text{by } 4(a). \end{aligned}$$

(*b*) is proved dually.

Theorem 11.2.2. *For any* $A \in S$, (*a*) $A \cup 1 = 1$ *and* (*b*) $A \cap 0 = 0$.

$$\begin{aligned} A \cup 1 &= 1 \cap (A \cup 1) && \text{by axioms } 1(b) \text{ and } 4(b),\\ &= (A \cup A') \cap (A \cup 1) && \text{by } 5(a),\\ &= A \cup (A' \cap 1) && \text{by } 3(a),\\ &= A \cup A' && \text{by } 4(b),\\ &= 1 && \text{by } 5(a). \end{aligned}$$

(*b*) is proved dually.

Theorem 11.2.3. *The Absorption Laws.*

For any A and $B \in S$,

$$(a)\ A \cup (A \cap B) = A;$$

$$(b)\ A \cap (A \cup B) = A.$$

$A \cup (A \cap B) = (A \cap 1) \cup (A \cap B)$ by axiom 4(*b*),

$= A \cap (1 \cup B)$ by 3(*b*),

$= A \cap 1$ by theorem 11.2.2(*a*) and axiom 1(*a*),

$= A$ by 4(*b*).

(*b*) is proved dually.

Theorem 11.2.4. *The uniqueness of the complement.*

There is only one element A' satisfying the conditions of axiom 5.

Suppose X and Y are such that $A \cup X = A \cup Y = 1$ and $A \cap X = A \cap Y = 0$. We will prove that $X = Y$.

$X = 1 \cap X$ by axioms 1(*b*) and 4(*b*),

$= (A \cup Y) \cap X$

$= (A \cap X) \cup (Y \cap X)$ by 1(*b*) and 3(*b*),

$= 0 \cup (Y \cap X)$

$= Y \cap X$ by 4(*a*) and 1(*a*).

Similarly $Y = X \cap Y = Y \cap X$ by 1(*b*). Hence $X = Y$.

Theorem 11.2.5. *The uniqueness of the zero and unity.*

There is only one element satisfying the condition of axiom 4(a), and only one satisfying the condition of axiom 4 (b).

We prove the result for 0: the uniqueness of 1 follows by duality.

Suppose $A \cup X = A \ \forall A \in S$. Then $0 \cup X = 0$. But $0 \cup X = X$ by 1(*a*) and 4(*a*) and hence $X = 0$.

Theorem 11.2.6. *A is the complement of A', i.e. $(A')' = A$.*

$A \cup A' = 1$ and $A \cap A' = 0$ give, by axiom 1,

$A' \cup A = 1$ and $A' \cap A = 0$ and so A is a complement of A'.

But the complement of A' is unique by theorem 11.2.4, and the result follows.

Theorem 11.2.7. *0 and 1 are complements. I.e.* (*a*) $0' = 1$ *and* (*b*) $1' = 0$.

By theorem 11.2.2, $0 \cup 1 = 1$ and $1 \cap 0 = 0$, or by 1(*b*) $0 \cap 1 = 0$. Hence 1 is a complement of 0 and is the unique complement by theorem 11.2.4.

It follows, either dually or by theorem 11.2.6, that 0 is the complement of 1.

Theorem 11.2.8. *The De Morgan Laws.*

$$(a)\ (A \cup B)' = A' \cap B';$$
$$(b)\ (A \cap B)' = A' \cup B'.$$

$(A \cup B) \cup (A' \cap B') = (A \cup B \cup A') \cap (A \cup B \cup B')$ by axiom 3(*a*),

$$= 1 \cap 1$$

by 2(*a*) and other axioms and theorem 11.2.2(*a*),

$$= 1 \text{ by } 4(b).$$

Also $(A \cup B) \cap (A' \cap B') = A' \cap B' \cap (A \cup B)$

by axioms 1(*b*) and 2(*b*), omitting the brackets for triple combinations, $= (A' \cap B' \cap A) \cup (A' \cap B' \cap B)$

by 3(*b*), $= 0 \cup 0$

by various axioms and theorems, including axiom 5(*b*),

$$= 0 \text{ by } 4(a).$$

Hence $A' \cap B'$ is a complement of $A \cup B$ and is unique by theorem 11.2.4. The first result follows and the second is proved dually.

We have now proved or assumed all the basic laws except those for inclusion. It is convenient to state them together, and we do this below.

The Commutative Laws

$$A \cup B = B \cup A;$$
$$A \cap B = B \cap A.$$

The Associative Laws

$$A \cup (B \cup C) = (A \cup B) \cup C;$$
$$A \cap (B \cap C) = (A \cap B) \cap C.$$

The Distributive Laws

$$A \cup (B \cap C) = (A \cup B) \cap (A \cup C);$$

$$A \cap (B \cup C) = (A \cap B) \cup (A \cap C).$$

The Idempotent Laws

$$A \cup A = A;$$

$$A \cap A = A.$$

The Absorption Laws

$$A \cup (A \cap B) = A;$$

$$A \cap (A \cup B) = A.$$

The laws for 0 *and* 1

$$A \cup 0 = A;$$

$$A \cap 1 = A.$$

$$A \cup 1 = 1;$$

$$A \cap 0 = 0.$$

The laws for complements

$$A \cup A' = 1;$$

$$A \cap A' = 0.$$

$$(A')' = A.$$

$$0' = 1;$$

$$1' = 0.$$

The De Morgan Laws

$$(A \cup B)' = A' \cap B';$$

$$(A \cap B)' = A' \cup B'.$$

Inclusion

The relation $A \subseteq B$, which we have met in set theory and statement algebra, is defined in terms of the operations and elements already defined, as follows.

Definition. *If A and B are elements of a Boolean algebra S, we say that $A \subseteq B$ (A is* contained *or* included *in B) if and only if* $A \cap B' = 0$.

Theorem 11.2.9. *Inclusion satisfies the following laws:*

(*a*) *The Reflexive Law.* $A \subseteq A \; \forall A \in S$;

(*b*) *The Anti-symmetric Law.* $A \subseteq B$ *and* $B \subseteq A \Rightarrow A = B$;

(*c*) *The Transitive Law.* $A \subseteq B$ *and* $B \subseteq C \Rightarrow A \subseteq C$.

(*a*) $A \cap A' = 0$ by axiom 5(*b*).

(*b*) Suppose $A \cap B' = 0$ and $B \cap A' = 0$. It follows from the De Morgan Laws that $(A \cap B')' = A' \cup B = 1$, and so A' and B are complements. Thus $B = (A')' = A$.

(*c*) If $A \cap B' = 0$ and $B \cap C' = 0$ we see that

$$\begin{aligned} A \cap C' &= A \cap C' \cap (B \cup B') \quad \text{since} \quad B \cup B' = 1, \\ &= (A \cap C' \cap B) \cup (A \cap C' \cap B') \\ &= 0 \cup 0 \quad \text{by hypothesis,} \\ &= 0. \end{aligned}$$

If $A \subseteq B$ and $A \neq B$ we sometimes write $A \subset B$ and say that A is *strictly* contained in B. We also sometimes write $A \subseteq B$ as $B \supseteq A$ and say B *contains* A (in line with set theory).

Let us consider the dual statement. This is that $A \cup B' = 1$. But this is equivalent to $(A \cup B')' = 0$ or, by the De Morgan Laws, that $A' \cap B = 0$, i.e. that $B \subseteq A$. Hence the dual of $A \subseteq B$ is $A \supseteq B$.

Theorem 11.2.10. (*a*) $A \subseteq B$ *and* $A \subseteq C \Rightarrow A \subseteq B \cap C$;

(*b*) $A \subseteq B \Rightarrow A \subseteq B \cup C$ *for any* C;

(*c*) $A \subseteq B \Leftrightarrow B' \subseteq A'$.

(*a*) $A \cap B' = A \cap C' = 0$ and so $A \cap (B' \cup C') = 0$ by the Distributive Laws.

Hence by the De Morgan Laws $A \cap (B \cap C)' = 0$ and so $A \subseteq B \cap C$.

(*b*)
$$\begin{aligned} A \cap (B \cup C)' &= A \cap (B' \cap C') \quad \text{by De Morgan Laws,} \\ &= (A \cap B') \cap C' \\ &= 0 \quad \text{by hypothesis.} \end{aligned}$$

(*c*)
$$\begin{aligned} A \subseteq B &\Leftrightarrow A \cap B' = 0 \Leftrightarrow (A')' \cap B' = 0 \\ &\Leftrightarrow B' \cap (A')' = 0 \Leftrightarrow B' \subseteq A'. \end{aligned}$$

11.3. Simple consequences

We show first that the two algebras in the first section are in fact Boolean algebras.

Theorem 11.3.1. *The subsets of a given set I form a Boolean algebra under union, intersection and complement, with the empty set as the zero and the universal set I as the unity.*

We merely require to verify the axioms on p. 315.

(1) Immediate from the definition.

(2) Also immediate: $A \cup (B \cup C)$ and $(A \cup B) \cup C$ are both the set of elements in at least one of A, B and C, and $A \cap (B \cap C)$ and $(A \cap B) \cap C$ are both the set of elements in all three of A, B and C.

(3) (*a*) If $x \in A \cup (B \cap C)$ then $x \in A$ or $x \in B$ *and* $x \in C$

$$\Rightarrow x \in A \text{ or } B \quad \textit{and} \quad x \in A \text{ or } C$$

$$\Rightarrow x \in (A \cup B) \cap (A \cup C).$$

Therefore $\quad A \cup (B \cap C) \subseteq (A \cup B) \cap (A \cup C).$

Also $\quad x \in (A \cup B) \cap (A \cup C) \Rightarrow x \in A \text{ or } B$

$$\textit{and} \quad x \in A \text{ or } C$$

$$\Rightarrow x \in A \quad \text{or} \quad x \in B \textit{ and } C$$

$$\Rightarrow x \in A \cup (B \cap C).$$

Therefore $\quad A \cup (B \cap C) \supseteq (A \cup B) \cap (A \cup C).$

Hence $\quad A \cup (B \cap C) = (A \cup B) \cap (A \cup C).$

(*b*) is proved similarly.

(4) Since there are no elements in the empty set $\varnothing$, the set of elements in either A or $\varnothing$ or both is merely A: hence $A \cup \varnothing = A$. Since every element is in I the set of elements in both A and I is A: hence $A \cap I = A$ and I is the unity. (Remember that we are considering only subsets of I, so that A is certainly completely in I.)

(5) By the definition of complement as the set of elements not in A (but in I of course) we see that all elements in I are in either A or A', and none are in both: hence

$$A \cup A' = I \quad \text{and} \quad A \cap A' = \varnothing.$$

Since the empty set and the universal set are the zero and

the unity respectively we will henceforth call them 0 and 1 in line with the general notation.

Note that if I is finite with n elements the Boolean algebra of its subsets has 2^n elements, since there are this many distinct subsets, as every member of I may or may not be included in a subset. If I has infinitely many elements we obtain an infinite Boolean algebra.

Theorem 11.3.2. *A set of statements that is complete in the sense that for any two statements in the set their disjunction and conjunction are in it, the negation of any statement of the set is in it, and the set contains statements that are always true and always false, is a Boolean algebra under these operations.*

(1) Clearly true, since 'either A or B or both' is logically equivalent to 'either B or A or both', and 'both A and B' is logically equivalent to 'both B and A'. (Remember that equality of statements means that they are logically equivalent.)

(2) Also clearly true. Both $A \cup (B \cup C)$ and $(A \cup B) \cup C$ are 'either A, B or C or more than one', and both $A \cap (B \cap C)$ and $(A \cap B) \cap C$ are 'all of A, B and C'.

(3) 'Either A or both B and C' is equivalent to '(either A or B) *and* (either A or C)'. 'Both A and (either B or C)' is equivalent to 'both A and B or both A and C'. (Note how difficult it is to word such composite statements unambiguously: the use of our algebraic symbols is much more precise, and we have even been reduced to using brackets above.)

(4) 'Either A or a statement that is never true' is logically equivalent to A (i.e. both are either true or false). 'Both A and a statement that is always true' is also equivalent to A.

(5) 'Either A or not A' is always true, and 'both A and not A' is always false.

The Associative Laws generalise to more than three elements in the usual way, and we can ignore the brackets when forming the union, or intersection, of more than two elements.

The Distributive Laws also generalise: for example,

$$A \cup (B \cap C \cap D) = (A \cup B) \cap (A \cup C) \cap (A \cup D)$$

and so on.

The Idempotent Laws lead to the fact that there is no theory of either powers or multiples of elements.

The Cancellation Laws, that $A \cup X = A \cup Y \Rightarrow X = Y$ or $A \cap X = A \cap Y \Rightarrow X = Y$ are not true (as may easily be seen in the algebra of sets by Venn diagrams), but if *both* hypotheses are true then the result follows.

Theorem 11.3.3. *The Cancellation Law.*

If both $A \cup X = A \cup Y$ and $A \cap X = A \cap Y$ then $X = Y$.

$X = X \cap (X \cup A)$ by the Absorption Law,

$= X \cap (Y \cup A)$ by hypothesis,

$= (X \cap Y) \cup (X \cap A)$

$= (X \cap Y) \cup (Y \cap A)$ by hypothesis,

$= Y \cap (A \cup X)$ by the Distributive and Commutative Laws,

$= Y \cap (A \cup Y)$ by hypothesis,

$= Y$ by the Absorption Laws.

11.4. Boolean functions and their normal forms

We may define functions in a Boolean algebra in a similar manner to those in ordinary algebra, but the work will be found to be much simpler. A *Boolean function* in the variables $X_1, X_2, \ldots, X_n$ is a finite expression involving $X_1, X_2, \ldots, X_n$ and the symbols $\cup$, $\cap$ and $'$. For simplicity we consider only those functions that do not involve constants (we may in fact consider a function involving a constant A as being obtained from a function including the extra variable Y by putting $Y = A$, and so our restriction is more apparent than real).

Let us first consider the Boolean functions in a single variable X.

Because of the Idempotent Laws we see that the only function involving only $\cup$ and $\cap$ is X itself. If we introduce complementation we obtain X', and then $X \cup X' = 1$ and $X \cap X' = 0$, but however we proceed we cannot obtain any other functions. Thus there are only four Boolean functions in a single variable X: viz. X, X', 1 and 0. The latter two contain constants indeed, but are written as short for $X \cup X'$ and $X \cap X'$, and so are admissible.

If we consider two variables X and Y the position is more complicated, and we can form various functions. But there are still only a limited number (16 in fact) and to see this let us consider the special case of the theory of sets. (This in fact is not so special as it seems, since any Boolean algebra is isomorphic to an algebra of sets, as we shall see in §11.7.) We investigate this by Venn diagrams.

Consider the Venn diagram showing the two sets (or variable elements of our Boolean algebra) X and Y in general position. This is shown in figure 27.

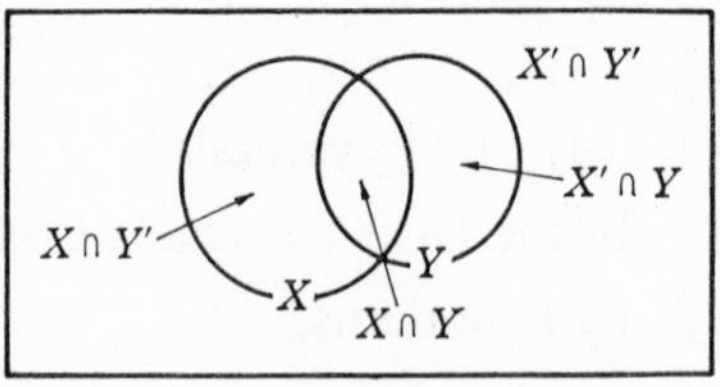

Fig. 27

The sets X and Y divide the universal set 1 into 4 regions, which may be written conveniently as $X \cap Y$, $X \cap Y'$, $X' \cap Y$ and $X' \cap Y'$. Now any Boolean function whatever must be able to be shown in our Venn diagram, without introducing any more sets (since the union, intersection and complement of any elements of the Venn diagram are shown in the diagram). Each of the four basic areas may or may not be included, and we thus obtain 16 possible functions, each of the form $Z_1 \cup Z_2 \cup \ldots \cup Z_m$ where $m \leqslant 4$ and each Z_i is one of the basic functions $X \cap Y$, $X \cap Y'$, $X' \cap Y$, $X' \cap Y'$.

A function put in this form is said to be in *disjunctive normal form*, and we have seen that any Boolean function in two variables can be expressed in this manner. This is true generally, for functions in any number of variables, and we proceed to show this. We may convince ourselves of the truth of the result, and may work out the normal form for particular examples, by considering Venn diagrams, but our proof will be independent of these, thus avoiding any begging of the question of whether or not a Boolean algebra may be represented as an algebra of sets.

Theorem 11.4.1. *Any Boolean function (without constants) in the n variables* $X_1, X_2, \ldots, X_n$ *may be expressed in* disjunctive normal form, *that is as the union of terms, each of the form* $\bigcap_{i=1}^{n} F_i(X_i)$ *where* $F_i(X_i)$ *is either* X_i *or* X_i' *with no two terms the same.*

We will give a process by which the reduction to the normal form in any case may be performed, and explain it briefly.

Stage 1. Any primes ′ outside brackets may be moved inside by use of the De Morgan Laws, so that we obtain a function involving merely the union and intersection of terms X_i and X_i'.

Stage 2. By the Distributive Law axiom 3(*b*) we may obtain the function as the union of terms, a typical one being called T say, each T being the intersection of various letters X_i and X_i'.

Stage 3. If any X_i appears twice in a term T we may ignore its repeated occurrence by the Idempotent Laws, and if both X_i and X_i' occur we replace their intersection by 0 and thus the whole term T by 0, and ignore it (unless the whole function is 0). Thus now any term T contains a variable or its complement at most once.

Stage 4. If any term T does not contain any term X_i or X_i' we may replace it as follows:

$$T = T \cap 1 = T \cap (X_i \cup X_i') = (T \cap X_i) \cup (T \cap X_i')$$

and obtain two terms containing X_i or X_i'. Repeat for all missing terms.

Stage 5. Finally we ignore any duplicate terms by the Idempotent Laws.

Example. We use X, Y, Z for convenience, rather than X_1, X_2, X_3.

$$\begin{aligned}
&(X \cup Y \cup Z) \cap [(X \cap Y) \cup (X' \cap Z)]' \\
&\quad = (X \cup Y \cup Z) \cap [(X \cap Y)' \cap (X' \cap Z)'] \\
&\quad = (X \cup Y \cup Z) \cap (X' \cup Y') \cap (X \cup Z') && \text{[Stage 1]} \\
&\quad = (X \cap X' \cap X) \cup (X \cap X' \cap Z') \cup (X \cap Y' \cap X) \\
&\qquad \cup (X \cap Y' \cap Z') \cup (Y \cap X' \cap X) \cup (Y \cap X' \cap Z') \\
&\qquad \cup (Y \cap Y' \cap X) \cup (Y \cap Y' \cap Z') \cup (Z \cap X' \cap X) \\
&\qquad \cup (Z \cap X' \cap Z') \cup (Z \cap Y' \cap X) \cup (Z \cap Y' \cap Z') && \text{[Stage 2]}
\end{aligned}$$

$$= (X \cap Y') \cup (X \cap Y' \cap Z') \cup (Y \cap X' \cap Z') \cup (Z \cap Y' \cap X) \qquad \text{[Stage 3]}$$

$$= (X \cap Y' \cap Z) \cup (X \cap Y' \cap Z') \cup (X \cap Y' \cap Z') \cup (X' \cap Y \cap Z') \cup (X \cap Y' \cap Z) \qquad \text{[Stage 4]}$$

$$= (X \cap Y' \cap Z) \cup (X \cap Y' \cap Z') \cup (X' \cap Y \cap Z'). \qquad \text{[Stage 5]}$$

The work in the above example is tedious to write out but is essentially trivial. It is probably easier in fact than investigating the problem by means of a Venn diagram, a process that is always open to us. Another method is by use of truth tables, as will be seen in the next section.

The main use of the normal form is to investigate the equality or otherwise of two functions. In order to test this we express both functions in the disjunctive normal form: if their normal forms are the same then they are clearly equal, and we proceed to show that if they are equal then they give the same normal form, i.e. the normal form for a given function in n variables is unique.

There are precisely 2^n possible terms to be included in the disjunctive normal form in n variables, for a typical term is of the form $\bigcap_{i=1}^{n} F_i(X_i)$ where $F_i(X_i) = X_i$ or X'_i. The normal form which contains all 2^n terms is called the *complete disjunctive normal form.*

Theorem 11.4.2. *If we give each of $X_1, X_2, \ldots, X_n$ the value 0 or 1 then there is just one term in the complete disjunctive normal form that is equal to 1 and all the rest are 0.*

For if $X_i = 0$, $X'_i = 1$, and so the term $\bigcap_{i=1}^{n} F_i(X_i)$ where $F_i(X_i) = X_i$ if we give X_i the value 1, and $F_i(X_i) = X'_i$ if we give X_i the value 0, has value $\bigcap_{i=1}^{n} 1 = 1$, and all other terms contain the intersection of elements with 0 and hence are 0.

Theorem 11.4.3. *If two functions in n variables are equal (i.e. the same function) then their disjunctive normal forms are identical (i.e. contain the same terms).*

If the functions are equal then they take equal values for any values of the variables X_i. Now take any set of values 0 or 1 for the n variables, as in theorem 11.4.2. The corresponding term as defined in theorem 11.4.2 is present in the normal form if and only if the functions have value 1, and not 0, for those values of the variables X_i. But the values of the two functions are the same, and hence any term (tested by the corresponding set of values) is either present in both forms or neither, and so the two forms contain the same terms.

Corollary 1. *The disjunctive normal form in n variables for any function in n variables is unique.*

Corollary 2. *A Boolean function in n variables is completely defined by its value when all possible values* 0 *or* 1 *are given to each variable, and may be written down in normal form by selecting the terms corresponding to the choices of* 0 *and* 1 *for the variables which give the value* 1 *to the function.*

In the example given previously, we see that the given function has value 1 when $X = Z = 1$ and $Y = 0$, or when $X = 1$ and $Y = Z = 0$, or when $Y = 1$ and $X = Z = 0$, but has value 0 for all other combinations of the values 0 and 1. Any other function with the same values must be equal, and will have the same disjunctive normal form.

Note that 0 and 1 must be counted as special functions in n variables, for any $n \geqslant 1$. 1 is written as the complete disjunctive normal form, and 0 is the disjunctive normal form with no terms. With these included we see that there are precisely 2^{2^n} distinct functions in n variables, since there are 2^n terms to be either included or not included. Thus we have made a complete classification of the possible Boolean functions in any finite number of variables.

Note that if a function does not contain a variable explicitly we must introduce it when we obtain the normal form: thus in three variables $X \cap Y$ is $(X \cap Y \cap Z) \cup (X \cap Y \cap Z')$. The 2^{2^n} Boolean functions in n variables include such cases, of course. Note also in this connection that the disjunctive normal forms are unique only when considered as being in a specific number of variables: the two normal forms $X \cap Y$, in two variables, and $(X \cap Y \cap Z) \cup (X \cap Y \cap Z')$, in three, are the same function.

The Conjunctive Normal Form

The above work may all be dualised, to obtain another type of normal form, the conjunctive normal form, where a function is expressed as the intersection of terms, each of the form $\bigcup_{i=1}^{n} F_i(X_i)$, where $F_i(X_i)$ is either X_i or X_i'. Corresponding to theorem 11.4.2 we have the result that there is just one term in the complete conjunctive normal form that is equal to 0 when any set of values 0 or 1 are applied to the variables, and the conjunctive normal form of any function contains merely those terms corresponding to the sets of values of the variables that give the value 0 to the function.

11.5. Truth tables

We saw in corollary 2 to theorem 11.4.3 that a Boolean function in n variables is completely determined by its values (0 or 1) when the variables take all possible combinations of the values 0 or 1. Thus to investigate a function we merely need to investigate these 2^n cases. For example, two functions are equal if and only if their values are equal in all 2^n cases, and their normal form may be written down from this set of values by the methods of the last section. It is convenient to display the investigation in the form of a table: such a table is known as a *truth table* because of its special importance in logic, as we shall see later in this section.

The procedure is best shown by working an example, and for this purpose we use the rather complicated example discussed in the last section.

Example. Investigate whether or not the Boolean functions

$$F = (X \cup Y \cup Z) \cap [(X \cap Y) \cup (X' \cap Z)]'$$

and

$$G = (X \cap Y') \cup (X' \cap Y \cap Z')$$

are equal (i.e. are the same function) and write down their disjunctive normal forms.

The functions are functions of three variables, and we have $2^3 = 8$ cases to consider. We do this in the truth tables in figures 28 and 29. These are worked out in steps, using the elementary results for union, intersection and complement of 0 and 1 ($1 \cap 1 = 1, 1 \cup 0 = 1, 1' = 0$, etc.).

We see that all eight of the values in the final columns correspond, and so the functions are indeed equal by corollary 2 of theorem 11.4.3. Their common disjunctive normal form is written down as the union of the three terms corresponding to the value 1 in the final column, each term being the intersection of the variables with value 1 and the complements of the variables with value 0, as in theorem 11.4.2. Hence the normal form is $(X \cap Y' \cap Z) \cup (X \cap Y' \cap Z') \cup (X' \cap Y \cap Z')$.

X	Y	Z	$X \cap Y$	$X' \cap Z$	$(X \cap Y) \cup (X' \cap Z)$	$[(X \cap Y) \cup (X' \cap Z)]'$	$X \cup Y \cup Z$	F
1	1	1	1	0	1	0	1	0
1	1	0	1	0	1	0	1	0
1	0	1	0	0	0	1	1	1
1	0	0	0	0	0	1	1	1
0	1	1	0	1	1	0	1	0
0	1	0	0	0	0	1	1	1
0	0	1	0	1	1	0	1	0
0	0	0	0	0	0	1	0	0

Fig. 28

X	Y	Z	$X \cap Y'$	$X' \cap Y \cap Z'$	G
1	1	1	0	0	0
1	1	0	0	0	0
1	0	1	1	0	1
1	0	0	1	0	1
0	1	1	0	0	0
0	1	0	0	1	1
0	0	1	0	0	0
0	0	0	0	0	0

Fig. 29

Our work so far with truth tables has been for general Boolean algebras, but they originated in the particular case of the algebra of statements. Let us now investigate them from this point of view.

We defined a statement as a sentence that is either true or false (but not both), and combined statements in various ways. For example the disjunction $A \cup B$ was a statement that is true if either A or B or both are true, and false if both A and B are false. We defined $A \cup B$ in terms of words to be 'either A or B or both', but could equally well have *defined* it by its truth value (false if both A and B are false, true otherwise) and this would have been a more satisfactory logical definition. Writing T for true and F for false we obtain the following table as defining $A \cup B$.

A	B	$A \cup B$
T	T	T
T	F	T
F	T	T
F	F	F

Fig. 30

We define $A \cap B$ similarly by figure 31, and the negation A' of A by figure 32.

A	B	$A \cap B$
T	T	T
T	F	F
F	T	F
F	F	F

Fig. 31

A	A'
T	F
F	T

Fig. 32

By means of repeated application of these tables we may investigate the behaviour of any statement formed by combining other statements A, B, etc. Two such complex statements are equal if they always have the same truth values (i.e. are logically equivalent). By drawing up tables similar to the above we may verify the Commutative, Associative and Distributive Laws. If we introduce 0 and 1 in the usual way, as statements that are always false and always true respectively we may also verify axioms 4 and 5, and thus show (independently of any intuitive meaning to be given to words, as was done in theorem 11.3.2), that the algebra of statements is a Boolean algebra.

We will leave the details to the reader, but will show the process for the Distributive Law (*a*): the verification is in the fact that the final columns in figures 33 and 34 are identical, showing that the two expressions have the same truth values for all truth values of *A*, *B* and *C*, and hence are logically equivalent and so 'equal'.

A	B	C	$B \cap C$	$A \cup (B \cap C)$
T	T	T	T	T
T	T	F	F	T
T	F	T	F	T
T	F	F	F	T
F	T	T	T	T
F	T	F	F	F
F	F	T	F	F
F	F	F	F	F

Fig. 33

A	B	C	$A \cup B$	$A \cup C$	$(A \cup B) \cap (A \cup C)$
T	T	T	T	T	T
T	T	F	T	T	T
T	F	T	T	T	T
T	F	F	T	T	T
F	T	T	T	T	T
F	T	F	T	F	F
F	F	T	F	T	F
F	F	F	F	F	F

Fig. 34

It is easy to see the similarity between the above tables showing truth values and the ones we previously used for general Boolean algebras. All we need to do is to replace the T and F by the particular statements 1 and 0: instead of saying for example that A is true we put $A = 1$, a statement that is always true, and similarly if B is false we put $B = 0$, a statement that is always false. Conjunction, disjunction and negation are then given in table form by the ordinary truth tables for intersection, union and complement—this is no accident, since we have shown that the algebra of statements is a particular case of a Boolean algebra.

We have seen how truth tables originated in the algebra of statements (the name itself implies as much), and how even there it is convenient to use 1 and 0 rather than the symbols T and F. We have also seen how, by using the idea of a normal form, we can study functions by means of truth tables, being interested only in the values 1 and 0 for the variables. This is usually the best way of investigating Boolean functions, and constructing those with required properties, particularly if the number of variables is small. Finally we note that the approach using the algebra of statements, where we are interested only in the truth or falsity of a statement, gives us an indication why we should expect that the consideration of values 1 and 0 for the variables is sufficient in the general case.

11.6. Implication and logical arguments

We have seen that the algebra of statements is a Boolean algebra, and of course one of the important uses of Boolean algebra (the original use in fact) is in its application to logic. We will not investigate this huge subject in any detail or depth, but will merely indicate how the study of the validity of a logical argument may be dealt with.

We need first to consider the idea of *implication*. Since we are dealing in this section with the special case of the algebra of statements (although we are free to use general results of Boolean algebra) we will denote our basic statements by P, Q, R, ..., and not A, B, C, ... and understand all such capital letters to stand for statements.

By $P \Rightarrow Q$ (P implies Q) we mean that if P is true then Q must be, but we say nothing about Q when P is false. In fact $P \Rightarrow Q$ is a function on P and Q which is true when P and Q are both true, false when P is true and Q false, and true when P is false:

P	Q	$P \Rightarrow Q$
1	1	1
1	0	0
0	1	1
0	0	1

Fig. 35

it is false *only* when P is true and Q false. We define it by the truth table in figure 35.

We can easily see that the function is equal to $P' \cup Q$.

Definition. *The function $P \Rightarrow Q$ (P implies Q) is defined to be the function $P' \cup Q$, where P and Q are any two elements in an algebra of statements.*

Note. There is no absolute need for any definition—we could work in terms of the already defined $\cup$, $\cap$ and $'$; but because of the importance of implication in logic it is convenient to introduce the symbol.

We sometimes write $P \Rightarrow Q$ as $Q \Leftarrow P$, and we say that $P \Leftrightarrow Q$ if $P \Rightarrow Q$ *and* $Q \Rightarrow P$. Hence $P \Leftrightarrow Q$ is the function $(P' \cup Q) \cap (Q' \cup P)$. The following theorem is as expected.

Theorem 11.6.1. *$P \Leftrightarrow Q$ is a true statement if and only if P and Q are both true or both false.*

The truth table for $P \Leftrightarrow Q$ is shown in figure 36 and the result follows at once.

P	Q	$P \Rightarrow Q$	$Q \Rightarrow P$	$P \Leftrightarrow Q$
1	1	1	1	1
1	0	0	1	0
0	1	1	0	0
0	0	1	1	1

Fig. 36

Related to $P \Rightarrow Q$ are three other functions: the *converse* $Q \Rightarrow P$, the *inverse* $P' \Rightarrow Q'$ and the *contrapositive* $Q' \Rightarrow P'$. Their truth tables are shown in figure 37 and we can see at once that $P \Rightarrow Q$ is equal to its contrapositive, and the converse and inverse are equal. But of course these two pairs of functions are not equal.

P	Q	$Q \Rightarrow P$	$P' \Rightarrow Q'$	$Q' \Rightarrow P'$
1	1	1	1	1
1	0	1	1	0
0	1	0	0	1
0	0	1	1	1

Fig. 37

The common mathematical terminologies used in proofs, such as 'if and only if', 'necessary and sufficient condition', may be expressed in terms of implication. For example 'P is a sufficient condition for Q' is $P \Rightarrow Q$, 'P is a necessary condition for Q' is the converse $Q \Rightarrow P$, and 'P if and only if Q' is $P \Leftrightarrow Q$.

Before we discuss the idea of a logical argument we must define a tautology.

Definition. *A* tautology *is a statement (usually a compound statement formed from a function of some basic statements) that is always true: i.e. its truth table has a column of all* 1's, *whatever the truth value of the basic statements.*

Examples. The function $P \cup P'$ (equalling 1) has truth value 1 whether the truth value of P is 1 or 0 and so the statement corresponding to it ('P or not P') is a tautology. Similarly $P \cup (P' \cap Q) \cup (P' \cap Q')$ is a tautology.

Note that a tautology is strictly speaking a statement, but we may without ambiguity use the word for the corresponding function—there is no real distinction between a function of the statements $P, Q, \ldots$ and the compound statement that it represents. For a function to be a tautology it must reduce to the function 1. Note also that a statement that happens to be true, for example '6 is an even number', is not a tautology in the logical sense—the term is confined to statements that are always *logically* true.

We may now investigate the validity or otherwise of logical arguments. In a logical argument we start with a finite set of statements $P_1, P_2, \ldots, P_n$, called *premises* and deduce another statement Q, called the *conclusion.* The argument is valid if the truth of Q follows from the truth of all P_i's, i.e. if the truth of all P_i's implies the truth of Q. This may be expressed symbolically as $(P_1 \cap P_2 \cap \ldots \cap P_n) \Rightarrow Q$, and for the argument to be valid this statement or function must be true whatever the truth values of each P_i or Q. Thus the statement

$$(P_1 \cap P_2 \cap \ldots \cap P_n) \Rightarrow Q$$

must be a tautology.

Definition. *The argument that yields the conclusion Q from the premises $P_1, P_2, \ldots, P_n$ is said to be* valid *if the statement*

$(P_1 \cap P_2 \cap \ldots \cap P_n) \Rightarrow Q$ is a tautology. Otherwise the argument is invalid.

Thus to investigate the validity of an argument we merely form the truth table for the function given above and check whether or not its last column consists entirely of 1's: if so then it is valid and otherwise it is invalid. Of course to check a suspected invalid argument we merely require one line of the table to give a 0, and this may often be spotted quickly. The procedure may seem needlessly complicated, but it is a certain way of testing the validity and, in an involved argument, the safest way.

We may sometimes short cut the truth table method by attempting to reduce the function $(P_1 \cap P_2 \cap \ldots \cap P_n) \Rightarrow Q$ to 1 using standard theorems, and if this can be done then the function is a tautology and the argument is valid. Yet another method is by a reduction of the argument to a series of simpler arguments, each known already to be valid.

All three methods will be used in worked exercise 3. Here we will demonstrate the validity of two simple types of argument and will discover an invalid argument.

Theorem 11.6.2. *Prove that the following are valid arguments.*

(*a*) *The law of modus ponens.*

Premises: P *and* $P \Rightarrow Q$.

Conclusion: Q.

(*b*) *The law of syllogism.*

Premises: $P \Rightarrow Q$ *and* $Q \Rightarrow R$.

Conclusion: $P \Rightarrow R$.

(*a*) The truth table for $[P \cap (P \Rightarrow Q)] \Rightarrow Q$ is given in figure 38, and the function is a tautology.

P	Q	$P \Rightarrow Q$	$P \cap (P \Rightarrow Q)$	$[P \cap (P \Rightarrow Q)] \Rightarrow Q$
1	1	1	1	1
1	0	0	0	1
0	1	1	0	1
0	0	1	0	1

Fig. 38

(*b*) The truth table for $F = [(P \Rightarrow Q) \cap (Q \Rightarrow R)] \Rightarrow (P \Rightarrow R)$ is given in figure 39, and the function is a tautology.

P	Q	R	$P \Rightarrow Q$	$Q \Rightarrow R$	$(P \Rightarrow Q) \cap (Q \Rightarrow R)$	$P \Rightarrow R$	F
1	1	1	1	1	1	1	1
1	1	0	1	0	0	0	1
1	0	1	0	1	0	1	1
1	0	0	0	1	0	0	1
0	1	1	1	1	1	1	1
0	1	0	1	0	0	1	1
0	0	1	1	1	1	1	1
0	0	0	1	1	1	1	1

Fig. 39

Alternatively

(*a*)
$$\begin{aligned} [P \cap (P \Rightarrow Q)] \Rightarrow Q &= [P \cap (P' \cup Q)] \Rightarrow Q \\ &= [(P \cap P') \cup (P \cap Q)] \Rightarrow Q \\ &= (P \cap Q) \Rightarrow Q \\ &= (P \cap Q)' \cup Q \\ &= P' \cup Q' \cup Q \\ &= P' \cup 1 \\ &= 1. \end{aligned}$$

(*b*)
$$\begin{aligned} F &= [(P' \cup Q) \cap (Q' \cup R)] \Rightarrow (P' \cup R) \\ &= [(P' \cup Q) \cap (Q' \cup R)]' \cup (P' \cup R) \\ &= (P' \cup Q)' \cup (Q' \cup R)' \cup (P' \cup R) \\ &= (P \cap Q') \cup (Q \cap R') \cup P' \cup R \\ &= (P \cup Q \cup P' \cup R) \cap (P \cup R' \cup P' \cup R) \\ &\qquad \cap (Q' \cup Q \cup P' \cup R) \cap (Q' \cup R' \cup P' \cup R) \\ &= 1 \cap 1 \cap 1 \cap 1 \\ &= 1. \end{aligned}$$

Example of an invalid argument

Premises: This animal is a horse only if it eats grass *and* this animal eats grass.

Conclusion: This animal is a horse.

Let P be the statement 'this animal is a horse' and Q be 'this animal eats grass'. Then the problem may be put in symbolic form thus:

Premises: $P \Rightarrow Q$ and Q.

Conclusion: P.

Thus we wish to discover whether or not the function $F = [(P \Rightarrow Q) \cap Q] \Rightarrow P$ is a tautology.

We suspect that F is not a tautology and so seek a line of the truth table that gives 0. Such a line must arise from $P = 0$ and we try $P = 0$, $Q = 1$. Then $(P \Rightarrow Q) = 1$ and so

$$(P \Rightarrow Q) \cap Q = 1,$$

giving the value 0 to F. Hence F is not a tautology and the argument is invalid.

11.7. The representation theorem

Any algebra of the subsets of a given set is a Boolean algebra, and it is a remarkable fact that the converse is also true in a certain sense: every Boolean algebra is isomorphic to an algebra of subsets (though not generally of *all* the subsets) of some universal set. (Two Boolean algebras are *isomorphic* if their elements may be put in 1–1 correspondence in such a way that $\cup$, $\cap$ and $'$ are all preserved.) In the finite case the theorem is rather stronger than the above: any finite Boolean algebra is isomorphic to the algebra of *all* subsets of some finite universal set. The proof of the theorem for the infinite case raises deep questions, and even for the finite case it is tedious, though comparatively elementary. We will not attempt to give the proof: the interested reader may find it in many standard works (for example *Sets, Logic and Axiomatic Theories* by R. R. Stoll).

The representation theorem classifies finite Boolean algebras completely: if the universal set contains n objects then the Boolean algebra has 2^n elements (there are precisely this number of subsets, since each of the n objects may or may not be present in any subset). Thus any finite Boolean algebra contains 2^n elements for some $n \geqslant 1$, and any two with the same number of elements are isomorphic.

11.8. Application to circuit theory

One of the most important applications of Boolean algebra at the present time is to electrical circuits and the design of computers. We will not discuss this large subject in any detail, but will confine ourselves to describing briefly one simple use in this field, that of the closure properties of a circuit involving switches in series and parallel.

Suppose we have a circuit which includes switches that may be either on or off. (We are interested only in the switch part of the circuit, and not in any other electrical components that may be present.) We denote switches by capital letters A, B, ..., and two switches that are always either both on or both off are denoted by the same letter. Let us denote the circuit consisting of the two switches A and B in parallel by $A \cup B$, and that consisting of them in series by $A \cap B$, and let A' be a switch that is always closed when A is open and open when A is closed. Then corresponding to any series-parallel circuit of switches is an algebraic expression involving $\cup$, $\cap$ and $'$, and any such expression may be interpreted by a series-parallel circuit.

We give a switch A the value 1 if it is closed and 0 if open, and similarly for a circuit, so that 1 represents a closed circuit, that is a circuit through which current may flow, and 0 represents one through which no current may flow. Then clearly $A \cup B$ is closed if either A or B is closed, and $A \cap B$ is closed only when both A and B are closed. Figure 40 shows the situation.

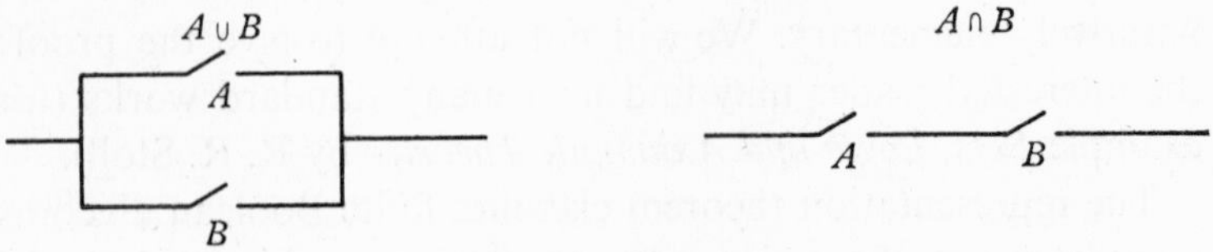

Fig. 40

Putting the possible states of A and B in a table and writing down the corresponding state of $A \cup B$, $A \cap B$ and A' we obtain, using the symbols 1 and 0, the tables in figure 41.

We see that these are precisely the truth tables for the corresponding expressions in a Boolean algebra.

A	B	$A \cup B$	$A \cap B$	A'
1	1	1	1	0
1	0	1	0	0
0	1	1	0	1
0	0	0	0	1

Fig. 41

We now say that two expressions are equal if they represent equivalent circuits, that is if the corresponding circuits are either both open or both closed for any given positions of the switches A, B, ... Finally we denote by 1 a switch that is always closed and by 0 one that is always open.

Since the truth tables are the same as for a Boolean algebra, we may easily verify (as we indicated in §11.5) that the axioms of a Boolean algebra are satisfied and that the system we have introduced is in fact such an algebra: it is akin to the algebra of statements. To fix the ideas we consider the first Distributive Law without reference to the truth tables.

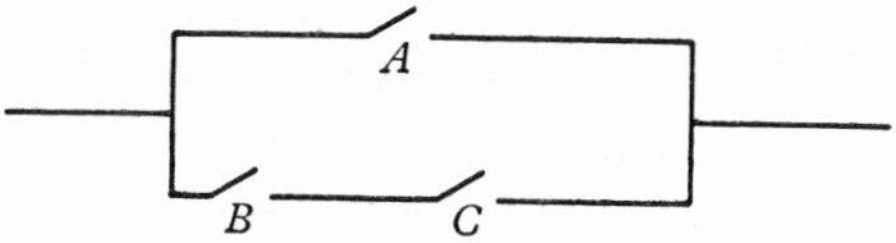

Fig. 42

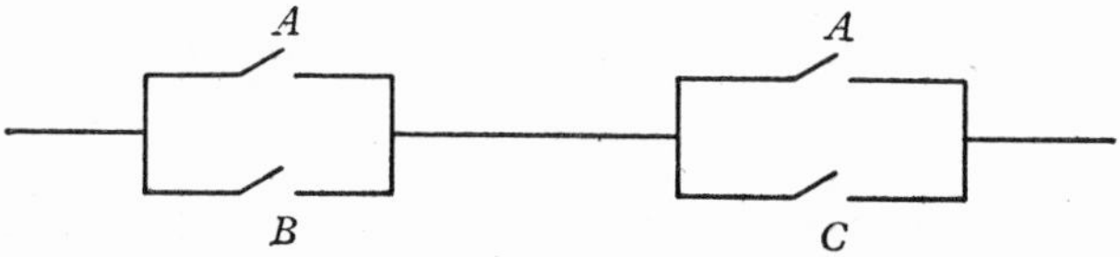

Fig. 43

Figure 43 shows the circuit corresponding to $A \cup (B \cap C)$ and figure 43 shows that corresponding to $(A \cup B) \cap (A \cup C)$. Remembering that both switches A must open and close together in the second case, we see that in either circuit the circuit is closed (current may flow) if either A is closed or both B and C are closed (or both) and is otherwise open. Hence the circuits are equivalent and the expressions are equal, verifying the first Distributive Law.

We will not take the theory any further, but will show two simple applications of it.

Example 1. Simplify the circuit shown in figure 44 (i.e. produce a simpler circuit with the same closure properties).

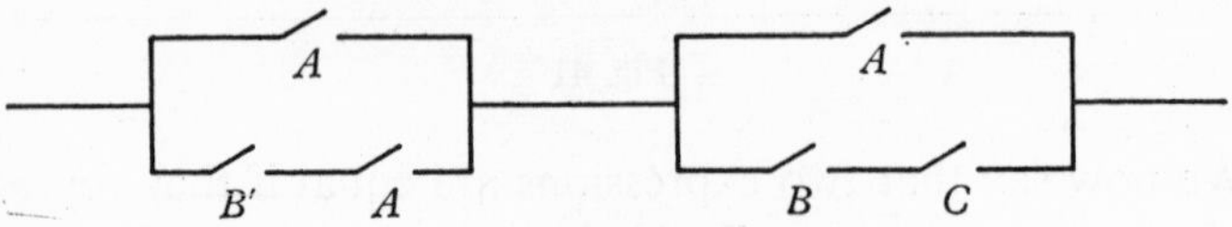

Fig. 44

The function corresponding to the circuit is

$$[A \cup (B' \cap A)] \cap [A \cup (B \cap C)].$$

This equals $A \cap [A \cup (B \cap C)]$ by the Absorption Laws and equals A by the Absorption Laws again.

Hence a simpler circuit is provided merely by the switch A.

Thus 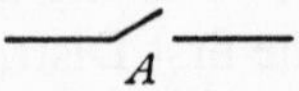

Example 2. Design a circuit for the operation of a light from three independent switches.

If the switches are A, B and C we require a circuit such that the change of state of any one switch changes the state of the circuit. If we take the circuit to be closed when all three switches are closed (this is arbitrary) we see that we require the function F representing the circuit to have the truth table shown in figure 45.

A	B	C	F
1	1	1	1
1	1	0	0
1	0	1	0
1	0	0	1
0	1	1	0
0	1	0	1
0	0	1	1
0	0	0	0

Fig. 45

The function F may be written down in disjunctive normal form as

$$F = (A \cap B \cap C) \cup (A \cap B' \cap C') \cup (A' \cap B \cap C') \cup (A' \cap B' \cap C)$$

which may be simplified to

$$F = \{A \cap [(B \cap C) \cup (B' \cap C')]\} \cup \{A' \cap [(B \cap C') \cup (B' \cap C)]\}$$

and the corresponding circuit is as shown in figure 46, where for example the switch B must throw, at the same time, both switches marked B one way and both switches marked B' the other, and, of course, if B is closed B' is open and vice versa.

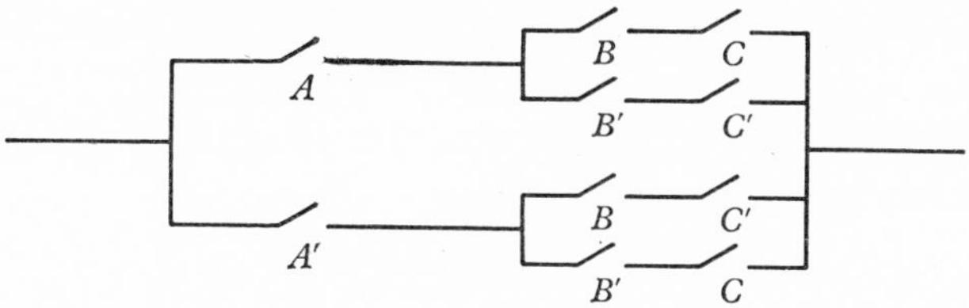

Fig. 46

11.9. Application to probability theory

Set theory is useful in the theory of probability, and we will indicate, again very briefly, why this is so.

Suppose we have an experiment with N possible and equally likely outcomes. We represent each outcome by an element of a universal set I, and consider any subset A of I. Then the probability of the experiment resulting in an element of A (we say the probability of the *event* A occurring) is $n(A)/N$, where $n(A)$ is the number of elements in A. This is denoted by $p(A)$.

For example suppose I draw a card from a pack of 52. Then $N = 52$. If A is the set of all spades, $n(A) = 13$ and the probability of drawing a spade is $13/52 = 1/4$.

The probability of A not occurring is clearly $1-p(A)$, which equals $p(A')$. The probability of either A or B occurring is $p(A \cup B) = n(A \cup B)/N$, and the probability of both A and B occurring is $p(A \cap B) = n(A \cap B)/N$. Thus if A is the set of spades and B the set of aces, the probability that I will draw either a spade or an ace is $n(A \cup B)/52 = 16/52 = 4/13$, and the probability that I will draw the ace of spades is

$$n(A \cap B)/52 = 1/52.$$

If A and B are disjoint so that $A \cap B = \varnothing$ we say that the events are *mutually exclusive*. Clearly in this case

$$n(A \cup B) = n(A) + n(B)$$

and so $p(A \cup B) = p(A)+p(B)$. Similarly the probability of several mutually exclusive events occurring is the sum of the probabilities of each separate one. Thus the probability that I will draw either a Knave, Queen or King is

$$1/13+1/13+1/13 = 3/13.$$

Theorem 11.9.1. $p(A \cup B) = p(A)+p(B)-p(A \cap B)$. (*The probability of either A or B occurring is the sum of the separate probabilities minus the probability that both will occur.*)

It is clear that $n(A \cup B) = n(A)+n(B)-n(A \cap B)$ and the result follows by dividing through by N.

Corollary.

$$p(A \cup B \cup C) = p(A)+p(B)+p(C)-p(B \cap C) -p(C \cap A)-p(A \cap B)+p(A \cap B \cap C).$$

For by the theorem,

$$\begin{aligned} p(A \cup B \cup C) &= p(A)+p(B \cup C)-p[(A \cap B) \cup (A \cap C)] \\ &= p(A)+p(B)+p(C)-p(B \cap C)-p(A \cap B) \\ &\quad -p(A \cap C)+p(A \cap B \cap A \cap C) \end{aligned}$$

and the result follows.

11.10. Lattices

In §11.2 we defined inclusion for a Boolean algebra in terms of intersection and complement, and in theorem 11.2.9 proved its basic properties. It is possible to go the other way—to start with inclusion as a fundamental concept and define union and intersection in terms of it. This approach brings in the ideas of *partially ordered sets* and *lattices*, which are important in their own right and not merely in connection with Boolean algebras.

Definition. S is a partially ordered set *if there is a relation between pairs of elements in S (not necessarily between every pair) written $A \subseteq B$ and satisfying the laws of theorem 11.2.9.*

We write $A \subseteq B$ also as $B \supseteq A$, and note that it is not necessary that for any pair either $A \subseteq B$ or $B \subseteq A$ (if this *is* the case then the set is *completely ordered*: an example is the reals).

(Note that we may reverse the elements in the definition and obtain the same properties for $\supseteq$.)

Examples. The integers, the rationals and the reals; the subsets of a given universal set (or any Boolean algebra); the subgroups of a given group; the positive integers with $p \subseteq q$ defined as meaning that p is a factor of q; are all partially ordered.

To define union and intersection in a partially ordered set we need the concept of *bound.*

Definition. *In a partially ordered set S an element U is a* least upper bound *of A and B if $U \supseteq A$ and $U \supseteq B$ and if $V \supseteq A$ and $V \supseteq B \Rightarrow V \supseteq U$. An element L is a* greatest lower bound *if $L \subseteq A$ and $L \subseteq B$ and if $M \subseteq A$ and $M \subseteq B \Rightarrow M \subseteq L$.*

Definition. *A* lattice *is a partially ordered set in which any two elements A and B possess a least upper bound and a greatest lower bound. These are denoted by $A \cup B$ and $A \cap B$ respectively.*

Examples. All the examples given above are lattices. For the integers, rationals or reals $A \cup B$ is the greater of A and B and $A \cap B$ is the lesser. For a Boolean algebra they have the obvious meanings, for the subgroups of a group $A \cap B$ is the ordinary intersection (a subgroup by theorem 1.8.5) and $A \cup B$ is the subgroup generated by the elements of A and B, while for the example of the positive integers $p \cup q$ is the L.C.M. and $p \cap q$ the H.C.F. of p and q.

It can be seen at once from the Anti-symmetric Law that $A \cup B$ and $A \cap B$ are unique. Several of the expected laws follow.

Theorem 11.10.1. *In a lattice S the least upper bound and greatest lower bound possess the following properties:*

(*a*) $A \cup B = B \cup A$ *and* $A \cap B = B \cap A$,

(*b*) $A \cup (B \cup C) = (A \cup B) \cup C$,

$A \cap (B \cap C) = (A \cap B) \cap C$,

(*c*) $A \cup A = A$ *and* $A \cap A = A$,

(*d*) $A \cup (A \cap B) = A$ *and* $A \cap (A \cup B) = A$.

(*a*) Immediate since the definition is symmetrical.

(*b*) Let $U = A \cup (B \cup C)$ and $V = (A \cup B) \cup C$. Then $V \supseteq A \cup B$ and $A \cup B \supseteq A$ and B. Hence by the Transitive Law $V \supseteq A$ and B, and also $V \supseteq C$. Thus by the definition $V \supseteq B \cup C$ and so, again by the definition, $V \supseteq U$. Similarly $U \supseteq V$ and so $U = V$. Similarly for the law for intersection.

(*c*) $A \supseteq A$ by the Reflexive Law and certainly

$$V \supseteq A \Rightarrow V \supseteq A,$$

so the result that $A \cup A = A$ follows. Similarly for $A \cap A$.

(*d*) $A \cap B \subseteq A$ and so $A \supseteq A \cap B$ and $A \supseteq A$. If $V \supseteq A \cap B$ and $V \supseteq A$ we certainly have that $V \supseteq A$ and so A is the least upper bound of $A \cap B$ and A. Similarly for the other law.

The Distributive Laws do not apply to general lattices.

Example. Consider the lattice of subgroups of the Vierergruppe V. Let $A = \{e, a\}$, $B = \{e, b\}$ and $C = \{e, ab\}$. Then $B \cap C = \{e\}$ and $A \cup (B \cap C)$ is the subgroup generated by e and a and is A. But $A \cup B$ is the subgroup generated by e, a and b and is V, as is $A \cup C$, so $(A \cup B) \cap (A \cup C) = V$ and not A. Thus the first Distributive Law is not true in this case.

Unity and zero in a lattice

If an element 1 exists with the property that $1 \supseteq A$ for all A in a lattice S, 1 is called a *unity* for S, and is clearly unique by the Anti-symmetric Law. Similarly a *zero* is an element 0 such that $0 \subseteq A \ \forall\ A \in S$ and is also unique if it exists.

Examples. The integers, rationals and reals have no unity and no zero, although lattices of subsets (for example all reals x such that $1 \leqslant x \leqslant 2$) might have. For subgroups of a group G, G is the unity and $\{e\}$ the zero.

Theorem 11.10.2. *If a lattice possesses a unity* 1, $A \cap 1 = A$ *for any* A, *and if the lattice possesses a zero* 0, $A \cup 0 = A$ *for any* A.

Certainly $A \subseteq 1$ and $A \subseteq A$. Hence $A \subseteq A \cap 1$. But

$$A \cap 1 \subseteq A$$

by definition and so $A \cap 1 = A$. Similarly for 0.

We now obtain our final theorem.

Theorem 11.10.3. *If S is a lattice that possesses a unity and a zero, is* Distributive (*i.e. both Distributive Laws hold*) *and is* complemented (*i.e. to any $A \in S \exists A'$ such that $A \cup A' = 1$ and $A \cap A' = 0$*) *then S is a Boolean algebra.*

Referring to the axioms of a Boolean algebra given in §11.2, the first two are satisfied by theorem 11.10.1, the Distributive Law is satisfied by hypothesis, as is axiom 5, and axiom 4 follows by theorem 11.10.2. Hence S is a Boolean algebra.

Note. The truth of one Distributive Law implies the truth of the other. (See worked exercise 4.)

Worked exercises

1. Prove that in a Boolean algebra $(A \cap B) \cup (A \cap B') = A$.

$$(A \cap B) \cup (A \cap B') = [A \cup (A \cap B')] \cap [B \cup (A \cap B')]$$

by the Distributive Laws and Associative Laws,

$$= [A \cup (A \cap B')] \cap (B \cup A) \cap (B \cup B')$$

by the Distributive Laws again,

$$= A \cap (B \cup A) \cap 1$$

using the Absorption Laws,

$$= A \cap (B \cup A)$$

$$= A \quad \text{by the Absorption Laws.}$$

In the theory of sets the result is illustrated by the Venn diagram in figure 47.

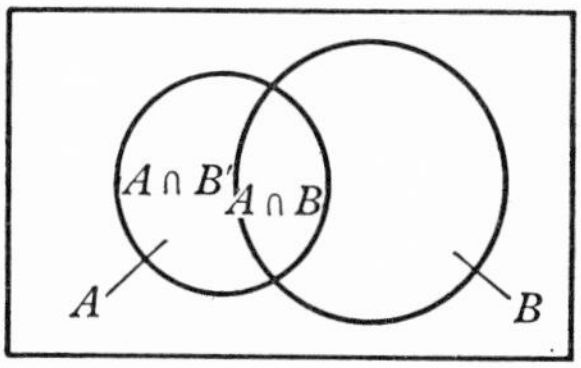

Fig. 47

2. Prove the De Morgan Laws by the truth table method.

The truth tables for $(A \cup B)'$, $A' \cap B'$, $(A \cap B)'$ and $A' \cup B'$ are shown

A	B	$A \cup B$	$(A \cup B)'$
1	1	1	0
1	0	1	0
0	1	1	0
0	0	0	1

Fig. 48

A	B	A'	B'	$A' \cap B'$
1	1	0	0	0
1	0	0	1	0
0	1	1	0	0
0	0	1	1	1

Fig. 49

in figures 48–51, and we see that the first two are identical, and so are the last two, thus proving the De Morgan Laws.

A	B	$A \cap B$	$(A \cap B)'$
1	1	1	0
1	0	0	1
0	1	0	1
0	0	0	1

Fig. 50

A	B	A'	B'	$A' \cup B'$
1	1	0	0	0
1	0	0	1	1
0	1	1	0	1
0	0	1	1	1

Fig. 51

3. Prove that the following argument is valid:

Premises: Henry is a dog,
If Henry is a dog then he has a long tail.
If Henry has a long tail then he has curly hair.

Conclusion: Henry has curly hair.

Let P, Q, R be the statements:

P: 'Henry is a dog'
Q: 'Henry has a long tail'
R: 'Henry has curly hair'.

Then the above premises and conclusion may be put in the form:

Premises: P; $P \Rightarrow Q$; $Q \Rightarrow R$.
Conclusion: R.

We will show that this argument is valid by each of the three methods mentioned on p. 335.

Method 1. *The Truth-table method*

We form the truth table for the function

$$F = [P \cap (P \Rightarrow Q) \cap (Q \Rightarrow R)] \Rightarrow R$$

and show it is a tautology.

P	Q	R	$P \Rightarrow Q$	$Q \Rightarrow R$	$P \cap (P \Rightarrow Q) \cap (Q \Rightarrow R)$	F
1	1	1	1	1	1	1
1	1	0	1	0	0	1
1	0	1	0	1	0	1
1	0	0	0	1	0	1
0	1	1	1	1	0	1
0	1	0	1	0	0	1
0	0	1	1	1	0	1
0	0	0	1	1	0	1

Fig. 52

By figure 52 the function is a tautology and so the argument is valid.

Method 2

We reduce the above function F to 1, showing it to be a tautology and the argument to be valid.

$$\begin{aligned} F &= [P \cap (P' \cup Q) \cap (Q' \cup R)]' \cup R \\ &= P' \cup (P \cap Q') \cup (Q \cap R') \cup R \\ &= (P' \cup P \cup (Q \cap R') \cup R) \cap (P' \cup Q' \cup (Q \cap R') \cup R) \\ &= 1 \cap (P' \cup Q' \cup Q \cup R) \cap (P' \cup Q' \cup R' \cup R) \\ &= 1 \cap 1 \cap 1 \\ &= 1. \end{aligned}$$

Method 3. *By using simpler arguments*

By the law of modus ponens we know that $[P \cap (P \Rightarrow Q)] \Rightarrow Q$ is valid, and that $[Q \cap (Q \Rightarrow R)] \Rightarrow R$ is also valid, i.e. that

$$[P \cap (P \Rightarrow Q)]' \cup Q = P' \cup (P \Rightarrow Q)' \cup Q = 1$$

and $$[Q \cap (Q \Rightarrow R)]' \cup R = Q' \cup (Q \Rightarrow R)' \cup R = 1,$$

and so $$\begin{aligned} 1 &= P' \cup (P \Rightarrow Q)' \cup Q \cup Q' \cup (Q \Rightarrow R)' \cup R \\ &= P' \cup (P \Rightarrow Q)' \cup (Q \Rightarrow R)' \cup R \\ &= [P \cap (P \Rightarrow Q) \cap (Q \Rightarrow R)]' \Rightarrow R. \end{aligned}$$

4. Prove that in a lattice if $A \cup (B \cap C) = (A \cup B) \cap (A \cup C)$ then

$$A \cap (B \cup C) = (A \cap B) \cup (A \cap C),$$

i.e. that the truth of one Distributive Law implies that of the other.

We know that the Commutative, Associative, and Absorption Laws are true in any lattice by theorem 11.10.1 and so, using these freely, we see that

$$\begin{aligned} A \cap (B \cup C) &= [A \cap (A \cup C)] \cap (B \cup C) \quad \text{by the Absorption Laws,} \\ &= A \cap [(A \cup C) \cap (B \cup C)] \\ &= A \cap [(A \cap B) \cup C] \\ &= [(A \cap B) \cup A] \cap [(A \cap B) \cup C] \quad \text{by the Absorption Laws,} \\ &= (A \cap B) \cup (A \cap C). \end{aligned}$$

(Similarly if the second Distributive Law is true then the first is also true.)

5. Find a valid conclusion from the following premises:

(*a*) babies are illogical;

(*b*) nobody is despised who can manage a crocodile;

(*c*) illogical persons are despised.

(This example is taken from Lewis Carroll's *Symbolic Logic*. It was solved by set theory in volume 1, p. 45.)

Define the following statements.

P: 'This person can manage a crocodile'.
Q: 'This person is a baby'.
R: 'This person is despised'.
S: 'This person is logical'.

Then (*a*) is $Q \Rightarrow S'$, (*b*) is $P \cap R = 0$ and (*c*) is $S' \Rightarrow R$.

Now $P \cap R = 0$ is equivalent to $R' \cup P' = 1$ (De Morgan) or $R \Rightarrow P'$. Hence we have the premises $Q \Rightarrow S'$, $S' \Rightarrow R$ and $R \Rightarrow P'$ and so, by a double application of the law of syllogism (theorem 11.6.2) a conclusion $Q \Rightarrow P'$ is valid. In words the conclusion is 'if this person is a baby he cannot manage a crocodile' or, more simply, 'babies cannot manage crocodiles'.

Unless otherwise stated the capital letters in exercises 11A and 11B refer to elements of a general Boolean algebra, and the exercises should not be proved by intuitive methods such as the use of Venn diagrams.

Exercises 11 A

1. If A is the statement 'this rabbit is white' and B is the statement 'all monkeys are good-tempered' write down the statements $A \cup B$, $A \cap B$, A', B'.

2. Prove by induction the generalised De Morgan Laws:

$$(A_1 \cup A_2 \cup \ldots \cup A_n)' = A_1' \cap A_2' \cap \ldots \cap A_n',$$
$$(A_1 \cap A_2 \cap \ldots \cap A_n)' = A_1' \cup A_2' \cup \ldots \cup A_n'.$$

3. Verify that the definition of $A \subseteq B$ as $A \cap B' = 0$ has the usual interpretation in the algebra of sets (i.e. $x \in A \Rightarrow x \in B$).

4. Repeat **3** for the algebra of statements (i.e. the truth of A implies the truth of B).

5. Prove the Consistency theorem that $A \subseteq B \Leftrightarrow A \cap B = A \Leftrightarrow A \cup B = B$.

6. Defining $A - B$ as $A \cap B'$ prove

(i) $A-(A-B) = A \cap B$,
(ii) $B \cup (A-B) = A \cup B$,
(iii) $A-(B \cup C) = (A-B) \cap (A-C)$,
(iv) $A-(B \cap C) = (A-B) \cup (A-C)$,
(v) $A-B = B-A = 0 \Leftrightarrow A = B$.

Interpret the definition of $-$ and these results in the algebra of sets.

7. Prove that (i) $A \cap B \subseteq A$, (ii) $A \subseteq A \cup B$.

8. Prove Poretsky's Law, that

$$B = (A \cap B') \cup (A' \cap B) \text{ if and only if } A = 0.$$

9. Write down the 16 possible Boolean functions in two variables (in any simple form).

Express the Boolean functions in **10–12** in (i) disjunctive and (ii) conjunctive normal form by the method of theorem 11.4.1.

10. $[(X \cap Y) \cup (X' \cap Y')]'$.

11. $[X \cup (Y \cap Z)]' \cap [X' \cap (Y \cup Z)]'$.

12. $(X \cup Y') \cap (Y \cup Z') \cap (Z \cup X')$.

13. Express the Boolean function in **10** in disjunctive normal form in the three variables X, Y and Z.

Construct truth tables for the Boolean functions in **14–16** and write down their disjunctive and conjunctive normal forms:

14. $[(X \cap Y') \cup (X' \cap Y)]'$.

15. $X \cup (X \cap Y' \cap Z) \cup (X' \cup Z)'$.

16. $(X' \cap Y \cap Z) \cup [X \cap Y' \cap (Z \cup X')]' \cup [(X' \cup Y) \cap (X \cup Z')]'$.

17. Using the normal forms already found, simplify the functions in **10–12** and **14–16** where possible.

18. Construct truth tables for $A \cup B \cup C$ and $A \cap B \cap C$.

19. Prove the Idempotent Laws by the truth-table method.

20. Prove the Absorption Laws by the truth-table method.

21. Give combination tables for $\cup$ and $\cap$ for the Boolean algebra with 2 elements and for that with 4 elements. (*Hint.* Use the Representation theorem.)

22. Prove that $X \cup A = Y \cup A$ *and* $X \cup A' = Y \cup A' \Rightarrow X = Y$.

23. Prove that $X \cap A = Y \cap A$ *and* $X \cap A' = Y \cap A' \Rightarrow X = Y$.

24. Prove that

$$(A \cap B) \cup (B \cap C) \cup (C \cap A) = (A \cup B) \cap (B \cup C) \cap (C \cup A).$$

In **25–27** let A, B, C be the statements.

A: 'It is cold to-day'.
B: 'The sun is shining'.
C: 'I am outside'.

25. Write out the statements:

(i) $(A \cap B)'$, (ii) $A' \cap B$, (iii) $A' \cup B'$.

26. Write in symbols: (i) 'the sun is not shining and I am outside,' (ii) 'either I am not outside and it is not cold to-day, or the sun is shining and I am outside'.

27. Write out the negations of the statements in **26.**

28. Write in implication form the statements:

(i) P if Q, (ii) P only if Q,

(iii) P is a necessary and sufficient condition for Q.

29. What is the negation of $P \Rightarrow Q$?

30. If P is the statement 'n is divisible by 4' and Q is 'n is even' we may write $P \Rightarrow Q$ as 'if n is divisible by 4 then n is even'. Write in similar form the converse, inverse, contrapositive and negation of $P \Rightarrow Q$. Which of these are in fact true in this case?

Check that the arguments in **31–33** are valid and illustrate by constructing suitable practical examples:

31. Premise: $P \cap Q$.
Conclusion: P.

32. Premise: P'.
Conclusion: $P \Rightarrow Q$.

33. Premises: $P \cup Q$ and P'.
Conclusion: Q.

34. Draw the circuits corresponding to the Boolean functions:

(i) $(A \cup B) \cap (A' \cup B')$, (ii) $(A \cup B \cup C) \cap [(A \cup B) \cap (A' \cap C')']$.

35. What Boolean functions correspond to the circuits in figures 53 and 54?

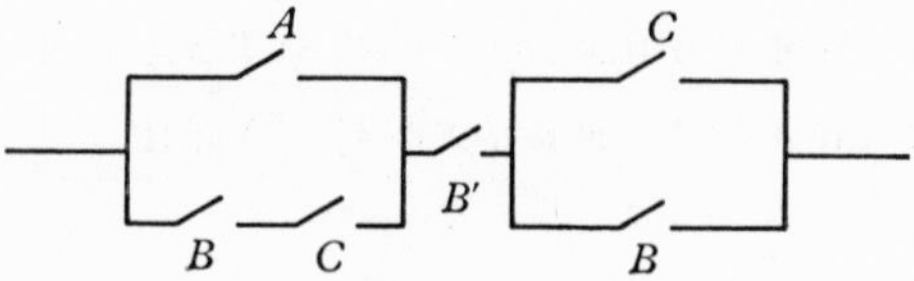

Fig. 53

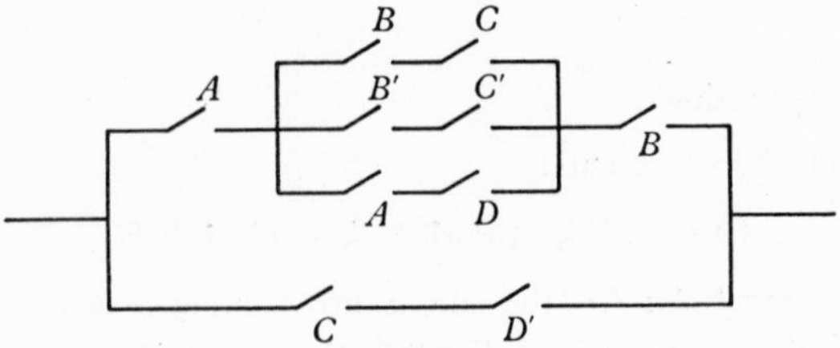

Fig. 54

36. Illustrate the Absorption Laws in terms of circuit theory.

37. Simplify the circuit in figure 55.

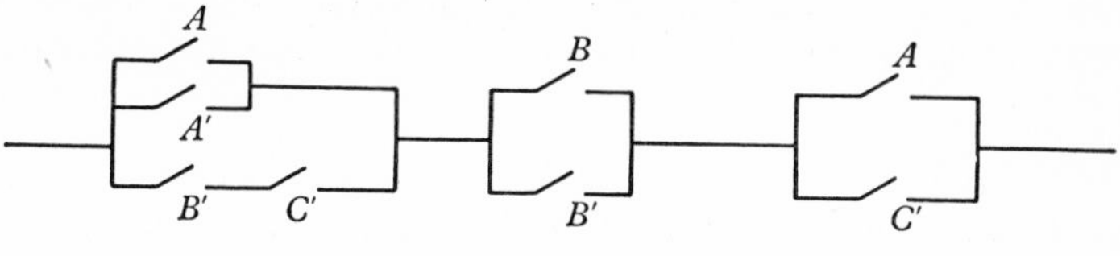

Fig. 55

38. If I throw 2 dice, one red and one green, what is the probability that (i) I throw a red 6; (ii) I throw a green 6; (iii) I throw a double 6; (iv) I throw at least one 6?

Verify the result of theorem 11.9.1 for this case.

39. Prove that a set is partially ordered if and only if there exists a relation $A \subseteq B$ satisfying

(*a*) $A \subseteq B$ *and* $B \subseteq A$ if and only if $A = B$,

(*b*) $A \subseteq B$ *and* $B \subseteq C \Rightarrow A \subseteq C$.

Exercises 11B

1. If a Boolean function F in n variables has r terms in its disjunctive normal form show that F' has $2^n - r$ terms, and that the conjunctive normal form of F also has $2^n - r$ terms. Show also that the conjunctive normal form of F may be obtained from the disjunctive normal form of F' by interchanging $\cup$ and $\cap$ and replacing each variable by its complement (and vice versa).

2. Show that the combination tables in figure 56 define an algebra with 2 elements 0 and 1 which satisfies axioms 1, 4, 5 and 3(*b*) but *not* 3(*a*).

$\cup$	0	1
0	0	1
1	1	0

$\cap$	0	1
0	0	0
1	0	1

Fig. 56

Construct a similar example which satisfies axioms 1, 4, 5 and 3(*a*) but not 3(*b*). (This shows that axioms 3(*a*) and 3(*b*) are independent of the others. We may similarly show that the pairs of axioms 1, 4, 5 are independent and that axioms 1, 3, 4, 5 give a set of 8 independent axioms for a Boolean algebra. As we have already stated axiom 2 may be deduced from these.)

3. Prove by the truth-table method that

$$[(A \cap B') \cup (B \cap C)] \cap [(A' \cap B') \cup (A \cap C) \cup (B \cap C)] = (B \cup A) \cap C.$$

Dualise this result. Express the function on each side of the above equation in disjunctive and conjunctive normal forms. Repeat for the dual equation.

4. Change the function

$$(X \cap Y \cap Z) \cup (X' \cap Y' \cap Z) \cup (X' \cap Y \cap Z') \cup (X \cap Y' \cap Z')$$

from disjunctive normal form to conjuctive normal form.

5. We write the statement 'for all x in a set X, P is true' as '$\forall x \in X:P$' and we write 'for some x in X, P is true' as '$\exists x \in X:P$'. Prove that

$$\text{(i) } (\forall x \in X:P)' = \exists x \in X:P', \quad \text{(ii) } (\exists x \in X:P)' = \forall x \in X:P'.$$

6. Check whether or not the following arguments are valid:

(i) Premises: (*a*) if the statement 'if to-day is Tuesday implies that the bus will not run then I will be late for work' is true then I will jump in the river and (*b*) I will be late for work.

Conclusion: I will jump in the river.

(ii) Premises: (*a*) my dog is not ill if my cat is not old and (*b*) my cat is old.

Conclusion: My dog is ill.

7. Prove that any circuit with n independent switches $X_1, X_2, \ldots, X_n$ is equivalent to a circuit of the form shown in figure 57, where each parallel part of the circuit contains n switches, one of each of either X_i or X_i', $i = 1, \ldots, n$, and there are $\leqslant 2^n$ parallel parts, all different.

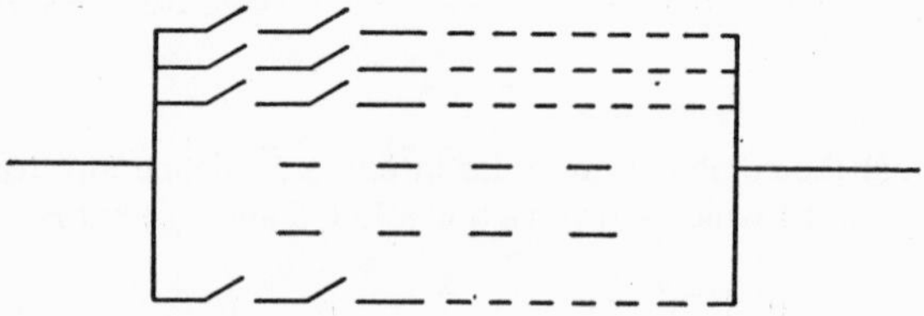

Fig. 57

8. Design a series-parallel circuit for the operation of a light with two independent switches B and C and a third master switch A, so that the light is always off when A is open but when A is closed each of B and C will switch the light on and off independently of the other.

9. I draw a card from each of two packs. Let A be the event of drawing a spade from the first pack and B that of drawing a spade from the second pack. Express the following events in terms of A and B and find their probabilities:

(i) drawing two spades,
(ii) drawing at least one spade,
(iii) drawing just one spade,
(iv) drawing a spade from the first pack and a card not a spade from the second,
(v) drawing two cards, neither of which are spades.

10. Prove that any Boolean algebra is a lattice.

11. Prove that in *any* lattice (not merely a Distributive one)

$$A \cup (B \cap C) \subseteq (A \cup B) \cap (A \cup C)$$

and
$$A \cap (B \cup C) \supseteq (A \cap B) \cup (A \cap C).$$

12. Prove that in a Boolean algebra

$$(A \cap (B \cup C) = B \cup (A \cap C)) \Leftrightarrow A \supseteq B.$$

13. In a Boolean algebra S we define $A+B$ as $(A \cap B') \cup (A' \cap B)$. ($A+B$ is called the *symmetric difference* of A and B.) Prove that S is a commutative ring with unity under $+$ and $\cap$.

14. Let R be a ring with unity such that $x^2 = x \; \forall \; x \in R$. By considering $(x+y)^2$ show that $xy+yx = 0$ and deduce that $2x = 0$ and hence that R is commutative. Define $x \cup y = x+y-xy$ and $x \cap y = xy$.

Prove that R is a Boolean algebra under $\cup$ and $\cap$ with the zero and unity of R as its 0 and 1, and $1-x$ as the complement of x.

12

FURTHER RESULTS

12.1. Introduction

In the preceding chapters we have covered most of the elementary theory of abstract algebra, and have indeed on occasions included some fairly deep and difficult properties (for example in the later parts of chapter 4 and in chapter 8). But of course there is an immense amount of work that we have not touched. Much of this is very difficult. We cannot hope to explain in a short chapter even what it is about, but there is some which, while it is not possible (nor desirable in a book at this level) for us to develop it rigorously and fully, is nevertheless amenable to brief indications of its scope and importance. In this chapter we will look at some such topics. Proofs will be omitted, and we will not even give references, since they may be found in the standard works on the various types of structure. There is of course no attempt at completeness in any sense—we merely select various ways in which the work we have done develops and try to show the reader some more of the power and beauty of the methods of the subject.

Most of the work continues the theory of groups, rings and fields.

12.2. The isomorphism theorems

When studying the structure of a group it is necessary to consider its subgroups, particularly its invariant subgroups (so that we can form the quotient group, which we expect to be a simpler group than the original). We are thus led into considering various subgroups of a given group, with their intersections and 'products', and it turns out that there are some interesting isomorphisms.

Let us first suppose that N is an invariant subgroup of a group G, and that H is any subgroup of G. Then $H \cap N$ is certainly a subgroup of H, and it can be proved that it is an *invariant* subgroup of H. We can also prove that the Frobenius product NH is a subgroup of G (remember that NH is the set of elements nh for $n \in N$ and $h \in H$) and it is immediate that N is an invariant subgroup of NH. The first theorem states that the quotient groups $H/(H \cap N)$ and NH/N are isomorphic.

Theorem 12.2.1. *The First Isomorphism theorem.*

If N is an invariant subgroup of G and H is any subgroup of G, then (i) *$H \cap N$ is an invariant subgroup of H;*

(ii) *NH is a subgroup of G and N is an invariant subgroup of NH;*

(iii) $H/(H \cap N) \cong NH/N$.

The situation may be illustrated as in figure 58, where a line connects a group with a subgroup of it, and the line is doubled if the subgroup is invariant. The theorem is sometimes known as the *Parallelogram theorem.*

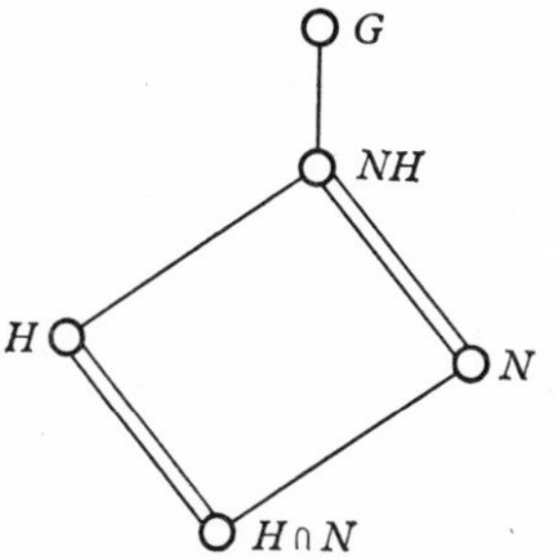

Fig. 58

Fig. 59

The second theorem deals with the case where H is also an invariant subgroup of G and contains N as a subset, so that in fact N is an invariant subgroup of H. The theorem then is as follows.

Theorem 12.2.2. *The Second Isomorphism theorem.*

Let N and H be invariant subgroups of G and let N be a subgroup of H.

Then (i) *H/N is an invariant subgroup of G/N;* (ii) *the quotient group $(G/N)/(H/N)$ is isomorphic to G/H.*

Conversely, any invariant subgroup of G/N may be written in the form H/N where H is some invariant subgroup of G containing N.

Figure 59 shows the situation of the three groups G, H and N.

There is a third isomorphism theorem, much more complicated than the first two, but important because it is used in the proof of the Jordan–Hölder theorem to be given in the next section.

Theorem 12.2.3. *Zassenhaus' lemma.*

If H_1, N_1, H_2, N_2 are subgroups of G, and if N_1 and N_2 are invariant subgroups of H_1 and H_2 respectively, then

(i) *$(H_1 \cap N_2)N_1$ is an invariant subgroup of $(H_1 \cap H_2)N_1$ and $(H_2 \cap N_1)N_2$ is an invariant subgroup of $(H_2 \cap H_1)N_2$;*

(ii) *the two corresponding quotient groups are isomorphic.*

We obtain the rather complicated diagram in figure 60, the double lines marked l_1 and l_2 being those corresponding to the invariant subgroups of the theorem, the double lines m_1 and m_2 being hypotheses, and the other double lines being obtained by using theorem 12.2.1.

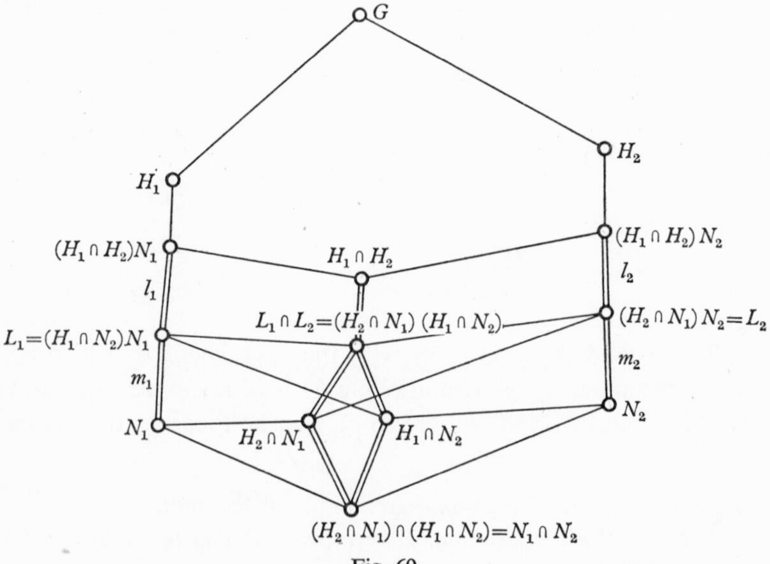

Fig. 60

12.3. Composition series: the Jordan–Hölder theorem

We recollect from §6.10 that a group with no proper invariant subgroups is called a *simple* group. From theorem 12.2.2 we see that if N is an invariant subgroup of G then G/N is simple if and only if no invariant subgroup H exists, properly containing N and properly contained in G. In this case we say that N is a *maximal invariant subgroup* of G.

If G is finite there must exist a maximal invariant subgroup N (there may be more than one). If N is not the trivial group $\{e\}$ then it must also possess a maximal invariant subgroup N_1, and so on, so that we obtain in the finite case a series $G \supset N \supset N_1 \supset \ldots \supset N_r \supset \{e\}$, where each subgroup is a maximal invariant subgroup of the preceding one, though not necessarily of G itself. All the quotient groups G/N, N/N_1, ..., N_{r-1}/N_r and N_r are simple groups.

Such a series is called a *composition series* for G, and gives an important method of, so to speak, decomposing G. But it would lose much of its importance if it were not unique in some sense.

The composition series itself is not unique—a group G may possess two that are not only distinct in that the subgroups are not identical, but are such that the subgroups are not even isomorphic. But there is uniqueness in the sense given in the next theorem, the famous Jordan–Hölder theorem.

Theorem 12.3.1. *The Jordan–Hölder theorem.*

If a group G possesses two composition series

$$G \supset N \supset N_1 \supset \ldots \supset N_r \supset \{e\} \quad \textit{and}$$

$$G \supset M \supset M_1 \supset \ldots \supset M_s \supset \{e\}$$

then $r = s$ *and the quotient groups*

$$G/N,\ N/N_1,\ \ldots,\ N_{r-1}/N_r,\ N_r \quad \textit{and}$$

$$G/M,\ M/M_1,\ \ldots,\ M_{s-1}/M_s,\ M_s$$

are isomorphic in pairs, though not necessarily in the same order.

By using theorem 12.2.3 the proof is not very difficult. The result is of the greatest importance, in that it gives us an essentially unique method of expressing a group as, loosely speaking, the product of the smaller quotient groups. (We are speaking loosely since G is *not* in general the direct product of the quotient groups. This is not true even for the simple case where G has a maximal invariant subgroup N which is simple, as we showed in §6.9.) Although the quotient group are determined uniquely by G, G is not determined uniquely by the quotient groups.

In a finite group the orders of the quotient groups are called the *composition indices* of G, and it is an immediate corollary to the Jordan–Hölder theorem that these are uniquely defined by G, though not in any particular order.

A finite group G must possess a composition series, but there is no guarantee that an infinite group does, although if this is the case the Jordan–Hölder theorem applies and any two series have isomorphic quotient groups. It can be shown that the criteria given in the next theorem are necessary and sufficient for the existence of a composition series.

Theorem 12.3.2. *A necessary and sufficient condition for a group G to possess a composition series is that it satisfies the following two criteria (automatically satisfied if G is finite):*

(*a*) *The descending chain condition.*

If $G \supseteq H_1 \supseteq H_2 \supseteq H_3 \supseteq \ldots$ where each H_i is an invariant subgroup of the preceding one, then $\exists N$ such that

$$H_n = H_N \, \forall n > N,$$

i.e. no infinite sequence with proper inclusions can exist.

(*b*) *The ascending chain condition.*

If H is a subnormal subgroup of G, in the sense that

$$H \subset N_1 \subset N_2 \subset \ldots \subset N_r \subset G$$

where each subgroup is an invariant subgroup of the following one, and if $H_1 \subseteq H_2 \subseteq H_3 \subseteq \ldots$ where each H_i is an invariant subgroup of H, then $\exists N$ such that $H_n = H_N \forall n > N$.

12.4. Sylow groups

An important problem in group theory lies in discovering some or all of the subgroups of a given group. For cyclic groups the process is straightforward since we showed in §13.8 of volume 1 that there is a unique subgroup of order r where r is any factor of the order of the cyclic group, and that these are the only subgroups. For general finite groups we know by Lagrange's theorem that the order of any subgroup is a factor of the order of the group, but there is no general method of discovering all the subgroups corresponding to any given factor, nor even if any exist.

Some theorems which go some way to helping us were discovered by the Norwegian mathematician Sylow in 1872. They are not difficult to prove though a little tedious.

Theorem 12.4.1. *Sylow's theorem.*

If G is a finite group and if p is any prime factor of the order n of G, suppose that p^μ is the highest power of p which divides n. Then G has at least one subgroup of order p^μ.

Thus any group of order 144 (say) possesses subgroups of order $2^4 = 16$ and $3^2 = 9$.

A corollary to Sylow's theorem is the following.

Theorem 12.4.2. *If p is any prime factor of the order of a finite group G, then G has at least one element of order p.*

This means of course that G possesses a cyclic subgroup of order p (the subgroup of powers of the element in question).

We cannot always tell how many subgroups there are of order p^μ in theorem 12.4.1. But we can go part way.

Theorem 12.4.3. *If there are k subgroups of order p^μ, with notation as in theorem* 12.4.1, *then $k \equiv 1 \pmod{p}$ and k is a factor of the order n of G.*

By using theorem 12.2.2 we can in fact prove a good extension of Sylow's theorem.

Theorem 12.4.4. *Any group of order p^μ possesses at least one invariant subgroup of order p^ν for any ν such that $0 < \nu < \mu$, and hence with the notation of theorem* 12.4.1 *G must possess a subgroup of order p^ν (though this need not be invariant in G).*

12.5. The basis theorem for Abelian groups

The classification of all groups is an impossible task, even when we restrict ourselves to finite groups, and a complete classification is not known even for all groups of a given order (save for certain simple orders). For Abelian groups the situation is much simpler, and finite Abelian groups may in fact be completely classified.

***Theorem* 12 5.1.** *The Basis theorem.*

Every finite Abelian group may be expressed as the direct product of cyclic groups, each of order p^μ for a prime p and some positive integer μ.

Thus all finite Abelian groups are merely direct products of cyclic groups. If we wish to find all Abelian groups of a given order, say 360, we express the order as a product of primes $2^3 . 3^2 . 5$. Then an Abelian group of order 360 must be the direct product of C_5 with cyclic groups of order a power of 3 with those of order a power of 2, such that the total powers of 3 and 2 are 3^2 and 2^3 respectively. Thus we obtain the following Abelian groups of order 360:

$$C_5 \times C_9 \times C_8, \quad C_5 \times C_3 \times C_3 \times C_8, \quad C_5 \times C_9 \times C_4 \times C_2,$$
$$C_5 \times C_3 \times C_3 \times C_4 \times C_2, \quad C_5 \times C_9 \times C_2 \times C_2 \times C_2,$$
$$C_5 \times C_3 \times C_3 \times C_2 \times C_2 \times C_2.$$

A similar analysis may be used for any order.

Cyclic groups corresponding to different primes may in fact be combined: $C_{p^\mu} \times C_{q^\nu} \cong C_{p^\mu q^\nu}$.

If we select a generator from each of the cyclic groups of theorem 12.5.1 we obtain a set of elements $f_1, f_2, \ldots, f_r$ such that each element of the group may be represented uniquely in the form $f_1^{\alpha_1} f_2^{\alpha_2} \ldots f_r^{\alpha_r}$, where $0 \leqslant \alpha_i \leqslant p_i^{\mu_i}$ where $p_i^{\mu_i}$ is the order of the cyclic group corresponding to f_i. The set of elements $\{f_i\}$ form a *basis* for the group.

A similar classification holds for infinite Abelain groups, provided they possess a finite set of generators.

***Theorem* 12.5.2.** *If G is an Abelian group with a finite set of generators, then*

$$G \cong C_{p_1{}^{\mu_1}} \times C_{p_2{}^{\mu_2}} \times \ldots \times C_{p_r{}^{\mu_r}} \times Z \times \ldots \times Z,$$

where Z is the infinite cyclic group.

12.6. Galois theory: the insolubility of the general quintic

In chapter 8 we considered a little of the theory of algebraic extensions of fields. This idea leads to important developments in the theory of fields and enables us to consider the deeper properties of the roots of polynomial equations. The theory is difficult and we will merely indicate some of its features.

If a polynomial P has coefficients in a field F we can adjoin the roots of P and form an extension field of F within which P can be factorised into linear factors. For example if $P = x^2 - 2$ where F is the rational field we can form the extension field consisting of the real numbers of the form $a + b\sqrt{2}$ where a and b are rational. The 'smallest' such field is that generated by F and the roots of P and is called a *root field* or *splitting field* of P: it can be proved to exist and to be unique up to isomorphism.

Now if we take the automorphisms of the root field of P which leave every element of F unchanged we obtain a group, which is called the *Galois group* of the polynomial equation $P = 0$ over the field F, and it is upon properties of this group that the deeper properties of the polynomial equation $P = 0$ rest. The Galois group is in fact a subgroup of the group of permutations of the distinct roots of $P = 0$.

We now introduce the concept of a *soluble* group. A group G is soluble if its composition indices (as defined in §12.3) are all prime: it follows that the quotient groups of the composition series are all cyclic of prime order. It may then be shown that a polynomial equation $P = 0$ is soluble by radicals if and only if its Galois group is soluble. (By 'soluble by radicals' we mean that its roots can be written in terms of a finite sequence of elements in the coefficient field F using the processes of addition, subtraction, multiplication, division and the extracting of roots of elements; this is equivalent to the fact that the roots must lie in an extension field of F obtained by successively adjoining roots of elements.)

Now it can be shown that there exists at least one polynomial equation of degree n over the rationals whose Galois group is the symmetric group S_n. (There will in fact be many such polynomials.) Theorem 6.10.2 proved that the alternating group of degree 5 is simple. If we denote this by A_5 we see that since any subgroup of index 2 is invariant $S_5 \supseteq A_5 \supseteq \{e\}$ is a composition series for S_5 with composition indices 2 and 60 $(=\frac{1}{2}5!)$, and 60 is not prime. Hence S_5 is not soluble and so any polynomial equation with it as its Galois group is not soluble by radicals.

Thus there exists at least one quintic equation which is not soluble by radicals.

For any $n > 5$ the alternating group A_n is also simple and the same argument shows that the general n-tic equation for $n > 5$ is therefore not soluble by radicals. For $n = 1, 2, 3$ or 4 the equation is of course soluble and general formulae may be given.

12.7. Ruler and compass constructions

The construction of a line segment from a given unit segment by the classical methods of ruler and compasses may be reduced to an algebraic problem, for it can be shown that the ruler and compass method is equivalent to the construction of algebraic expressions using the four rules of addition, subtraction, multiplication and division together with the extraction of square roots. Thus a segment of length x can be constructed if and only if it lies in a field obtained from the rational field by adjoining square roots. Such an extension field must have degree a power of 2, and any of its members must satisfy an irreducible equation over the rationals of degree a power of 2. Thus the only lengths that can be constructed are those given by roots of irreducible polynomial equations of degree a power of 2, over the rationals.

Let us consider briefly the three classical ruler and compass constructions.

Duplication of the cube

The problem is to construct a segment of length $\sqrt[3]{2}$, which is a root of the irreducible cubic $x^3 - 2 = 0$. Since this has degree 3, which is not a power of 2, the solution is impossible by means of ruler and compasses.

Trisection of the angle

Given an angle 3θ we wish to construct θ. But this is equivalent to constructing the length $\cos\theta$ given $\cos 3\theta$, and we know that $4\cos^3\theta - 3\cos\theta = \cos 3\theta$. Thus we wish to construct a solution of $4x^3 - 3x = \cos 3\theta$, for a general value of $\cos 3\theta$. It can be shown (and is easily believable) that this equation is irreducible for some values of $\cos 3\theta$ (in fact for most values) and hence, since it is again a cubic, the construction is impossible in general, although it may of course be performed in certain special cases (e.g. when $\theta = 15°$).

Squaring the circle

This depends on the construction of $\sqrt{\pi}$, which in turn depends on the construction of π. But it was shown by Lindemann in 1884 that π is transcendental, i.e. it is not a root of any polynomial equation with rational coefficients. Hence it is certainly not a root of an equation of degree a power of 2, and so the construction is impossible.

12.8. Symmetric polynomials

A polynomial in $x_1, x_2, \ldots, x_n$ is *symmetric* if it is invariant under any permutation of the x_i's.

The set of symmetric polynomials forms a subring of the polynomial ring.

The *elementary symmetric polynomials* are the polynomials Σx_i, $\Sigma x_i x_j$, ..., $x_1 x_2 \ldots x_n$, which are of course the coefficients of the powers of x in the polynomial $(x-x_1)(x-x_2)\ldots(x-x_n)$. The fundamental theorem is as follows.

Theorem 12.8.1. *Any symmetric polynomial in* $x_1, x_2, \ldots, x_n$ *is expressible uniquely as a polynomial in the elementary symmetric polynomials.*

Since the converse is obviously true, the symmetric polynomials are precisely the polynomials in the elementary symmetric polynomials. The uniqueness shows that the elementary symmetric polynomials are algebraically independent over R, and so the subring of symmetric polynomials is in fact the ring $R[P_1, P_2, \ldots, P_n]$ where the P_i's are the elementary symmetric polynomials.

For example if $n = 2$ all symmetric polynomials are expressible uniquely as polynomials in x_1+x_2 and $x_1 x_2$, and the subring of symmetric polynomials is $R[(x_1+x_2), x_1 x_2]$.

12.9. Modules

The study of the structure of rings is a difficult subject, and we could do very little in chapter 3. The work is helped by the introduction of the concept of a *module*, which is a composite structure involving a ring R and a commutative group M.

Definition. *If R is a ring and M a commutative group, M is a* right R-module *if for any $x \in M$ and $r \in R$ there is an element $x \circ r \in M$ such that*

(i) $(x+y) \circ r = x \circ r + y \circ r$;

(ii) $x \circ (r+s) = x \circ r + x \circ s$;

(iii) $x \circ (rs) = (x \circ r) \circ s$.

Let us see what this implies. For a fixed r and variable x condition (i) states that the mapping $x \to x \circ r$ is an endomorphism of M (i.e. a homomorphism of the Abelian group M into itself). But we saw in §3.9 that the endomorphisms of any Abelian group form a ring, and thus, corresponding to any element r in R, we have an endomorphism, which is an element of the ring $E(M)$ of endomorphisms of M. Conditions (ii) and (iii) show that this correspondence is a ring-homomorphism of R into $E(M)$. Conversely any such ring-homomorphism leads to a right R-module as defined above.

The ring R is itself a right R-module if we define $x \circ r$ to be the ring product xr (clearly satisfying the three conditions above). Thus we obtain a ring-homomorphism of R into the ring of group-endomorphisms of the additive group of R.

If R is a field then an R-module is a vector space over R provided we have the additional condition that $x \circ 1 = x$, i.e. that the element 1 corresponds to the identity endomorphism of M.

If N is a subgroup of M such that $x \circ r \in N$ for all $x \in N$ and $r \in R$ we call N a *submodule* of M, and note that when we consider R as an R-module then any ideal is a submodule.

Before going much further we must lay down two conditions similar to the chain conditions for groups of theorem 12.3.2.

The descending chain condition for modules

If $N_1 \supseteq N_2 \supseteq N_3 \supseteq \ldots$ where the N_i are submodules of a module M, then M is said to satisfy the descending chain condition if for some integer r we have $N_r = N_{r+1} = \ldots$.

The ascending chain condition for modules

If $N_1 \subseteq N_2 \subseteq N_3 \subseteq \ldots$ where the N_i are submodules of a module M then M is said to satisfy the ascending chain condition if for some integer r we have $N_r = N_{r+1} = \ldots$.

One way of considering the structure of a ring is to look at the structure of its ideals, and it turns out to be convenient to restrict ourselves to commutative rings that satisfy the ascending chain condition where the submodules correspond of course to ideals. Such rings are called *Noetherian* (after the great mathematician Emmy Noether). It can be shown that the ring of polynomials in n indeterminates over any field is Noetherian.

If we define a *primary ideal* to be an ideal I such that $xy \in I \Rightarrow$ either $x \in I$ or $y^r \in I$ for some positive integer r, we can prove that every ideal in a Noetherian ring is a finite intersection of primary ideals, and that such a representation is in a certain sense unique (see the theory of *radicals* in any standard work on ring theory for a detailed explanation of 'certain sense').

12.10. The Krull–Schmidt theorem

In §12.5 we saw that any finitely generated Abelian group may be expressed as the direct product of cyclic groups. Clearly the expression of a group as a direct product is a useful method of studying its structure, and we extend the discussion to general groups.

A group is said to be *indecomposable* if it cannot be expressed as the direct product of two non-trivial subgroups: if it can then it is called *decomposable*. We might expect that any group can be expressed as the direct product of indecomposable subgroups (strictly speaking of groups that are isomorphic to subgroups, since if $G = A \times B$ we know that A and B, while not being strictly speaking subgroups of G, are isomorphic to subgroups of G). We require however that the group satisfies a descending chain condition for invariant subgroups (remember that if $G = A \times B$ then A and B are isomorphic to *invariant* subgroups of G).

The descending chain condition for invariant subgroups

G is said to satisfy the descending chain condition for invariant subgroups if given a sequence $H_1 \supseteq H_2 \supseteq H_3 \supseteq \ldots$ where each H_i is an invariant subgroup of G, there exists N such that $H_N = H_{N+1} = \ldots$. (This condition is weaker than that given in theorem 12.3.2.)

If this condition is satisfied, we are able to show that an indecomposable direct factor of G exists, and that the expression of G as a direct product of indecomposable subgroups must stop.

Theorem 12.10.1. *If G satisfies the descending chain condition then it can be expressed as a direct product of a finite number of non-trivial indecomposable groups.*

We now wish to show that the decomposition is unique. This requires in addition to the descending chain condition an ascending chain condition.

The ascending chain condition for invariant subgroups

G is said to satisfy the ascending chain condition for invariant subgroups if given a sequence $H_1 \subseteq H_2 \subseteq H_3 \subseteq \ldots$ where each H_i is an invariant subgroup of G, there exists N such that $H_N = H_{N+1} = \ldots$. (This again is weaker than that given in theorem 12.3.2.)

If G satisfies both chain conditions then there exists a unique decomposition of G, as is shown by the following famous theorem.

Theorem 12.10.2. *The Krull–Schmidt theorem.*

If G satisfies both chain conditions and if

$$G = H_1 \times H_2 \times \ldots \times H_r \quad \textit{and} \quad G = K_1 \times K_2 \times \ldots \times K_s$$

are two decompositions of G into indecomposable subgroups, then $r = s$ *and, for a suitable renumbering of the* K_i*'s,* $H_j \cong K_j$ *for all* $j, 1 \leqslant j \leqslant r$.

We may apply the above to module theory.

Theorem 12.10.3. *If M is a right R-module that satisfies the descending chain condition then it can be expressed as a direct sum (using the additive notation and defining direct sum of R-modules in the obvious way by* $(x, y) \circ r = (x \circ r, y \circ r)$*) of a finite number of non-trivial indecomposable modules.*

Theorem 12.10.4. *The Krull–Schmidt theorem for modules.*

If M is a right R-module that satisfies both chain conditions and if

$$M = N_1 \times N_2 \times \ldots \times N_r \quad \text{and} \quad M = L_1 \times L_2 \times \ldots \times L_s$$

are two decompositions of M into indecomposable submodules, then $r = s$ *and, for a suitable renumbering of the* L_i*'s,* $N_j \cong L_j$ *for all* $j, 1 \leqslant j \leqslant r$.

We cannot at once apply the above to rings, since we are using modules that are all R-modules, that is all referring to the same ring R. But by means of the theory for modules it can be shown that the following theorem is true. Note that this says nothing about uniqueness.

Theorem 12.10.5. *If R is a ring with unity and satisfying the two chain conditions for right ideals (a right ideal being a subgroup I with* $xr \in I$ *for* $r \in R$ *and* $x \in I$*, but not necessarily* $rx \in I$*) then R can be expressed as a direct sum of a finite number of non-trivial subrings.*

12.11. Representations of groups and rings

If A is any algebraic structure, whose automorphisms form a group, a *representation* of a group G is a homomorphism of G into this group of automorphisms: thus if A is a familiar structure we can study G with reference to a known group, particularly if the homomorphism is in fact 1–1, in which case the representation is said to be *faithful*.

As an example let A be a vector space V of dimension n with scalar field F. Then the automorphisms of A are merely the non-singular linear transformations of V and may be represented, by choice of a base, by the non-singular $n \times n$ matrices over F. A faithful representation of G then gives G as a subgroup of this group of matrices.

Another useful example is given by taking A as a set and considering an automorphism of A to be a permutation of the elements. We then represent G as a subgroup of a symmetric group (i.e. a permutation group).

We can apply the same sort of ideas to rings. If the endomorphisms of A form a ring then a *representation* of a ring R is a ring-homomorphism of R into this ring of endomorphisms: if it is 1–1 the representation is again said to be *faithful*.

As before if A is a vector space V the endomorphisms are given (by choice of a base) by the $n \times n$ matrices with coefficients in F, and a representation of R gives R in terms of a subring of this matrix ring: if the representation is faithful then R is isomorphic to a matrix ring.

If A is a commutative group we have already seen that a representation of a ring R corresponds to a module 'product' giving A as an R-module.

As the reader may guess from the above, the idea of a representation is exceedingly useful in the advanced theory of the structure of groups and rings, and has been developed in great detail. But it is too difficult for us to do more than mention the broad outlines of the definition and its scope.

12.12. Algebraic topology

Topology is the study of properties of figures, or *spaces*, that are invariant under continuous transformations (or deformations). In the study of topological spaces considered as a whole probably the most important question is to decide whether or not two such spaces are topologically equivalent; that is whether or not one can be continuously deformed into the other. It turns out that abstract algebra, particularly group theory, forms a powerful tool in answering this question. Briefly the situation is that we associate in some way a group with a given space and show that two topologically equivalent spaces possess the same associated groups: thus if we know that two spaces have different associated groups then they cannot be equivalent topologically. Unfortunately the converse is not true—two different spaces may have the same group associated with them.

There are many ways of linking algebraic structures with spaces—so much so that a whole new branch of algebra, called *homological algebra*, has been developed to cope with the special algebraic problems that arise in connection with topology. We will briefly indicate two lines of approach.

Homotopy—the fundamental group

For simplicity we will consider two-dimensional spaces and think of them as being embedded in the everyday three-dimensional space, although the theory of course applies to spaces of any dimension. We will call such spaces *surfaces*, to help the reader's intuition.

Take a fixed point P on a surface S and consider all paths (i.e. curves) starting and ending at P and lying completely on S. Two such paths are said to be *homotopic* if one can be transformed into the other by a continuous deformation *on* S. It can easily be shown that homotopy is an equivalence relation. Now form the product of two paths by tracing first one then the other. If two pairs of paths are homotopic then their products are also homotopic and product of homotopy classes of paths is well-defined. The classes under this definition form a group associated with the point P. This group is in fact independent of P (provided one can reach all points of S from P without leaving the surface) and is called the *fundamental group of* S.

Examples. The fundamental group of the surface of a sphere is the trivial group $\{e\}$. The fundamental group of the torus is $Z \times Z$, where Z is the infinite cyclic group. (Generators are given by paths α and β in figure 61 and the group is commutative since $\alpha\beta$ is homotopic to $\beta\alpha$.) The fundamental group of the real projective plane (obtained as shown in figure 62 as a circular surface with the opposite ends of any diameter identified) is the cyclic group C_2, and the fundamental group of the Klein bottle (represented by the surface in figure 63 with the edges identified as shown by the arrows) is $Z \times C_2$. Thus all the above examples are topologically inequivalent, since all have different fundamental groups.

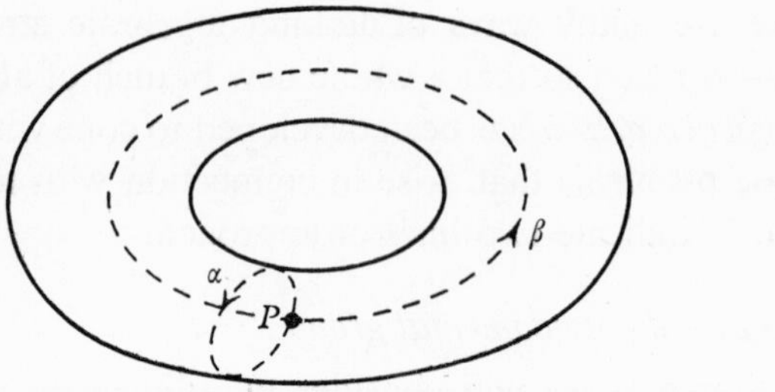

Fig. 61

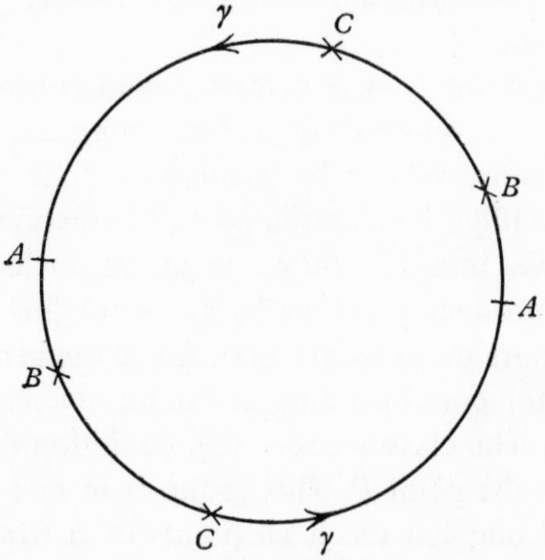

Fig. 62

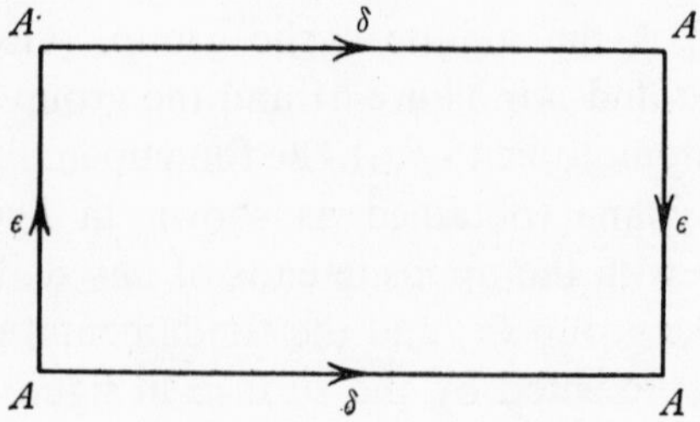

Fig. 63

Homology

Again we will restrict ourselves to the two-dimensional case of a surface. Suppose the surface S is divided into cells, such that each cell is topologically equivalent to the inside and circumference of a circle. (In the elementary simplicial complex theory the cells are triangles, but this is not necessary, although it makes the theorems easier to prove.)

Such a division will lead to the existence of curves on S, dividing the cells, and points (vertices) where the curves meet (think of the net of a polyhedron). If we add together (formally) any integer multiples of the various cells we obtain a 2-*chain*, if we do the same with the curves we obtain a 1-*chain*, and a similar process with the vertices gives a 0-*chain*. (For example if the cells are denoted by $c_1, c_2, \ldots, c_r$, a 2-chain is a formal sum $u_1 c_1 + u_2 c_2 + \ldots + u_r c_r$, where the u_i's are integers, positive, negative or zero.) The *boundary* of a cell is obtained by taking the 1-chain obtained by going round it (speaking loosely), and the boundary of a curve is the difference of its end-points. The boundary of a point is defined to be zero. The boundary of a chain is obtained by taking the same multiples of its components (thus the boundary of $\Sigma u_i c_i$ is $\Sigma u_i b(c_i)$, where $b(c_i)$ is the boundary of c_i). Note that negative multiples of the components are allowed—in fact each cell and curve must be orientated in an arbitrary but constant manner. A chain with boundary zero is called a *cycle*: thus we obtain 2-cycles, 1-cycles and 0-cycles (every 0-chain is of course a 0-cycle). The set of all p-chains clearly forms an Abelian group C^p, and the set of p-cycles will form a subgroup Z^p of C^p. A p-cycle that is itself a boundary of a $(p+1)$-chain is called a *bounding p-cycle* (in the case of S being a surface there will be no bounding 2-cycles). The bounding p-cycles also form a group B^p, a subgroup of Z^p, which is invariant since all groups are Abelian.

The quotient group Z^p/B^p is called the pth *homology group* of S: it consists of the p-chains with zero boundaries, modulo those that are boundaries themselves.

Thus corresponding to any surface we have the zero-th, first and second homology groups. In general for an n-dimensional space we will have homology groups of each order up to the nth.

The homology groups are topological invariants: if two spaces are either the same or topologically equivalent then they have identical homology groups. But as usual the converse is not true.

Examples. For a sphere the zero-th and second homology groups are the infinite cyclic group Z, and the first homology group is $\{e\}$. This may be seen by taking a single cell, with no curves and a single point as a vertex (i.e. a circle with the whole circumference identified). The cell has zero boundary and so is a 2-cycle, generating the second homology group. The point is not a boundary and so generates the zero-th homology group, whereas there are no 1-cycles and so the first homology group is $\{e\}$.

The torus may also be considered as a single cell, with zero boundary, whose second homology group is therefore Z. Similarly the zero-th homology group is Z, as may be seen by taking a single point A on the surface. But in order to open out the torus we must take two curves from A to A, corresponding to the paths α and β in figure 61, and these are 1-cycles. But no combination of these is a boundary (since the only 2-cell has zero boundary) and so the first homology group is $Z \times Z$.

For the real projective plane see figure 62. The only 2-cell has non-zero boundary and so there are no 2-cycles and the second homology group is $\{e\}$. If we call the curve that forms the upper half of the circumference γ, we see that γ and any multiple of γ are 1-cycles, and that 2γ is a boundary. Hence the first homology group is C_2. Since there is just one point A which is not a boundary the zero-th homology group is Z.

Finally consider the Klein bottle as shown in figure 63. There are no 2-cycles and the second homology group is $\{e\}$. Both δ and ϵ are 1-cycles and so Z^1 is $Z \times Z$. But 2ϵ is the boundary of the 2-cell and so B^1 is the subgroup generated by 2ϵ. Hence the first homology group Z^1/B^1 is $Z \times C_2$. As before the zero-th homology group is Z.

In the above examples the first homology groups have all been isomorphic to the fundamental groups, but this is by no means always the case. In fact, while the homology groups are always Abelian (by the definition) the fundamental groups may be non-Abelian. For example, the fundamental group of the double torus (a doughnut with two holes) is the group generated by four elements a, b, c, d with the single relation $aba^{-1}b^{-1}cdc^{-1}d^{-1} = e$ and is non-Abelian, while the first homology group is $Z \times Z \times Z \times Z$ and is of course Abelian.

The above is not the only homology theory: many others have been developed in order to refine the classification of topological spaces, and the whole subject affords an excellent example of the uses to which abstract algebra is being put in mathematics.

ANSWERS TO EXERCISES

Exercises 1A (p. 25)

1. No. **2.** Yes. **3.** No. **4.** Yes.

5. No. **6.** No.

10. C_5: all non-zero elements are generators.
C_8: 1, 3, 5, 7 are generators; 4 has order 2; 2, 6 have order 4.
C_{12}: 1, 5, 7, 11 are generators; 6 has order 2; 4, 8 have order 3; 3, 9 have order 4; 2, 10 have order 6.

11. (i) C_6; (ii) C_{10}.

12. (i) Vierergruppe; (ii) C_4; (iii) C_6.

14. (i) $\begin{pmatrix} 1 & 2 & 3 & 4 \\ 3 & 2 & 4 & 1 \end{pmatrix}$; (ii) $\begin{pmatrix} 1 & 2 & 3 & 4 \\ 2 & 4 & 3 & 1 \end{pmatrix}$; (iii) $\begin{pmatrix} 1 & 2 & 3 & 4 \\ 2 & 4 & 1 & 3 \end{pmatrix}$;

(iv) $\begin{pmatrix} 1 & 2 & 3 & 4 \\ 4 & 2 & 1 & 3 \end{pmatrix}$; (v) $\begin{pmatrix} 1 & 2 & 3 & 4 \\ 2 & 1 & 3 & 4 \end{pmatrix}$.

15. (i) Even; (ii) odd; (iii) odd.

17. (i) Vierergruppe; (ii) Vierergruppe; (iii) D_{10}.

20. Generators a, b; $a^4 = b^2 = e, ab = ba$.

21. No.

Exercises 2A (p. 53)

29. C_2. **30.** C_4. **31.** C_2. **32.** S_3.

Exercises 3A (p. 87)

12. n. **13.** 2. **14.** 3. **15.** 12.

19. 1(1), 5(2), 7(2), 11(2). **23.** $R_2 \oplus R_2$.

24. $Q \equiv 2x^2+2x+1$, $S \equiv 4$. **25.** $Q \equiv x^3+2x^2+3x$, $S \equiv x$.

26. $c = 1, 3, 5, 7$. **27.** $c = 0, 2, 4, 6, 8, 10, 12, 14$.

31. Zero-ring, additive structure C_2. **32.** R_3, unity 4.

33. R_2, unity 3. **34.** Zero-ring, additive structure C_2.

35. R_5, unity 6. **36.** R_2, unity 5.

44. Ring of integers. **45.** R_6.

47. Trivial group $\{e\}$. **48.** C_2.

Exercises 3B (p. 90)

4. Additive structure is $C_2 \times C_2$. **10.** $0, 1, x, 1+x$.

11. $a+bx+cx^2$, each of $a, b, c = 0, 1$, or 2.

Exercises 4A (p. 116)

16. 2, 4, 2, 4.

17. $x^2, x^2+x = x(x+1), x^2+1 = (x+1)^2, x^2+x+1$.

18. $x = 1, y = 3$. **19.** Vierergruppe $C_2 \times C_2$.

21. Kernel—when it is F; image—when it is $\cong F$.

23. $\{e\}$. **24.** C_2. **25.** $\{e\}$. **26.** $C_2 \times C_2$.

29. p, J_p.

Exercises 4B (p. 118)

1. $p(p-1)$. **2.** $s = -8, t = 53; 53$.

5. Rationals, F.

Exercises 5A (p. 144)

16. $\pm 2 \pm \sqrt{3}$, etc. **19.** The identity automorphism.

20. $x-1$. **21.** $x, x-1, x-2$. **22.** None. **23.** None.

25. x^2+1, x^2+x+2, x^2+2x+2 (plus associates).

30. $4 = 2.2 = (1+\sqrt{-3})(1-\sqrt{-3})$.

35. $q = 1-i, r = 0$. **36.** $q = 3+i, r = 1$.

37. $q = 2i, r = -2-3i$. (Other answers are possible in **35–37**.)

38. $(1+i)(2-i)$, or associates such as $(1-i)(1+2i)$.

39. $2+3i$. $s = -1, t = 2i$.

Exercises 5B (p. 146)

3. Prime factorisation is $(x+1)(x+2)(x^2+2)$.

4. (i) A prime. (ii) $(\sqrt{-2})(1-\sqrt{-2})$. (iii) $(1-\sqrt{-2})(1+2\sqrt{-2})$.

Exercises 6A (p. 176)

1. Reals under addition modulo 1. **2.** Integers.

3. $C_2 \times C_2$. **4.** $\cong$ polynomials. **5.** Reals $\times$ reals.

6. Reals. **7.** Reals under addition modulo 2π ($\cong$ complex numbers of modulus 1 under multiplication). **8.** C_5.

9. C_4. **10.** $C_2 \times C_2$. **13.** No. **14.** Yes; C_2.

15. No. **16.** Yes; $C_2 \times C_2$. **17.** Yes; D_6.

18. Yes; C_3. **22.** 2, 3, 4, 5, 6, 7, 9.

23. $\{e\}$, $\{a, a^3\}$, $\{a^2\}$, $\{b, ba^2\}$, $\{ba, ba^3\}$.

24. $\{(1234)\}$, $\{(2143), (3412), (4321)\}$; $\{(2314), (4132), (3241), (1423)\}$; $\{(3124), (2431), (4213), (1342)\}$, writing $(abcd)$ for $\begin{pmatrix} 1 & 2 & 3 & 4 \\ a & b & c & d \end{pmatrix}$.

26. $\{e, b\}$; $\{e, ba^2\}$.

27. There are 4, each consisting of all elements that leave one object unaltered. **32.** $\{3n\}$; $\{3n+r\}$, $r = 0, 1, 2$; C_3.

33. $\{(x, 0)\}$; $\{(x, a)\}$; reals

34. $\{(x, 0, 0)\}$; $\{(x, a, b)\}$; two-dimensional vectors.

35. $\{\lambda n\}$; $\{\lambda n+r\}$, $r = 0, 1, \ldots, (n-1)$; C_n.

36. $\{e, a^3, a^6, a^9, a^{12}\}$; $\{a^{3n+r}\}$, $r = 0, 1, 2$; C_3.

Exercises 6B (p. 177)

1. $\{e, a^2\}$; $C_2 \times C_2$; D_8; C_2.

Exercises 7A (p. 203)

2. No. **3.** Additive structure the Vierergruppe, multiplicative structure that of R_4 with 0, 1, $1+i$, i corresponding to 0, 1, 2, 3.

4. R_2. **5.** $\cong$ 3, with $x \leftrightarrow 1+i$, $1+x \leftrightarrow i$.

6. No. **7.** No. **8.** Zero-ring, additive structure C_2.

12. R_3.

13. Additive structure C_4. All products zero except

$$2.2 = 2.6 = 6.2 = 6.6 \text{ (all } = 4).$$

17. **3**$-[2]$, **4**$-[2, x]$, **5**$-[2, x^2]$, **8**$-[4]$. **19.** $x+1$.

21. (i) $\sum_i r_i a_i + \sum_i a_i r_i' + \sum_{ij} s_{ij} a_i s_{ij}'$; (ii) $\sum_i r_i a_i + \sum_i n_i a_i$;

(iii) as (i) $+\sum_i n_i a_i$.

28. **4** is both prime and maximal. **38.** Polynomials over integers, etc.

Exercises 7B (p. 206)

6. $[m+ni]$ (principal ideal domain).

8. (i) $[x^2+y^2-1]$; (ii) $[x^2+y^2-2, x-y]$.

9. R = ring of integers, $I = [n]$, etc.

Exercises 8 A (p. 222)

13. $a+b\sqrt[4]{2} \Leftrightarrow a+bi\sqrt[4]{2}$, $a+b\sqrt[4]{2} \Leftrightarrow a-bi\sqrt[4]{2}$.

14. $\{a+b\omega\sqrt[3]{2}\}$, where $\omega^2+\omega+1 = 0$.

Exercises 9 A (p. 267)

13. $\lambda_1 = 3, \lambda_2 = 5, \lambda_3 = -1$.

14. $\lambda_1 = 3, \lambda_2 = -1, \lambda_3 = -1$.

15. $\lambda_1 = 2, \lambda_2 = 1, \lambda_3 = -2$.

16. $x_3 = x_1+\frac{1}{2}x_2$.

37. No. **41.** D_6. **42.** 1. **44.** 2.

45. 3. **46.** 1. **47.** 2.

48. $\begin{pmatrix} -5 & 3 \\ 2 & -1 \end{pmatrix}$.

49. $\begin{pmatrix} 3 & 2 & -3 \\ 1 & 1 & -1 \\ -\frac{4}{3} & -\frac{4}{3} & \frac{5}{3} \end{pmatrix}$.

51. $(1, 1, -1)$.

52. $\left(-\frac{\lambda}{3}, 1-\frac{5\lambda}{3}, \lambda\right)$.

53. No solution. **54.** $(1-\lambda, \lambda, \lambda)$. **55.** $(1, 0)$.

56. No solution. **59.** $(\lambda, \lambda, 3\lambda)$.

60. $(\lambda, \mu, \mu-2\lambda)$.

Exercises 9 B (p. 270)

1. $\nu(\mathrm{A}) = n-r(\mathrm{A})$. (i) Line through O; (ii) Plane through O; (iii) Line through O in 3-dimensions (or plane if $r(\mathrm{A}) = 1$). (iv) Point O (or line if $r(\mathrm{A}) = 2$, plane if $r(\mathrm{A}) = 1$).

7. (iii) True. (i), (ii) and (iv) can be false.

12. (i) $ab'-a'b \neq 0$. (ii) $ab'-a'b = 0$ and $ac'-a'c \neq 0$. (iii) $ab'-a'b = 0$ and $ac'-a'c = 0$.

13. $\left(\frac{A-64}{B-1}, \frac{24B-7A+424}{B-1}, \frac{-13B+6A-371}{B-1}\right)$ if $B \neq 1$.
No solution if $B = 1$ and $A \neq 64$.
$y = \frac{1}{6}(53-7z)$, $x = \frac{1}{6}(13+z)$ if $B = 1$ and $A = 64$.

Exercises 10 A (p. 303)

9. (i) No. (ii) Yes. (iii) No.

17. $x_3-a_3 = 0$, $x_4-a_4 = 0$, for example.

18. No.

24. $D(0, 1, 1)$, $E(1, 0, 1)$, $F(1, 1, 0)$, $X(0, -1, 1)$, $Y(1, 0, -1)$, $Z(-1, 1, 0)$.

25. $AD{:}\,y-z = 0$, $BE{:}\,z-x = 0$, $CF{:}\,x-y = 0$, $EF{:}\; x-y-z = 0$, $FD{:}\,-x+y-z = 0$, $DE{:}\,-x-y+z = 0$.

26. $x+y+z = 0$. **27.** $(CA, EY) = -1$.

32. A rotation through θ anti-clockwise about O.

33. Reflections in Ox and Oy.

34. An enlargement about O by factor k. $k = \pm 1$.

39. $(2, \pm\sqrt{3}i)$. **40.** $(\pm\sqrt{\frac{27}{5}}\,i, \pm\sqrt{\frac{32}{5}})$.

43. $p^2; p^2+p; p+1; p$. **45.** All -1.

Exercises 10B (p. 307)

7. Arcs of great circles. **13.** S_3. **17.** $n+1$.

Exercises 11A (p. 348)

26. (i) $B' \cap C$; (ii) $(C' \cap A') \cup (B \cap C)$.

28. (i) $Q \Rightarrow P$; (ii) $P \Rightarrow Q$; (iii) $P \Leftrightarrow Q$. **29.** $P \cap Q'$.

35. $[A \cup (B \cap C)] \cap B' \cap (C \cup B)$;
$\{A \cap [(B \cap C) \cup (B' \cap C') \cup (A \cap D)] \cap B\} \cup (C \cap D')$.

37. $A \cup C'$. **38.** $\frac{1}{6}, \frac{1}{6}, \frac{1}{36}, \frac{11}{36}$.

Exercises 11B (p. 351)

4. $(X \cup Y \cup Z) \cap (X' \cup Y' \cup Z) \cap (X' \cup Y \cup Z') \cap (X \cup Y' \cup Z')$.

6. (i) Valid.
[Premises $\{(P \Rightarrow Q) \Rightarrow R\} \Rightarrow S$ and R, conclusion S.]
(ii) Invalid.
[Premises $Q' \Rightarrow P'$ and Q, conclusion P.]

9. (i) $A \cap B, \frac{1}{16}$; (ii) $A \cup B, \frac{7}{16}$; (iii) $(A \cap B') \cup (A' \cap B), \frac{3}{8}$;
(iv) $A \cap B', \frac{3}{16}$; (v) $A' \cap B', \frac{9}{16}$.

INDEX